GARBAGE IN THE CITIES

HISTORY OF THE URBAN ENVIRONMENT

Martin V. Melosi and Joel A. Tarr, Editors

GARBAGE IN THE CITIES

Refuse, Reform, and the Environment

REVISED EDITION

MARTIN V. MELOSI

UNIVERSITY OF
PITTSBURGH PRESS

Published by the University of Pittsburgh Press, Pittsburgh, Pa., 15260

Manufactured in the United States of America

Printed on acid-free paper

10 9 8 7 6 5 4 3 2 1

Library of Congress Cataloging-in-Publication Data

Melosi, Martin V., 1947-

Garbage in the cities : refuse, reform, and the environment / Martin V. Melosi.— Rev. ed.

p. cm. — (History of the urban environment)

Includes bibliographical references and index.

ISBN 0-8229-5857-0 (pbk. : alk. paper)

1. Refuse and refuse disposal—United States—History. I. Title. II. Series.

TD788.M45 2004

363.72'85'0917320973—dc22 2004015714

To Joel Tarr,
Colleague, Mentor, Friend

CONTENTS

TABLES AND FIGURES

TABLES

FIGURES

PREFACE

A colleague recently brought to my attention that Daniel C. Walsh, an adjunct professor at the Earth Institute at Columbia University, published an article in 2002 based on one hundred years' worth of New York City refuse collection records he uncovered at the city's archives. Walsh challenges many long-held assumptions about the things we dispose of and how much we throw away. The information on New York suggests that we probably discard half as much today as we did in 1940, largely because we do not have ash residue from burning coal, and packaging material has become lighter.[1] Somebody or something other than householders, of course, is now responsible for wastes produced from energy generation in the form of vapor, particulate pollution, waste heat, or solid residues. Packaging material has not so much decreased in volume as changed in form, ostensibly becoming lighter. Thus Walsh's revelations may not be so startling. They nonetheless urge us to pause and at least question what has happened to our waste habits over the years. Walsh is a non-historian helping us do historians' work.

In the more than twenty years since the first edition of *Garbage in the Cities*, garbage has not changed as much as garbage history has. The challenge to the historian is to adjust to those changes while writing his or her story. Many things have affected my perspective on the solid waste issue since 1981. Some new wrinkle always draws me back to the mundane, sometimes quirky, but unrelenting problem of waste. I am frequently reminded that the urge to discard things is basic to understanding a great deal about our culture. As social anthropologist Michael Thompson stated, "People in different cultures may value different things, and they may value the same things differently, but all cultures insist upon some distinction between the valued and the valueless."[2]

Some changes that have taken place over the last couple of decades suggest modifications in our waste-collection and disposal habits: Rhode Island enacted the nation's first statewide mandatory recycling law in 1986; McDonald's announced plans to stop using Styrofoam packaging for its foods due to consumer protests; the Environmental Protection Agency declared in 1996 that the nation had reached a 25 percent recycling rate and would set a new goal of 35 percent; the EPA asserted in 2000 that waste reduction and recycling could help limit global warming.

Some changes have been more symbolic than substantive. The garbage barge *Mobro* (1987) searched unsuccessfully in six states and three countries for a place to dump its cargo. Ultimately, it returned home to Islip, New York, where the refuse was incinerated in Brooklyn and the ash residue buried in a landfill near Islip. While the event was hardly comparable to the wanderings of Odysseus, the *Mobro* came to be viewed as a fitting reminder of a feared landfill disposal shortage in the Northeast and a potential garbage crisis throughout the country. The closing of Fresh Kills Landfill in Staten Island—the largest landfill in the world—in March 2001 was met with jubilation. Several months later, after the September 11 attacks, Fresh Kills was reopened to accept not only the debris from the collapse of the World Trade Centers 1 and 2, but also the remains of many Americans who had died there. Landfill had become cemetery.

On a more personal note, my involvement with preparing a National Historic Landmark designation for the Fresno Sanitary Landfill made me realize how deeply embedded our views of waste, wastefulness, and the places where we choose to discard that waste have ossified over the years. On August 27, 2001, Secretary of the Interior Gail Norton announced the designation of the Fresno landfill as a National Historic Landmark to be listed with some 2,300 sites. On August 28, the designation was "temporarily rescinded" and remains in limbo in 2004. The about-face came as a result of protests from environmentalists, opponents of the George W. Bush administration, and others who opposed the naming of a Superfund site (the landfill was placed on the National Priorities List in 1989) as a historic landmark. A spokesperson for Secretary Norton stated that the Superfund status "got lost in the translation of the signing-off process." Almost every major newspaper and even the electronic media picked up on the controversy. It mattered little that this was not the first Superfund site to be recognized as a historic landmark; that the site had merit as the first technology of its kind revolutionizing disposal and improving prevailing health conditions; or that historic landmarks are not designated to celebrate or to promote such properties, but to inform the public about them with respect to the heritage of the United States. To many people a dump is a dump, and there is little more that needs to be said.[3]

The original edition of *Garbage in the Cities* was the first historical treatment of its kind. The revised edition is intended to help place in perspective the important role of solid waste management in our urban and environmental history, as well as to offer some new insights gained in more than twenty years of change and reflection on one of the most curious, but important, parts of our daily lives. The introduction and chapters 1–6 have been slightly revised, and the notes include some new sources. Chapters 7 and 8 are new and reflect important changes in the field, and speak more directly about private waste companies.

ACKNOWLEDGMENTS

It has been a pleasure, and a little bit strange, revising a book that has been so meaningful to my career. The strange part has been reading and comprehending prose I wrote more than twenty-three or twenty-four years ago. Possibly even stranger is that I have yet to shake the title of "garbage historian." I wear it now, however, with much less trepidation—and with a little pride. What has not changed is my gratitude to those who helped me complete the original edition of *Garbage in the Cities,* and I offer my hearty thanks to those who made the revised edition possible.

My wife Carolyn still comes first. How she puts up with my work habits and always finds a way to give me moral support is unfathomable. Our daughters, Gina and Adria, who are out in the world now, must know that they are never beyond our thoughts.

Colleagues and friends who read and criticized the original manuscript, or who made useful suggestions, still deserve special recognition. They include Sara Alpern, Mary Clare Fabishak, Richard Fenton, Suellen Hoy, Dale Knobel, John Lenihan, Brad Rice, the late Mike Robinson, and Joel Tarr. Special thanks to Margaret Ingram and Noel Parsons of Texas A&M University Press: Margaret gave me excellent advice about the organization of the book, and Noel, among other things, came up with the title. Cynthia Miller of the University of Pittsburgh Press has been a friend and a supporter of my work for many years, and guided the revised edition to completion. I also wish to acknowledge the scores of people who I have met or corresponded with over the years—inside and outside of academia—who found this work useful in some way.

The staffs of the following libraries and other institutions were generous with their time: M. D. Anderson Library, University of Houston;

Sterling Evans Library, Texas A&M University; Perry-Castañeda Library, University of Texas, Austin; New York University Library; Engineering Societies Library, New York City; New York Public Library; Haven Emerson Public Health Library, New York City; Municipal Reference and Research Center, New York City; National Library of Medicine, Bethesda, Maryland; Library of Congress; Smithsonian Institution; and the Environmental Protection Agency, Washington, D.C.

The Rockefeller Foundation was the first—and a vital—source of financial support for the initial research project. I would also like to thank Texas A&M University and the University of Houston for additional research funds and clerical support.

The original volume was dedicated to an extraordinary teacher at the University of Montana, Jules Karlin, who inspired me to become a historian. This revised edition is for my friend and colleague, Joel Tarr, whose own work and constant support helped to shape my career as I became a professional historian. Writing can be a lonely task, but it is made worthwhile by all the people I depend on and trust.

GARBAGE IN THE CITIES

INTRODUCTION

Since human beings have inhabited the earth, they have generated, produced, manufactured, excreted, secreted, discarded, and otherwise disposed of all manner of waste. Among myriad types of rejectamenta, refuse—solid waste—has been one of the most abundant, most cumbersome, and potentially most harmful. Beginning with ancient civilizations, there has always been refuse. There has not always been a refuse problem, however, at least not one of the magnitude that has developed in modern times. Simply to equate poor sanitation with the age of a society is to overlook the major factors that produce a refuse problem with serious health and environmental repercussions.

Refuse is primarily an urban blight. Agrarian societies throughout history have successfully avoided solid waste pollution; cities and towns have faced the gravest dangers. Although varying in degree and intensity, the urban refuse problem is exacerbated by limited space and dense populations. A refuse problem must be understood by those affected by it to have negative effects on human life. The problem may be seen at first as merely a nuisance or annoyance, and only later as a health hazard or part of a broader environmental crisis. It is the modern industrial society, not the ancient society, that has experienced the most intense refuse problem. With the Industrial Revolution in Europe and the United States came the manufacture of material goods on a large scale and attendant pollutants.

With the emergence of modern metropolises, people concentrated in urban areas as never before. The modern urban-industrial society, however, also developed its own brand of environmental consciousness and civic awareness.

Garbage in the Cities focuses on the refuse problem in industrial and postindustrial America. Because of the nation's rapid growth and rising affluence, the magnitude of the waste production has been staggering in the nineteenth and twentieth centuries. The first six chapters of the book concentrate on the period between 1880 and 1920, beginning with a decade in which citizens first became interested in the "garbage nuisance," and ending soon after World War I, when the priorities of war distracted attention from almost every municipal problem. The American experience with refuse pollution was formative during this era. It was linked, in part, to the European experience, but it was also the result of a unique set of circumstances that produced the affluent, wasteful society whose material progress became the envy of the world. The last two chapters concentrate on the period after 1920, and explore the ways in which the refuse problem—and how Americans addressed it—evolved over time and continues to evolve.

In order to place in perspective the nature and extent of the American refuse problem, it is helpful first to trace the impact of waste on human society from ancient times through the Industrial Revolution in Europe. The historic connection between refuse and urbanization is apparent within this context, as is the significance of local circumstance and popular and institutional attitudes toward waste.

With the shift from hunting and gathering to food producing around 10,000 BC, human beings began to forsake the nomadic life for more permanent settlements, thus laying the groundwork for the first urban sites. In time, the demands of this new lifestyle produced many challenges, including the need for improved methods of waste disposal. On-site dumping and natural decomposition would never do; casual rural habits could not be tolerated in denser urban environs.

New ways of dealing with discards progressed slowly, however. In ancient Troy, wastes were left on the floors of homes or simply thrown into the streets. In parts of Africa, similar habits prevailed to the point where street levels rose and new houses had to be constructed on higher ground. As Lewis Mumford graphically stated, "For thousands of years city dwellers put up with defective, often quite vile, sanitary arrangements, wallowing in rubbish and filth they certainly had the power to remove, for the occasional task of removal could hardly have been more loathsome than walking and breathing in the constant presence of such ordure. If one had any sufficient explanation of this indifference to dirt and odor that are repulsive to many animals, even pigs, who take pains to keep themselves and their lairs clean, one might also have a clue to the slow and fitful nature of technological improvement itself, in the five millennia that followed the birth of the city."[1] This bleak portrayal suggests a lack of resolve by ancient civilizations to promote good sanitation. While the general state of uncleanliness was appalling in many locations, there were several examples to the contrary. Ancient Mayans in the New World placed their organic waste in dumps and used broken pottery and stones as fill. In the Indus River Valley city of Mohenjo-Daro (founded about 2500 BC), a precedent-setting experiment in central planning led to the construction of homes with built-in rubbish chutes and trash bins. The city also had an effective drainage system and a scavenger service. The residents of Harappa, in the Punjab in eastern Pakistan, equipped their homes with bathrooms and drains. Excavations of ancient Babylon, Greece, and Mesopotamia revealed drains, cesspools, and sewerage systems. Carthage and Alexandria also had well-constructed sewers. In the Egyptian city Heracleopolis (founded before 3000 BC), the wastes in the nonelite quarters were ignored, but in the elite and religious quarters, efforts were made to collect and dispose of all wastes, which usually ended up in the Nile. In Crete, a most advanced civilization in terms of sanitation, the homes of the Sea Kings had bathrooms connected to trunk sewers by 2100 BC, and by 1500 BC, the island had areas set aside for the disposal of organic wastes. Records of China dating from the second century BC indicate forces of sanitary police who were charged with removing animal and human carcasses and traffic police who oversaw, among other things, street sweeping in the major cities.[2]

Religion, as well as utilitarian and social conventions, played a major role in the establishment of sanitary practices in the ancient world. Most notable were the Jewish laws of cleanliness, likely derived from Minoan, Assyrian, Babylonian, Indian, and Egyptian origins. About 1600 BC, Moses wrote a code of sanitary laws that was perpetuated and enlarged upon through the centuries. Every Jew was expected to remove his own waste and bury it far from the living quarters. Later, the Talmud ordered the streets of Jerusalem to be washed daily, a severe law in such an arid region.

Like a number of other ancient cities conscious of health and sanitation needs, Jerusalem also had a sewer system and its own water supply as early as 800 BC.[3]

The achievements of Mohenjo-Daro, Harappa, and Jerusalem—as well as other cities—did not produce a universal standard of cleanliness in the ancient world. Into the classical period, refuse problems plagued even the high culture of Athens. In the fifth century BC, garbage and other accumulated waste cluttered the city's outskirts and threatened the Athenians' health. On balance, however, the Greeks made some important contributions to sanitation. About 500 BC, Greeks organized the first municipal dumps in the Western world. (The municipal dumps bordering the city also became sites for abandoning unwanted babies.) The Council of Athens began enforcing an ordinance requiring scavengers to dispose of wastes no less than one mile from the city walls. Athens also issued the first known edict against throwing garbage into the streets.[4]

Rome, because of its size and dense population, faced sanitation problems unheard of in Greece. The city was effective in dealing with water, sewerage, and some public health matters. The Cloaca Maxima—a large underground conduit—was an outstanding example of a drain used in a civilization more than two thousand years ago. And in addition to building the famous aqueduct system, the Romans supervised public baths, houses of prostitution, and wine-drinking establishments. They also regulated food vendors. By the end of the reign of Augustus Caesar in AD 14, Rome had an effective public health administration.

Although well organized by pre-nineteenth-century standards, refuse collection and disposal were deficient for Rome's needs. The volume of waste was staggering, yet municipal collection was restricted to state-sponsored events, such as parades and gladiatorial games. By law, property owners were responsible for adjacent streets, but enforcement of the law was lax. The wealthy employed slaves to collect and dispose of waste, and independent scavengers collected garbage and excrement to be resold as fertilizer. Open dumping remained the standard disposal practice, despite all of its obvious shortcomings. In a city of approximately one and one-quarter million people, the waste problem far exceeded the means to deal with it. Well before the Fall of Rome, the city became incredibly unhealthy and dirty. Ironically, as Rome experienced a population reduction to about twenty thousand in the thirteenth century, and as the rest of the Western world similarly deurbanized, the breakdown of sanitation services had a more localized impact.[5]

The persistent clichés that cast the medieval period as the Dark Ages with recurrent plagues suggest that Europe became a vast garbage dump after the Fall of Rome. Such generalizations are overstated. The population of Europe was scattered and was spared the massive waste problem Rome experienced in classical times. Despite the crudity of medieval dwellings and living conditions, sparsely populated areas did not have to contend with the refuse pollution experienced in the great cities of the past. With the rise of medieval cities, conditions were gradually improved. According to public health historian George Rosen, "All the institutions needed for a hygienic mode of life had to be created anew by the medieval municipalities. It was within this urban environment that public health, thought, and practice revived and developed further in the medieval world."[6]

All of the basic needs—safe water, sewerage, and so forth—had to be met by a new urban society. The collection and disposal of waste was a particularly difficult problem at a time when rural habits were being reintroduced into town life. Hogs, geese, ducks, and other animals shared the urban habitat with human beings. By the thirteenth century, the larger European cities were once again coming to grips with refuse. Cities began paving and cleaning their streets at the end of the twelfth century. Paris began paving its streets in 1184, when, according to contemporary accounts, King Philip II ordered the streets to be paved because he was annoyed by the offensive odors emanating from the mud in front of his palace. Augsburg became the first city in Germany to pave its streets, though not until 1415. Street cleaning at public expense came some time later—in 1609, for example, in Paris. In the German principalities, street-cleaning work was often assigned to Jews and to the servants of the public executioner. It was hardly an ennobling profession.[7]

Waste collection and dumping in medieval cities have a varied history. In 1388, the English Parliament banned waste disposal in public watercourses or ditches. Paris had a very unusual experience with the refuse problem. In 1131, a law was passed prohibiting swine from running loose in the streets after young King Philip, son of Louis the Fat, was killed in a riding accident caused by an unattended pig. The monks of the Abbey of Saint Anthony protested the law, and were granted a dispensation because their herds of swine were a major source of income. The controversy over allowing animals to run free raged on for years, however. Until the fourteenth century, Parisians were allowed to cast garbage out their windows, and although several attempts were made at effective collection and disposal, by 1400 the mounds of waste beyond the city gates were so high that they posed an obstruction to the defense of Paris. One ingenious regulation provided that whoever brought a cart of sand, earth, or gravel into the city had to leave with a load of mud or refuse. Little by little, the people of medieval Europe were becoming aware of waste as a health hazard. Public resistance to new regulations was strong, however, and primitive collection and disposal methods were widespread. No adequate solution was in sight. The steady transition of the medieval towns into modern cities, with multistory tenements, high concentrations of people and business establishments, and growing quantities of inorganic as well as organic wastes exacerbated the problem.[8]

Until the transition of Europe from a predominantly agrarian to an urban-industrial culture, the refuse problem remained much as it had been in the Middle Ages. Although the Renaissance brought a revival of classical art to Europe and heralded a new era of rationalism, early modern Europe did not undergo a sufficient physical or demographic change to influence the development of new methods to cope with waste. Change was gradual until the onset of the Industrial Revolution. Only in the major cities could the rudiments of a sanitation system be found. Most people continued to discard garbage and rubbish helter-skelter. In Edinburgh, regarded by many as the filthiest city in all of Europe, citizens cast garbage into the streets in the evening, hoping that the scavengers would collect it the next morning. In Naples, the breakwater sheltering moored vessels was so badly clogged by 1597 that city leaders almost decided to build a new breakwater rather than clean the old one. Cities continued to pass laws and ordinances against the most unsanitary practices, but to little avail. The plagues that invaded Europe between 1349 and 1750 provided some inducement for better sanitation, but responsibility largely remained an individual matter well into the nineteenth century.[9]

The Industrial Revolution, which originated in England in the 1760s, brought down the old order in Europe, replacing it with a new one characterized by vast economic expansion and rapid urbanization. The major physical consequence of the Industrial Revolution was the tremendous environmental change in the cities. As never before, urbanites were forced to confront massive pollution in many forms. In this context, the refuse problem emerged as a major blight.[10]

Historian Eric E. Lampard suggested that the Industrial Revolution was "a particular form of social change" and that its occurrence "transcends explanation in purely economic terms." Lampard argued that the first phases of the Industrial Revolution produced a kind of "disorder" rather than an instantaneous new order; the gradual nature of the change distressed and bewildered town and country people alike.[11] During the transition from a preindustrial to an industrial society, dislocations, distress, instability, and uncertainty of change shook the people to their roots. The transition from rural to urban, from agrarian to industrial, had a similar impact on the physical environment.

The effect of the Industrial Revolution on urban society was not all negative, but its imprint on the physical city was often grim. Mumford has written that "industrialism, the main creative force of the nineteenth century, produced the most degraded urban environment the world had yet seen; for even the quarters of the ruling classes were befouled and overcrowded."[12] Asa Briggs, in more measured but also critical words, observed, "The worst aspects of nineteenth-century urban growth are reasonably well known. The great industrial cities came into existence on the new economic foundations laid in the eighteenth century with the growth in population and the expansion of industry. The pressure of rapidly increasing numbers of people, and the social consequences of the introduction of new industrial techniques and new ways of organizing work, involved a sharp break with the past. The fact that the new techniques were introduced by private enterprise and that the work was organized for other people not by them largely determined the reaction to the break." He went on to say, "The priority of industrial discipline in shaping all human relations was bound to make other aspects of life seem secondary."[13] Neglect of the physical environment was to be expected in a society in which priorities were shaped by an "industrial discipline."

The demographic shift in England profoundly affected city growth and led to serious problems of overcrowding. The English were the world's first urbanized society. Twenty percent of the population lived in cities and towns of 10,000 or more by 1801, with one-twelfth of the people residing in London. By 1851, more than half of the English were city dwellers. At the beginning of the nineteenth century, only the Netherlands was more urbanized. During the reign of Queen Victoria (1837–1901), the population of Great Britain doubled, and the 1901 census indicated that 77 percent of the country's 36 million citizens lived in urban areas.

The inability to house such a growing population led to serious overcrowding and sanitary problems. In 1843, in one section of Manchester there was one toilet for every 212 people. "It was impossible," Lampard wrote, "for the nineteenth-century market-economy to house the growing, urbanizing, population in any but the most rudimentary way. Public and philanthropic efforts could do little more than advertise the 'problem.'" Although the housing crisis eased somewhat after the turn of the century, all types of structures, including cellars and other dank places, were converted for human habitation. In Liverpool, one-sixth of the population lived in underground cellars. As late as the 1930s, London had 20,000 basement dwellings considered unfit for occupation. Many dwellings had insufficient ventilation, inadequate privies, and little or no sunlight.[14]

The crush of people and the concentration of industry in and around cities produced living and working conditions of incredible deprivation, especially for the poor and the working class. The pages of Charles Dickens overflow with graphic images of the wretchedness of life in the industrial city. Stinking water, smoky skies, ear-shattering din, and filthy streets made living conditions grim. Conditions in the factory were no better. The factory was "a new kind of prison; the clock a new kind of jailer."[15] The lessons of good sanitation and public health learned over the years were forgotten or ignored. Nuisance laws were rarely enforced, public health laws went unheeded, and in some quarters cleanliness was all but forgotten.[16]

The life of the urban poor and the working class reveals the neglect of sanitation and proper collection and disposal of waste. It suggests but one dimension of the growing waste problem in industrial-urbanized societies. As Lampard noted, industrial-urban nations are "effluent" societies.[17] The growing production and consumption of goods made the scale and magnitude of the waste problem much greater than that encountered by previous cultures. Even if sanitary standards were improved to the point of rendering the unhealthy safe and the dirty clean, rising affluence, which brought still more production, would produce an ever-larger quantity of waste. The growth to maturity of an industrial society, therefore, was no guarantee that the refuse problem would decline, even though sanitary conditions might improve. The moderate rise in the standard of living and the improvement in living conditions in England by the time of the Great Exhibition in London in 1851 did not signal an end to the waste problem. Changes for the better simply meant that the most immediate unpleasant effects of the Industrial Revolution were subsiding.

In the mid-to-late nineteenth century, England could boast about reversing some of the most debilitating physical defects of the industrial city, especially poor sanitary and health conditions. The harshness of the industrial city could not be neglected forever. When the subtle became painfully obvious, when the affluent were touched by some of the same misfortunes as those of the suffering poor, something was done. Several forces converged to halt the downward spiral of the environment. One of the most important was the "service revolution." City services had been established over time to meet the most pressing needs: fire and police protection, water supplies, and even waste collection—largely by scavenging. Their growing size and the extent of their problems required industrial cities to supply many citywide services that had previously been provided selectively by volunteers or paid agents. Several scholars have argued that, along with the rise of laissez-faire capitalism, the nineteenth century also experienced a kind of "municipal socialism," that is, a demand for services provided by the city rather than the individual. Although some scholars have exaggerated the range and quantity of services provided by this municipal socialism, the needs of the large, heterogeneous industrial city did force a rethinking of ways in which those needs could be met. One of the results of the new emphasis on citywide services was the development of rudimentary public works and public health agencies or departments.[18]

Another, and perhaps the most essential, factor in bringing about the first effort to improve sanitation in the industrial city was the emergence of modern public health science. Surveys undertaken by the Poor Law Commission, first in London and then throughout England, evaluated the health of the working population. In 1842, the commission published *Report on the Sanitary Condition of the Labouring Population in Great Britain* under the primary authorship of barrister-turned-sanitarian Edwin Chadwick. The document was well researched, well argued, and widely disseminated, and painted a vivid picture of urban blight and the lack of sanitary conditions throughout the country. The most significant feature of the report was the conclusion that disease, especially communicable disease, was related in some way to filthy environmental conditions (the exact connection would not become clear until the inception of the germ theory of disease after 1880). The establishment of the Sanitary Commission in 1869 and subsequent enactment of public health laws provided the foundation for environmental sanitation that led to a reduction in urban disease. With the advent of the "sanitary idea" and ultimately with modern science and information-gathering procedures brought to bear on public health, conditions in industrial England began to improve. Similar programs in other parts of Europe and in the United States signaled a new "age of sanitation." The emergence of bacteriological science and the rise of the germ theory of disease led to the discrediting of environmental sanitation as the sole means of curbing communicable diseases. Nonetheless, these first steps offered immediate, and in some cases dramatic, relief from some of the ravages of the urban environment. The industrial city had not been brought under control, but at least the most obvious environmental hazards were being confronted.[19]

While Europe was in the midst of its Industrial Revolution, the United States was emerging as a new nation. Many of the difficult lessons learned in the industrial cities of Europe had little applicability in the colonial society of North America. Some aspects of the European experience with sanitation problems were transmitted to the New World, but not in ways that would help Americans avoid those problems. The evolution of American society established a different context for dealing with health and sanitation.

Preindustrial America was a highly decentralized society, but from the beginning it had some form of urban life. Indeed, cities and towns played central roles in establishing American traditions, in fostering a strong economy, and in providing staging areas for territorial expansion. The importance of American cities and towns was disproportionate to their size. From the early seventeenth century until the eve of America's own industrial revolution in the mid-nineteenth century, the total urban population remained small, as did the physical size of the cities. The first federal census of 1790 showed that city dwellers represented only 5.1 percent of the population, and only two cities exceeded 25,000. By 1840 the urban population had increased to 10.8 percent, and only New York exceeded 250,000. Between 1790 and 1840, however, the number of cities increased from 24 to 131.[20]

What distinguished the American experience with sanitation problems from the European experience during a comparable period of growth were factors of space and magnitude. In the American colonies, the abundance of land and natural resources such as water supplies mitigated massive sanitation problems even in cities and towns. Since no American city reached the size of London during that period, the need to deal with health and sanitation problems on a grand scale did not materialize. That is not to say that American cities were free of refuse and poor sanitation—only that any parallels between the two societies must be drawn with an understanding of local conditions.[21]

American cities periodically experienced appalling sanitary and health problems. Carl Bridenbaugh wrote that in colonial times the casting of rubbish and garbage into the streets was "a confirmed habit of both English and American town-dwellers." In the condition of streets, however, "colonial villages vied with, but never equaled, the filthiness prevalent in contemporary English towns," though the swine roaming the streets scavenging for food and causing obstructions to people and horses were reminiscent of scenes in most European villages.[22] In eastern cities, where crowding became a chronic problem as early as the 1770s, the streets reeked with waste, wells were polluted, and deaths from epidemic disease mounted rapidly. Even in the burgeoning cities in the West and South, problems were sometimes legion. As late as the 1860s, Washingtonians dumped garbage and slop into alleys and streets, pigs roamed freely, slaughterhouses spewed nauseating fumes, and rats and cockroaches infested most dwellings—including the White House. No wonder the infant mortality rate was very high in the capital city.[23]

Because of the time differential, preindustrial American cities benefited earlier and more quickly from sanitary sciences than did their counterparts in Europe. The connection between filth and disease was dogma. In colonial cities, removal of waste and street cleaning were considered effective ways of preserving public health as well as eliminating nuisances. By the mid-nineteenth century, several cities had established boards of health and had passed ordinances against indiscriminate dumping of refuse and the free roaming of animals.[24]

These measures alone were not enough to curb the problem of waste or to maintain consistently high standards of sanitation. Environmental sanitation alone could not protect cities from epidemics; until the development of the science of bacteriology, cities were constrained to deal with them. Ordinances were not reinforced with adequate inspections, surveillance, or policing to ensure the compliance of citizens. City leaders' concern for cleanliness was not always matched by their constituents' concern.[25]

The quality of sanitation in preindustrial America was determined largely by local circumstances. Some city leaders had the foresight to place a high priority on city cleanliness, while others ignored the problem. Epidemics ravaged several cities, while others were spared because of their relative isolation or because of attention to comprehensive sanitary measures. While rudimentary public works systems emerged in several of the larger or more progressive communities, individuals or private scavengers handled the waste problems in many towns and villages. Little progress was made in establishing clear lines of responsibility for collection and disposal of refuse, except in New York and Boston, and even in New York advances were slow in coming. The burghers of New Amsterdam had been among the first to pass laws against casting waste into the streets (1657), but the condition of the streets remained the responsibility of the householders. In the late eighteenth and early nineteenth centuries New York City established municipal control over several sanitary services, but jurisdictional disputes between state and local governments and between city and individuals continued. The time for comprehensive community-wide programs and general environmental reform had not yet arrived. Americans must have found it difficult to comprehend the massive pollution problems confronting industrial London, if they heard about them at all. They must have found it even more difficult to anticipate that similar problems would threaten them in the not-too-distant future.[26]

The impact of the Industrial Revolution on American cities was no less severe than its impact on European cities had been. Like Europe, the United States experienced an array of environmental problems—smoke, noise, and tainted water—if not for the first time, at the very least in greater intensity than before. Also like Europe, the United States was confronted with a waste problem in its industrial era that had two distinct dimensions. One was linked to the physical distress caused by overcrowding, poor sanitation, and primitive methods of collection and disposal, especially in the late nineteenth and early twentieth centuries; the other was tied to the rising affluence of the middle class, abundance of resources, and consumerism, which persisted well into America's postindustrial era in the late twentieth century. Of course, it is best to keep in mind that forces other than industrialization also influenced various forms of pollution and the waste problem, including agricultural cultivation, transmittal of epidemic diseases, a variety of technical choices in the nonindustrial sectors of society, and urban processes not attributable to industrial activity.

Industrialism in the United States, however, went hand in hand with the transformation of the country into an urban nation. As early as 1820, there was a significant link between urban development and industrial growth. From 1840 to 1920, the urban population grew from 1,845,000 to over 54 million. This represented at least 29 percent growth in each decade and as much as 92.1 percent growth between 1840 and 1850. The number of urban areas also grew at a fast pace—from 131 in 1840 to 2,722 in 1920, extending across the country.[27]

During this period of remarkable growth, especially between 1870 and

1920, the industrial city was a dominant urban form. Densely populated, physically expansive, and economically vital, the industrial city was characterized by outward expansion, an ever-rising skyline, and specialized land use in the form of well-defined business and residential districts. Relatively new cities with strong industrial economies such as Cleveland, Pittsburgh, and Milwaukee began experiencing rapid population growth and economic prosperity. Industrialization also transformed, and attracted factories to, many older commercial or preindustrial cities such as Boston, Baltimore, Philadelphia, and New York.[28]

Industrial cities paid a high price for their rapid population growth and economic dynamism. They experienced crowded tenement districts, chronic health problems, billowing smoke, polluted waterways, traffic congestion, unbearable noise, and mounds of putrefying garbage. Americans were unprepared to deal with the extent of these pollutants and the rapid transformation of the United States into an urbanized nation.

By 1885, the need for a plentiful and inexpensive source of energy to run factories and to heat homes led to the extensive use of coal. Bituminous (soft) coal was the most widely used; only a small portion of it was consumed in the generating of power and heat, and most of the residue went directly into the air, encrusting buildings, clothing, and the lungs of city dwellers.[29]

The concentration of factories in and around cities added to environmental problems. Iron and steel mills, textile mills, and chemical plants were often constructed near waterways, mainly because of the large quantities of water needed for steam conversion and for chemical solution manufacturing. Waterways also proved useful for disposing of wastes. A 1900 study suggested that 40 percent of the pollution load on American rivers was industrial in origin. Manufacturers contributed substantially to land pollution as well, dumping heaps of rubbish, garbage, slag, ashes, and scrap metal on available vacant land. Slaughterhouses and other animal-processing industries dumped animal wastes in open pits or on vacant lots; tanning companies polluted waterways by washing hides in them. The noise produced by large factories could deafen workers and disrupt surrounding residential neighborhoods.[30]

The numbers of factories and the dense concentration of industries around cities turned many nuisances into full-fledged environmental disasters. By 1899, 40 percent of the 500,000 industrial establishments in the country were factories. At least three-fourths of American manufacturing was concentrated in New England, the mid-Atlantic states, and the north-central states. By 1900, thirty-four of the forty-four states were manufacturing more than 50 percent of their goods in urban areas; in eighteen states more than 75 percent of the products came from urban factories. Industrial specialization added to this high concentration. For instance, the highly polluting iron and steel industry was concentrated in the greater Pittsburgh area, and, not surprisingly, the amount of smoke was stifling. Chicago, Saint Louis, and Kansas City led the nation in slaughtering and meatpacking, and the citizens suffered a great deal from the festering wastes and noxious odors.[31]

Sizeable human concentrations in the industrial cities exacerbated the environmental problems. The rapid population growth of the United States and its cities during this period is well known. Between 1850 and 1920, while the world population increased by 55 percent, the population of the United States soared by 357 percent. The most phenomenal growth occurred in the cities, primarily because of immigration and rural-to-urban migration. During this period, nearly 32 million people entered the United States, most of them from southern and eastern Europe. By 1910, 41 percent of American city dwellers were foreign-born. About 80 percent of the new immigrants settled in the Northeast. Migration from rural areas of the country was also impressive. Although statistics are scant, a conservative estimate is that 15 million rural people moved to the cities between 1880 and 1920. During those years, the rural population fell from 71.4 percent to 48.6 percent.[32]

Statistics can measure the magnitude of these shifts in population, but they cannot measure the human dimension of the attendant environmental problems. Many city dwellers in the industrial cities lived and worked in oppressive social and physical surroundings. It is almost impossible to comprehend the overcrowding in some cities. From 1820 to 1850 the average block density in lower Manhattan increased from 157.5 to 272.5 persons. In 1894, New York City's Sanitary District A averaged 986.4 people an acre in thirty-two acres, which translated to 300,000 people in a space of five or six blocks. Bombay, India, the second-most crowded area in the world, had 759.7 people an acre; Prague, the European city with the worst slums, had 485.4 people an acre.[33]

Jane Addams, in *Twenty Years at Hull-House,* recalled the seeming disregard for the crowded and inferior living conditions of those years, "The mere consistent enforcement of existing laws and efforts to their advance often placed Hull-House, at least temporarily, into strained relations with its neighbors. I recall a continuous warfare against local landlords who would move wrecks of old houses as a nucleus for new ones in order to evade the provisions of the building code, and a certain Italian neighbor who was filled with bitterness because his new rear tenement was discovered to be illegal. It seemed impossible to make him understand that the health of the tenants was in any wise as important as his undisturbed rents."[34]

Jacob Riis, in *How the Other Half Lives,* wrote, "Thousands were living in cellars. There were three hundred underground lodging-houses in the city when the Health Department was organized. Some fifteen years before that [about 1852] the old Baptist Church in Mulberry Street, just off Chatham Street, had been sold, and the rear half of the frame structure had been converted into tenements that with their swarming population became the scandal even of that reckless age."[35]

Appalling stories of overcrowding, like David Brody's example about the thirty-three Serbian workers and their boss who lived in a five-room house, or the common practice of keeping farm animals in basements and even slaughtering them there, were all too familiar.[36]

In such surroundings, health problems, disease, and high mortality rates were to be expected. Typhoid spread throughout New Orleans from sewage standing in unpaved streets. In 1873, Memphis lost nearly 10 percent of its population to yellow fever. Mortality figures for "Murder Bay," a black district in Washington, D.C., not far from the White House, were twice as high as those for white neighborhoods. The residents of that slum lived in ghastly surroundings, picking their dinners out of garbage cans and dumps. By 1870, conditions in New York City had deteriorated so badly that infant-mortality rates were 65 percent higher than those of 1810. Correlations between living conditions and disease in tenements led to some understanding of the debilitating effects of a bad environment on health, but improvements would not come quickly.[37]

Even the elite were not completely insulated from the environment of the industrial city. The crush of human beings, the concentration of factories, and the expansion of the city affected everyone. Even the wealthy banker had to endure the trip from his country estate to his downtown office. The new urban environment challenged every Jeffersonian notion of individuality and self-reliance. Yet the need to confront the most immediate environmental problems was at hand, and urbanites needed to meet this challenge.

The first efforts to resolve the environmental crisis were directed piecemeal at the most obvious concerns. Thus the needs for sources of pure water and adequate sewerage received top priority because they affected citizens collectively and were vital for good health. To the credit of many cities, effective programs to tap pure water sources and construct modern sewer systems were underway by the 1870s. Efforts to control smoke pollution and excessive noise lagged far behind and did not gain momentum until the mid-1890s. Such problems were more elusive, more difficult to gauge and measure, and even more difficult to monitor and control. Smoke came to symbolize material progress and the economic activity vital to the growth of industrial cities and the nation. Noise also seemed to indicate a society on the move. Only when the skies remained black with soot and the din made it difficult to think, eat, or sleep did the reformers gain public and official support.[38]

The refuse problem gained public recognition as an environmental issue soon after the efforts to assure clean water and adequate sewerage in the early 1880s, and just before the first attempts to abate smoke and excessive noise in the mid-1890s. At first it was considered a mere nuisance, but by the 1890s the garbage problem was recognized as a major pollutant of the industrial era. Between 1880 and 1920, American cities began coping effectively with the immediate threats caused by refuse but failed to confront the more fundamental problems associated with the production of wastes. Not until the 1960s and 1970s did Americans link the resolution of the refuse problem to American affluence and the consumption of goods. Yet, by the onset of the twenty-first century, Americans were still trying to learn how to *manage* the solid waste problem, not *solve* it.

ONE

OUT OF SIGHT, OUT OF MIND

The Refuse Problem in the Late Nineteenth Century

In the late nineteenth century, urban America discovered the garbage problem. In an 1891 issue of *Harper's Weekly,* an observer noted, "As the world grows older it becomes not only conscious of new problems which it has to solve, but it becomes more keenly conscious of the importance of old ones which it has only imperfectly met."[1] The refuse problem attained such massive proportions in the industrial United States that even the most insensitive city dweller could no longer ignore it. Heaps of garbage, rubbish, and manure cluttered alleys and streets, putrefied in open dumps, and tainted the watercourses into which refuse was thrown. Nineteenth-century Americans began to realize that mounds of garbage, rubbish, and other discards were not simply eyesores but unnecessary encumbrances and potential health hazards.

The budding environmental consciousness that made urbanites sensitive to impure water supplies, poor drainage and sewerage, and smoky skies also influenced their thinking about solid wastes. As the garbage nuisance came to be seen as a serious environmental problem, the impulse for reform acquired broad dimensions. The traditional acceptance of individual responsibility for refuse collection and disposal made way for an acknowledgment of community responsibility. Some measure of civic protest against the most serious sanitation problems and civic involvement in promoting municipal cleanliness were seen as necessary to persuade officials to act in the best interests of the community. Engineers reevaluated prevailing collection and disposal methods and frequently offered technical and administrative alternatives meant to provide solutions to the problem. Little by little, nineteenth-century Americans confronted the complexities of a nagging environmental difficulty and opened the door for more sophisticated reform measures in the twentieth century.

The extent of the refuse problem in the years after the Civil War was an indicator of the rapid urban growth of the nineteenth century and the dense concentration of people at the core of the cities. At a meeting of the American Public Health Association (APHA) in 1879, the Reverend Hugh Miller Thompson described a dumping ground in New Orleans sited on the edge of the swamp, "Thither were brought the dead dogs and cats, the kitchen garbage and the like, and duly dumped. This festering, rotten mess was picked over by rag-pickers and wallowed over by pigs, pigs and humans contesting for a living in it, and as the heaps increased, the odors increased also, and the mass lay corrupting under a tropical sun, dispersing the pestilential fumes where the winds carried them."[2]

In 1891, the head of the Boston Street Department bemoaned how the city's streets, once "models of cleanliness," were rapidly deteriorating. "The reason why the streets had grown more filthy from year to year," he said, "was easily discovered. The system of cleaning in vogue, while it answered for twenty years ago, had been entirely outgrown. Notwithstanding the enormous growth of the city, the system has never been changed to keep pace with this growth."[3]

Contemporary statistics help the modern reader to visualize the staggering quantities of refuse that brought the problem to public attention (it should be noted, however, that the information on refuse collection from the nineteenth century was based on estimates and indicated only the amount of refuse actually collected, not the great volumes that remained in the streets and alleys). For example, Boston authorities estimated that in 1890 scavenging teams collected approximately 350,000 loads of garbage, ashes, street sweepings, and other debris. In Chicago during the same period, about 225 street teams gathered approximately 2,000 cubic yards of refuse daily. Seasonal variations in the amounts and kinds of wastes complicated the picture. In Manhattan at the turn of the century, scavengers averaged 612 tons of garbage daily; during July and August the volume increased to 1,100 tons daily. Transportation of wastes to dumping sites was easier in the summer months, and the warm weather made mandatory the frequent collection of rapidly putrefying offal. Also in the summer, larger quantities of produce and dairy products were available to urbanites. For example, residents of New York City consumed approximately 750,000 watermelons during the warmer months of each year in the 1890s. Something had to be done with the rinds.[4]

Surveys conducted during the first two decades of the twentieth century offer some of the earliest reliable statistical evidence of the mounting refuse problem in the United States. Table 1 shows the amount of garbage (organic waste) collected in eighteen major cities in 1916. In most of the major cities, the amount of refuse collected yearly ranged from one-half to three-quarters ton per capita.[5] Between 1900 and 1920, each citizen of Manhattan, Brooklyn, and the Bronx annually produced about 160 pounds of garbage, 1,231 pounds of ashes, and 97 pounds of rubbish. One expert estimated that each American contributed approximately 100 to 180 pounds of garbage, 300 to 1,200 pounds of ashes, and 50 to 100 pounds

TABLE 1

Tons of Garbage Collected in Eighteen Major Cities, 1916

City (Population)	*Garbage Collected (Tons)*	*Garbage Collected per Capita (Tons)*
Baltimore, Md. (593,000)	37,915	0.064
Boston, Mass. (781,628)	52,650	0.067
Bridgeport, Conn. (172,113)	19,897	0.116
Cincinnati, Ohio (416,300)	40,692	0.098
Cleveland, Ohio (674,073)	59,708	0.089
Columbus, Ohio (220,000)	20,393	0.093
Dayton, Ohio (155,000)	16,621	0.107
Detroit, Mich. (750,000)	72,785	0.097
Grand Rapids, Mich. (140,000)	8,678	0.062
Indianapolis, Ind. (271,758)	23,267	0.086
Los Angeles, Calif. (600,000)	51,062	0.085
New Bedford, Mass. (118,158)	10,162	0.086
New York, N.Y. (5,377,456)	487,451	0.091
Philadelphia, Pa. (1,709,518)	101,678	0.059
Pittsburgh, Pa. (579,090)	73,758	0.127
Rochester, N.Y. (275,000)	30,782	0.112
Toledo, Ohio (220,000)	23,971	0.109
Washington, D.C. (400,000)	46,293	0.116

Source: Rudolph Hering and Samuel A. Greeley, *Collection and Disposal of Municipal Refuse* (New York, 1921), pp. 13, 28.

of rubbish yearly. By comparison, European city dwellers produced substantially less refuse than their American counterparts. A 1905 study indicated that fourteen American cities averaged 860 pounds of mixed rubbish per capita per year, while in eight English cities the amount was 450 pounds per capita, and in seventy-seven German cities, 319 pounds.[6] Franz Schneider Jr., research associate of the Sanitary Laboratory of the Massachusetts Institute of Technology, imaginatively calculated that if the entire year's refuse of New York City was gathered in one place, "the resulting mass would equal in volume a cube about one eighth of a mile on an edge. This surprising volume is over three times that of the great pyramid of Ghizeh, and would accommodate one hundred and forty Washington monuments with ease."[7]

Horses, the major means of individual and commercial transportation and an essential source of energy for factories in the late nineteenth century, rivaled human beings in creating waste problems. At the turn of the century, there were (in a conservative estimate) 3 to 3½ million horses in American cities. Historians Clay McShane and Joel Tarr estimated that the horse population was urbanizing at a rate that was 50 percent greater than the human population. In New York City alone, there were about 130,000 horses; in Chicago, something over 74,000. Engineers estimated that the normal, healthy city horse produced between 15 and 35 pounds of manure and about a quart of urine daily, most of which ended up in the streets. Cumulative totals of manure produced by horses were staggering: 26,000 horses in Brooklyn produced about 200 tons daily; 12,500 horses in Milwaukee produced 133 tons. Even when provisions were made for carting off the manure and depositing it beyond the city limits, a considerable amount spilled from the wagons and remained in the streets until street cleaners made their rounds. On unpaved and paved streets the manure was ground up, producing fine powder that blew into the faces of passers-by. The manure in Brooklyn, one observer noted, "was removed promptly by farmers in the early spring; but as summer advanced and it was not needed for the crops, it was allowed to accumulate."[8]

The proliferation of horsecars for mass transit in the 1850s exacerbated the problem. By 1890, approximately 84,000 horses and mules were pulling animal-powered streetcars. Of the total trackage in that year, 4,062 of 5,783 miles were serviced by horsecars. The horses' discharges not only cluttered the streets and corroded the metal streetcar tracks but also threatened the health of city dwellers. Stables were notorious breeding places for disease. Unbearable as well was the stench coming from the stables and manure pits that many horsecar companies maintained within the city limits as sources of additional income. Since the life expectancy of a city horse was only about four years (due in large measure to overwork and abuse), carcasses were plentiful and had to be eliminated. New York City scavengers removed 15,000 dead horses from the streets in 1880. As late as 1916, when motor vehicles dominated the streets, Chicago scavengers still had to remove 9,202 carcasses.[9]

The arrival of the electric streetcar and the automobile appeared to be the panacea for the horse problem. The Automobile Chamber of Commerce, whose interests were clearly linked to promoting motor vehicles, declared that the replacement of Chicago's approximately 80,000 horses with motor trucks and automobiles would not only improve the city's sanitation problem but also save the city a million dollars a year in street cleaning. Whether these figures could be trusted is a matter for speculation, but street cleaners enthusiastically heralded the arrival of the automobile as the means of their salvation. Little did they realize what the transition from animal power to mechanical power signaled for the city's physical environment. Trading manure and horse carcasses for hydrocarbons, noxious fumes, and waste heat from internal-combustion engines was no bargain. Horses, however, did not immediately disappear from the urban scene. Although the process of substitution of animal power with machine power proceeded more quickly in the United States than in Europe, horses continued to be important in functions involving freight and deliveries in some locations until the mid-twentieth century.[10]

With so much organic waste accumulating in the cities, giving off foul odors, and attracting flies and rats, urbanites could not avoid recognizing a connection between refuse and health hazards. Contemporary sanitary science seemed to confirm the connection. In the mid-nineteenth century, experiments in England and the United States demonstrated that there was some relationship between communicable diseases and putrefying wastes. The efforts of Edwin Chadwick in the 1840s led the way for new sanitation laws in England and inspired American health officials to consider refuse collection and disposal from the perspective of improving health conditions. The prevailing wisdom at mid-century was that disease was caused by environmental factors. The "miasmatic," or filth, theory of disease dominated American thinking on sanitation into the 1890s. According to the theory, gases emanating from putrefying matter or sewers were the cause of contagious diseases, and city cleanliness, proper drainage and sewerage, and adequate ventilation of buildings would improve the health of city dwellers. "Environmental sanitation," which included the proper removal of solid wastes, seemed to offer an immediate and effective solution to many city health problems. According to medical historian Howard D. Kramer, "A sanitary program based on these beliefs won considerable popular support and scored several major victories." Among those victories were the New York Metropolitan Health Law (1860); the creation of state boards of health, beginning with the Massachusetts board in 1869; the founding of the American Public Health Association in 1872; and the establishment of the short-lived National Board of Health in 1879.[11]

Environmental sanitation as the means of ridding the cities of death and pestilence had some serious drawbacks, however. In the 1880s, the discovery of specific pathogenic organisms—bacteria—enabled public-health officers and sanitarians to understand the actual causes of many contagious diseases. By the turn of the century, the "germ theory" of disease replaced older notions about the relationship between filth and disease and led to the implementation of bacteriological laboratories as the chief means of controlling epidemics. The views of "anticontagionists," advocates of the miasmatic theory, ultimately fell into disrepute, and environmental sanitation came to be considered far less important as a means of controlling disease.[12]

Municipal cleanliness through environmental sanitation, though it did not fully explain the nature of disease transmittal, was a worthwhile goal with a beneficial purpose, especially in light of the deplorable state of public health in nineteenth-century American cities.[13] It inspired a rudimentary environmental awareness, akin to modern ecological concepts, which gave impetus to refuse-management reform and provided a rallying point for attracting more advocates. Contemporary accounts confirm the growing perception of refuse as a health problem of serious consequence by the 1880s. One observer considered clean streets an important "sanitary objective."[14] Referring to street conditions in Boston, an engineer noted that, street cleaning "having been found to have so large an influence on the health and mortality of a community, a mere occasional attempt to clear up what street litter we cannot climb over is not sufficient."[15] In 1892, in the wake of a cholera epidemic in New York City, *Engineering News* asserted:

> When we consider that the sanitary wellbeing of two millions of people may depend upon the manner in which this refuse is disposed of, it is seen that some intelligent solution of this complex problem

> must be reached, and our city authorities can not afford to allow matters to relapse into old ruts as soon as the present cholera agitation is over. It is no case for cheese-paring or economy, and when it is once realized that a large sum of money must be spent for this public life insurance it is certain that men can be found to point out the means.[16]

If the magnitude of the urban refuse problem did not convince city officials that effective sanitation measures were needed, the rising concerns over the health question did. As one writer noted about conditions in New York City in 1894, "It is quite in order that New York should be grappling with the garbage problem at this time, for almost every other large city in the civilized world is in a similar predicament. . . the garbage problem is the one question of sanitation that is uppermost in the minds of local authorities."[17] City officials also realized that the complexities of urban life in the late nineteenth century made collection and disposal of refuse by private citizens impractical. In 1893, a special sanitary committee in Boston asserted, "The means resorted to by a large number of citizens to get rid of their garbage and avoid paying for its collection would be very amusing were it not such a menace to public health. Some burn it, while others wrap it up in paper and carry it on their way to work and drop it when unobserved, or throw it into vacant lots or into the river . . . the destruction of garbage by individual householders in any large city is too dangerous an experiment to be seriously considered by any intelligent community."[18]

But cities in post–Civil War America were only beginning to take charge of their own affairs and demand "home rule" from rural-dominated state legislatures. That made the offering of necessary services slow in coming.[19]

Of course, even before the Industrial Revolution, several of the largest communities had provided rudimentary sanitation services or at least had contracted out some of the work; however, the question of who was ultimately responsible for collection and disposal of refuse was not yet decided by the 1880s and 1890s. Because of the impracticality of private action, the major cities especially were compelled to choose between available alternatives. They could either contract the service by taking bids from private scavenging companies or establish a municipal service. The contract system was initially more popular because it required little or no capital outlay by the city, while still allowing for a modicum of municipal supervision. Advocates of the contract system, moreover, feared that municipally operated services bred political corruption. An editorial in *Engineering News* in 1890 suggested that for New York City, "[a] little less Tammany v. County Democracy and a little more organized muscle would work wonders in cleaning our streets."[20] The contract system was also touted as an incentive for free enterprise in the cities. New York City's Department of Street Cleaning reintroduced the contract system in 1890 after a long period of municipal control. Soon afterward, the following appeared in *Engineering News:*

> Properly administered and stringently enforced, a contract means the immediate discharge of all labor that does not turn money into the contractor's pocket by the vigorous use of muscle; the man must work or the master cannot meet his engagements and liabilities, and as long as the contractor is made to feel that his contract price will be his sole return for service, and that he will be strictly held to the specifications of that contract, the public can depend upon him to manage the labor and hold the Commissioner responsible for the class of work performed. Under this condition of affairs responsibility can be fixed, and as every citizen would be an inspector, the citizen knows who to howl at if things go wrong.[21]

Faith in the contract system was not shared by all city officials and concerned citizens. In the 1890s, especially, the contract system came under severe criticism as reformers looked to municipal ownership of utilities and municipal operation of services to cure many of the city's physical and social ills. The attack on the contract system was particularly strong in Chicago. As the head of the Chicago Board of Health stated in his 1892 report:

> One sentence will almost express it—there are few if any redeeming qualities attached [to the contract system]. No matter what guards are placed around it, the system remains vicious. If the contractor intends to approach his duty, the men he employs are not to be depended upon; he can not follow each one; the result is bad service. You ask the remedy. I reply, let the city purchase its own plant and do the work; at the end of the year you have at least the plant to show for the investment—at present nothing but the remembrance of how this man or that man has neglected your alley.

The Civic Federation of Chicago charged that the contractors collecting the garbage were swindling the city and completing only half the work promised.[22] While in Chicago the debate over the contract system focused on the questions of efficiency and economy of service, in other cities the health issue dominated the controversy. In its 1895 report, the Board of Health of Newton, Massachusetts, contended that, since it was the responsibility of the city to care for its citizens' health, sanitary services should not be left to the mercy of profit-motivated contractors.[23]

Municipal conditions varied so greatly that a consensus in the nineteenth century seemed impossible. Future debates over the best approach would focus on the relative cost of each system, the effectiveness of the methods utilized, and the degree of municipal control. For the present, however, previous practice dictated the method employed. The 1880 United States census, the first census in which comprehensive urban statistics were compiled, reveals no national trend favoring one system over the other. In only 48 of 199 cities surveyed was there a municipal system for collection and disposal of garbage and ashes. Only 38 cities employed the contract system, while 59 cities left the responsibility to private parties.[24] As figure 1 indicates, however, larger cities (those with populations over 30,000) were more likely to have a formal, citywide system of collection than were smaller cities (those with populations under 30,000). Furthermore, smaller cities were more likely to favor private collection than were larger cities. These findings tend to reinforce the contention that urban growth and expansion had a major impact on the extension of services.[25]

There was a much more significant trend toward municipal responsibility for street cleaning than there was for refuse collection and disposal. Data from the 1880 census indicate that 140 of the 199 cities surveyed (or 70 percent) had municipal street-cleaning services.[26] That is largely attributable to the accessibility of streets to all citizens. These "arteries" (a popular contemporary term for streets as parts of the "urban organism"), which allowed human beings, animals, and goods to move from one place to another, had to be free of obstacles. The question of ultimate responsibility for street cleaning was more easily determined because the streets had no clear territorial limits and thus transcended individual responsibility. In addition, during the late nineteenth century, streets were undergoing a major change in use. As historian Clay McShane noted, streets traditionally served "vital neighborhood and family social uses. Pushcart vendors brought their wares to urban housewives, whose mobility was limited by the slow, expansive transportation system of the era. Surviving lithographs

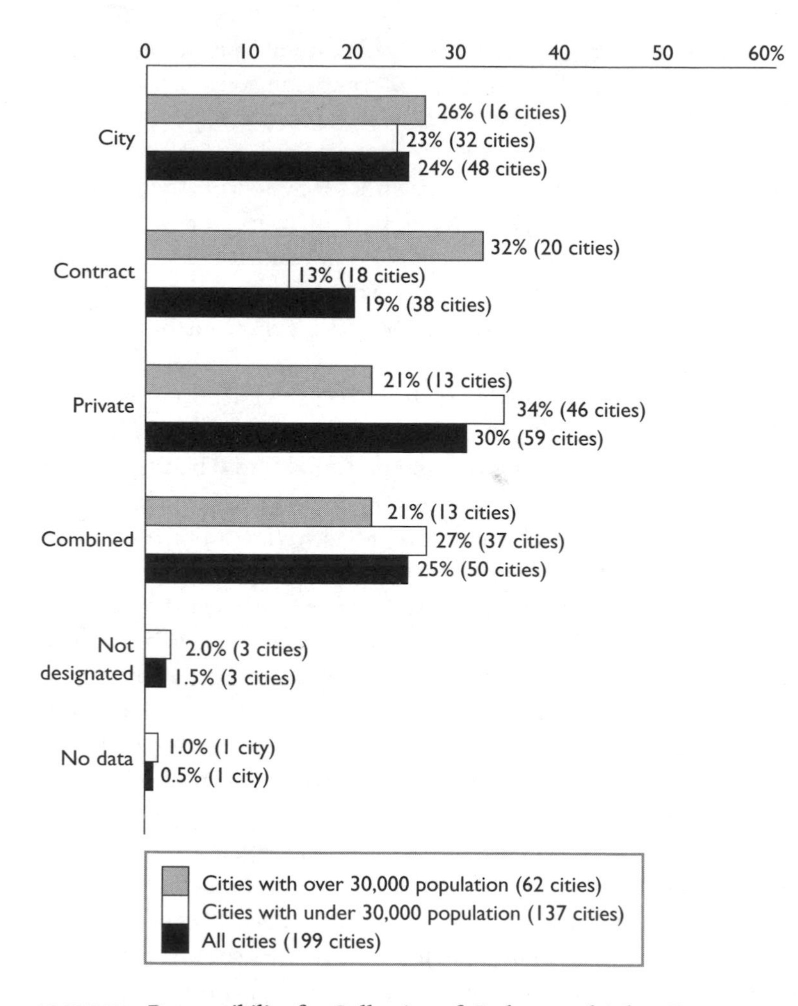

FIGURE 1. **Responsibility for Collection of Garbage and Ashes, Percentage and Number of Cities, 1880.**

Source: U.S. Department of the Interior, Census Office, *Report on the Social Statistics of Cities, Tenth Census, 1880,* compiled by George E. Waring Jr. (Washington, D.C., 1886).

and photos show great herds of children playing in the streets, generally the only available open spaces." By 1900, however, streets were designed to cope with the heavy volume of traffic or provide a means for removing the vast quantities of horse manure.[27] Efforts to pave more streets with materials which could withstand the burdens of heavy transportation use may have permitted city bureaucrats to lose sight of the important original purpose of city streets but brought attention to the need for an effective city street-cleaning program.

In the growing urban bureaucracies after mid-century, environmental sanitation initially became the province of municipal health authorities.[28] The 1880 census reveals that at least 94 percent of the cities surveyed had a board of health, a health commission, or a health officer. Of these authorities, 46 percent had some direct control over the collection and disposal of refuse.[29] As table 2 shows, larger cities were more likely to give their health boards and commissions direct control over these services.[30]

There is no evidence, however, that late-nineteenth-century health boards were capable of effective reform of collection and disposal practices. For instance, few sanitary authorities in 1880 operated without overt political interference. Physicians and sanitarians did not dominate many boards of health. In fact, some boards had no physicians as members. Without the communities' best-qualified public health experts in positions of immediate authority over health matters, considerations other than health did influence the boards' actions. Moreover, with the limited funds that most cities provided to their boards—most lacked operating budgets or salaried employees and received funds only in times of epidemics or other emergencies—few could establish long-range programs for the protection of health or for environmental sanitation. The boards of health were susceptible to the whims of the city councils or state legislatures that dominated them. Most often, their effectiveness was limited to cataclysmic events and epidemics, with meager resources for anything else.

Public health officers and sanitarians, nevertheless, dominated the thinking about collection and disposal practices in the United States during the 1880s and 1890s. In 1887, citing the unsatisfactory state of collection and disposal methods throughout the nation, the APHA appointed

TABLE 2

Responsibility for Refuse Collection and Disposal, 1880

	Percentage		
Cities with Sanitary Authorities, by Population (Number of Cities)	*Full or Some Authority*	*Little or No Authority*	*No Data*
Under 30,000 (127)	38	61	1
30,000–99,999 (40)	55	45	
Over 100,000 (20)	75	25	

Source: U.S. Department of the Interior, Census Office, *Report on the Social Statistics of Cities, Tenth Census, 1880,* compiled by George E. Waring Jr. (Washington, D.C.), 1886.

a Committee on Garbage Disposal, headed by the eminent sanitarian, Dr. S. S. Kilvington. The major task of the committee was to determine the extent of the refuse problem in the United States. The committee spent ten years gathering statistics from every major city in the country, examining European methods, and analyzing American practices.[31] Efforts such as these demonstrated the degree to which the health question provided refuse reform with a sense of direction and purpose.

Warnings by sanitarians and public health officers about the potential health hazards associated with accumulating waste soon began raising public consciousness about the need for better sanitation. Publications as diverse as *Harper's Weekly, Munsey's Magazine, Scientific American,* and *Engineering News* featured articles deploring the "garbage problem" or "filthy streets." A writer for *Harper's Weekly* asked: "What shall be done with the garbage? This is one of the great problems in the administration of modern cities."[32] A commentator in *Scientific American* stated: "The disposal of the refuse in cities, while it has been a problem in the sanitation of our larger towns, is yet to be solved. There is probably not a city of any size in the United States where the disposal of wastes is satisfactory or conducted in such a manner as to meet the demands of cleanliness and hygiene."[33]

Citizens' neglect of sanitation matters was a popular theme in the newspapers and magazines. Another writer for *Harper's Weekly* asserted: "The average citizen, accustomed to endure nuisances as a humpback carries his deformity, saunters along sublimely indifferent to foul smells, obstructed sidewalks, etc."[34] John S. Billings, however, founder of the Army Medical Library and *Index Medicus,* believed that public apathy was making way for civic awareness. "Quite recently," he stated in 1893, "there seems to be a growing interest in sanitary matters in our cities, and people are asking whether the city is in good condition to resist the introduction or spread of cholera, and to what extent it is worth while to expend money to secure pure water, clean streets, odorless sewers, etc."[35]

Increasingly during the 1880s and 1890s, protest against inadequate refuse collection and disposal was becoming a primary function of many citizens' groups and civic organizations. They too relied heavily on the health argument to justify their dissent. In 1881, the Citizens' Committee of Twenty-one of New York City circulated petitions, held mass meetings, and met with city officials to change the method of street cleaning, since the committee believed that the current system "was one better calculated to advance political interests than to secure cleanliness and health."[36] At about the same time, the Citizens' Association of Chicago was calling for similar changes. By 1893, organized protests against unsanitary conditions had multiplied greatly. The Municipal Order League of Chicago (primarily a women's organization) had formed in 1892, according to the Board of Heath, "to assist the civil authorities, not to theorize, publicly suggest impracticable and impossible socialisms." During its first year, the league distributed twenty thousand printed cards to be hung in kitchens throughout the city; the cards contained suggestions for the disposal of wastes and outlined local ordinances about refuse. "Improvement associations" formed in neighborhoods throughout Chicago to promote civic responsibility with respect to waste disposal. The *Chicago Herald* sponsored a Public Improvement Bureau. Several times a week the *Herald* printed blank forms on which citizens were to write complaints. The forms were collected and forwarded to the appropriate city department.[37]

In the struggle for sanitation reform, the Ladies' Health Protective Association of New York City (LHPA) was possibly the most influential civic group in the country. Organized in 1884, the association was the outgrowth of the efforts of fifteen women of the exclusive Beekman Hill area to force the removal of a large, stench-ridden manure pile from their neighborhood. Ultimately the LHPA undertook several projects, including school and slaughterhouse sanitation, street-cleaning reform, and improvement in refuse disposal methods.[38] Although influenced by national trends in public health, the LHPA was a community organization without extensive medical or technical expertise—something that also could be said of most other civic groups. For that reason, the LHPA often stated its protests in aesthetic rather than scientific or medical terms. As an organization dominated by middle- and upper-middle-class women who perceived their reform efforts as extensions of their roles as housewives and mothers, the LHPA couched its protests in domestic terms, "Even if dirt were not the unsanitary and dangerous thing we know that it is, its unsightliness and repulsiveness are so great, that no other reason than the superior beauty of cleanliness should be required to make the citizens of New York, through their vested authorities, quite willing to appropriate whatever sum may be necessary, in order to give to themselves and to their wives and daughters, that outside neatness, cleanliness and freshness, which are the natural complement and completion of inside order and daintiness, and which are to the feminine taste and perception, simply indispensable, not only to comfort but to self-respect."[39]

Probably because of, rather than in spite of, its layman's outlook about

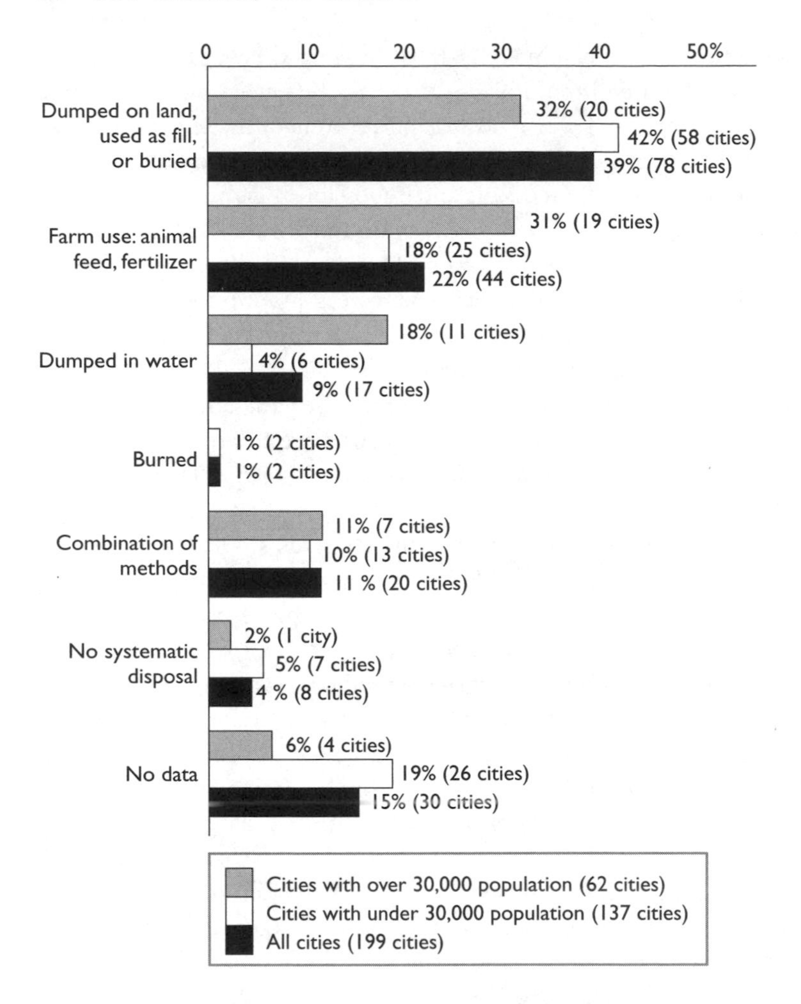

FIGURE 2. **Garbage Disposal Methods, Percentage and Number of Cities, 1880.**
Source: U. S. Department of the Interior, Census Office, *Report on the Social Statistics of Cities, Tenth Census, 1880,* compiled by George E. Waring Jr. (Washington, D.C., 1886).

the problem of waste, the LHPA put sanitation into terms that many citizens and political leaders could understand and appreciate. The association's successful efforts in lobbying for improved sanitary measures in New York City largely grew out of its ability to popularize the health concerns of sanitarians and public health officers. Similar groups in other cities followed the lead of the LHPA.[40] The heightened awareness of sanitary problems in the 1880s and 1890s ultimately led to criticism of the

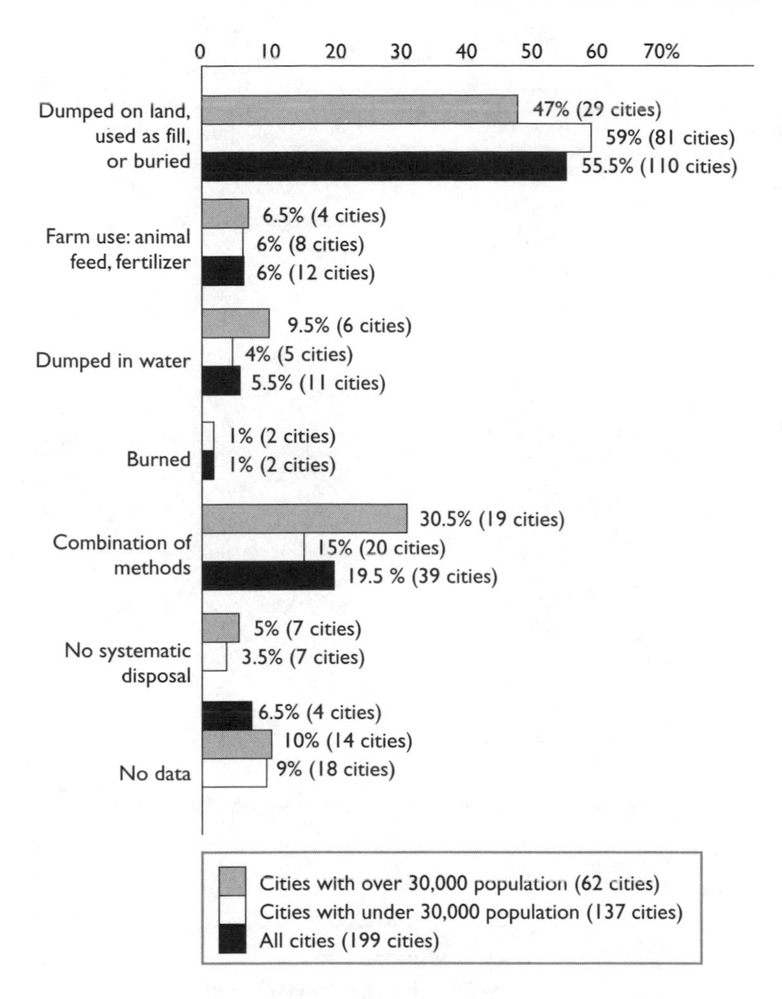

FIGURE 3. **Street Sweeping Disposal Methods, Percentage and Number of Cities, 1880.**

Source: U.S. Department of the Interior, Census Office, *Report on the Social Statistics of Cities, Tenth Census, 1880,* compiled by George E. Waring Jr. (Washington, D.C., 1886).

inadequate collection and disposal practices of the time. The methods that cities relied on were so primitive that the best one could expect was frequent removal of the garbage and trash from the immediate range of human senses. A philosophy of "out of sight, out of mind" prevailed. Much of the refuse was simply removed from one location to create a nuisance or health hazard in another. Disposal methods fell into two categories: indiscriminate discharging and utilitarian application. In the first category,

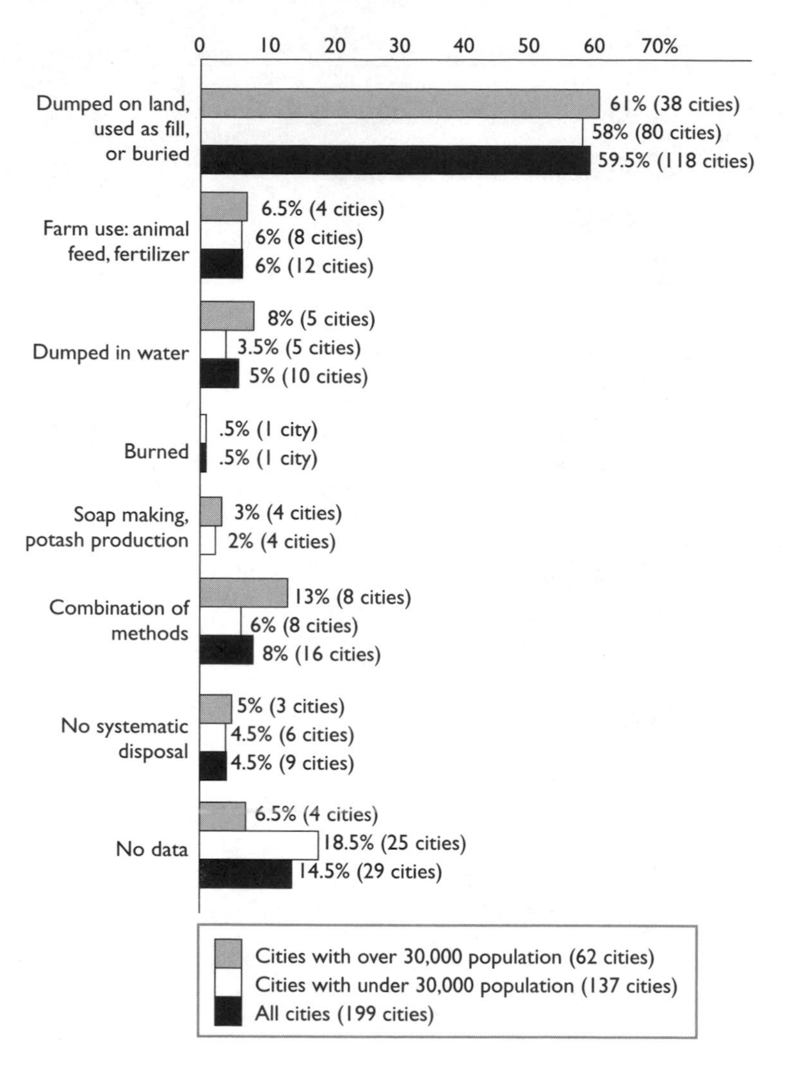

FIGURE 4. **Ash Disposal Methods, Percentage and Number of Cities, 1880.**
Source: U.S. Department of the Interior, Census Office, *Report on the Social Statistics of Cities, Tenth Census, 1880,* compiled by George E. Waring Jr. (Washington, D.C., 1886).

dumping of refuse on open land or into water was the most common practice; the second category included use of organic matter for fertilizer, animal feed, road surfaces, and landfill (See figures 2, 3, and 4).

All of these methods had one commonality: few precautions were taken to ensure that refuse disposal was sanitary. Even utilitarian practices, which offered a means of solving the problem by recycling some wastes, frequently created new sanitation problems. Untreated organic material used for filling potholes in roads and other depressions created serious health hazards. Waste used for agricultural purposes—fertilizing or animal feed—often was handled without proper safeguards. The transfer of garbage from the city to the farms sometimes required an intermediary stop at a "swill yard" within the city limits, where the waste was unloaded for farmers to cart away. These swill yards were dangerous breeding grounds for disease and, like open dumps, emitted a revolting stench. The common practice of feeding garbage to swine and other animals also became questionable; studies indicated that the meat or products of animals fed on garbage might not be fit for human consumption. Investigations in the mid-1890s in New England revealed that cases of trichinosis among hogs fed with garbage increased from 3 to 17 percent over a three-year period and that the annual mortality from hog cholera increased alarmingly.[41] The major cause of the health problems was not the feeding of organic matter to animals but rather the unsophisticated manner in which refuse was transported and utilized.

In the nineteenth century, the quantity of refuse was so great that cities could not give it away, let alone sell it to farmers, as had been the earlier custom. The cost of transporting refuse from the cities to the countryside and the rapid decomposition of organic materials meant that only farms close to urban centers could use the wastes for feed or fertilizer. Rapid urban growth coupled with constriction of rural areas further complicated the situation.[42]

In addition to organic wastes—garbage, manure, human excrement, dead animals—there were also tons of coal and wood ashes, street sweepings, wastepaper, cans, old shoes, and other assorted rejectamenta. Some waste, such as kitchen and restaurant garbage, had to be collected more frequently than rubbish. A scavenger could hardly employ the same method for the collection of manure that he used for wastepaper or discarded clothing. Cities had to contend with very basic and often perplexing questions in disposing of their refuse. Should solid wastes be separated in the homes and business establishments for ease of disposal? Was some of the material salvageable or recyclable? Was disposal of separated or mixed refuse more economical? How could the city or a contractor best employ human labor in the collection and disposal of waste? What about mechanical devices? Were there sufficiently effective techniques for the disposal of the various kinds of waste? The questions were as difficult to pose as to answer. Unfortunately, city officials had little success with either questions or answers throughout much of the nineteenth century.

If convenient utilitarian methods of disposal were unavailable, most cities ignored the more complex alternatives and resorted to dumping their refuse wherever space allowed. The open dumps and the watercourses became the most expedient, but hardly the best, places for disposal. Before the turn of the century, however, many city officials were forced to confront the cost of expedience. For example, many cities, especially those not situated along waterways, dumped refuse on vacant lots or near the "least desirable" neighborhoods, that is, those occupied by the poor, the working class and/or ethnic and racial minorities. Protests from the unfortunates who lived near the putrefying mounds often went unheard or were ignored. As the concentration of people in the inner cities became acute in the 1880s and 1890s, and as more dumps were created to meet the needs of the rapidly multiplying population, widespread outcries were heard, and these outcries could not easily be ignored. City officials were also faced with a new problem: rapidly multiplying commercial and residential building programs and high land values made it difficult to acquire new dumping sites. As the health officer of Washington, D.C., stated in his 1889 report, "Appropriate places for [refuse] are becoming scarcer year by year, and the question as to some other method of disposal . . . must soon confront us. Already the inhabitants in proximity to the public dumps are beginning to complain. . . . I can not urge too strongly upon the Commissioners the necessity for action in this direction. The waste that is taken from yards and dwelling places must be provided for, and that provision should not be longer delayed."[43]

By far the most pernicious method of disposal was the dumping of refuse into lakes, rivers, harbors, and even the open sea. The logic of such a method was painfully simple, as the Reverend Thompson noted about New Orleans's use of the Mississippi, "At 4 p.m. each day a tug picks up the scows, tows them two miles down the river below the city, where the garbage is dropped into the stream, and disappears into the devouring jaws of gar, pike, codfish and the other greedy denizens of the great stream, which attend in countless numbers at their daily dinner hour. What is spared by them is whirled away into the waters, and not a trace of any part of the offensive matters can be discovered four miles below."[44]

In 1886, New York City dumped 1,049,885 of its 1,301,180 cartloads of refuse into the ocean and for many years continued to rely on ocean dumping as a primary means of disposal. In Chicago the lack of "convenient and suitable" dumping grounds led the city to dispose of much of its waste in Lake Michigan, three miles from the mouth of the Chicago River.[45]

Such a disposal method had many repercussions. Cities that dumped their refuse in rivers had to contend with constant complaints from cities downstream. New York dumped so much of its waste into the Atlantic that the approaches to the harbor were often clogged, and the public and private beaches on the New Jersey shoreline looked like cesspools. It was not uncommon for swimmers to be nudged by mattresses and old shoes, which certainly gave them a fright. Besides the obvious nuisances, pollution created untold damage to marine ecology. As one commentator lamented in an 1894 issue of *Harper's Weekly,* "All in all . . . dumping at sea is not a cheap method of garbage disposal. Even if it were cheap, however, it is not a commendable or even a permissible method of disposal. It has been adopted as a makeshift by New York as by many other seaport cities because it seemed easiest in the beginning, but it can nowhere be regarded as a finality."[46]

Street cleaning fared better than refuse removal in the nineteenth century, but its effectiveness varied widely from city to city. As historian Lawrence H. Larsen wrote, "Although Americans living in the nineteenth century boasted of many aspects of their growing urban mosaic, hardly any talked in glowing terms about the cleanliness of their streets."[47] By 1880, the herds of swine that had been used as inexpensive scavengers were gone from most cities. The system that replaced them, however, hardly solved the problem of dirty streets. The 1880 census suggests that approximately 84 percent of the cities surveyed relied on hand sweeping to keep the streets clean.[48] This method could have been effective if carefully maintained, but unfortunately, many cities did not clean their streets regularly, except for the major business thoroughfares. Few cities bothered to clean their streets in the winter. Although most cities had antilittering ordinances, citizens rarely obeyed them, and policemen rarely felt compelled to enforce them.[49] It was easy to blame the dirty streets on the most convenient scapegoats, as *Popular Science Monthly* did in 1891: "In those cities and parts of cities where the people of the laboring class and the poor are crowded in tenement-houses, and where a considerable part of the population is foreign-born and from countries where personal and public cleanliness have not been enforced by proper police regulations, it is no trifling task to secure cleanliness of the streets."[50]

City officials claimed that inadequate funding or physical problems made clean streets impossible. The 1887 annual report of the Boston Board of Health asserted that the three major causes of inadequate street sanitation were poor paving, the continual digging up of streets for pipe laying, and an insufficient work force.[51] There was considerable substantiation for these claims. Street paving was primitive in the nineteenth century. Before the widespread use of asphalt, sweeping was difficult. In 1880, more than half of urban streets were unpaved (see table 3), and as late as 1909, only one-third of city streets were asphalt.[52]

Street department budgets rarely kept up with the rapid growth of

TABLE 3

Comparison of Paved to Unpaved Streets and Alleys in Fifty of the Largest U.S. Cities, 1890

City	*Total Length of Streets and Alleys (Miles)*	*Total Length of Paved Streets and Alleys (Miles)*	*Percentage of Paved Streets and Alleys*
Albany, N.Y.	140	55	39.3
Allegheny, Pa.	138	73	52.9
Atlanta, Ga.	200	29	14.5
Baltimore, Md.	780	459	58.8
Boston, Mass.	408	408	100.0
Brooklyn, N.Y.	653	375	57.4
Buffalo, N.Y.	372	194	52.2
Cambridge, Mass.	79	24	30.4
Camden, N.J.	100	31	31.0
Chicago, Ill.	2,048	629	30.7
Cincinnati, Ohio	503	284	56.5
Cleveland, Ohio	462	69	14.9
Dayton, Ohio	325	49	15.1
Denver, Col.	756	0	0.0
Detroit, Mich.	614	160	26.1
Fall River, Mass.	106	3	2.8
Grand Rapids, Mich.	149	89	59.7
Indianapolis, Ind.	400	234	58.5
Jersey City, N.J.	165	52	31.5
Kansas City, Mo.	383	51	13.3
Los Angeles, Calif.	800	84	10.5
Louisville, Ky.	314	183	58.3
Lowell, Mass.	104	19	18.3
Memphis, Tenn.	90	35	38.9
Milwaukee, Wis.	419	249	59.4
Minneapolis, Minn.	1,025	27	2.6

continued

the industrial cities, and allocations for street cleaning had to compete with the costs of street construction and maintenance—both vital to urban growth. In forty major cities surveyed in the 1890 census, the average percentage of the street budget earmarked specifically for cleaning was 17.5 percent, while the rest of the budget was allotted for construction and maintenance. The extremes, however, were great. Newark spent 60 percent of its street budget on cleaning, while Memphis spent slightly over 1 percent.[53]

As the number of streets multiplied, meager street-cleaning forces were

TABLE 3
Continued

City	*Total Length of Streets and Alleys (Miles)*	*Total Length of Paved Streets and Alleys (Miles)*	*Percentage of Paved Streets and Alleys*
Nashville, Tenn.	251	147	58.6
Newark, N.J.	186	48	25.8
New Haven, Conn.	140	32	22.9
New Orleans, La.	635	109	17.2
New York, N.Y.	575	358	62.3
Omaha, Nebr.	508	52	10.2
Paterson, N.J.	207	55	26.6
Philadelphia, Pa.	1,151	750	65.2
Pittsburgh, Pa.	356	143	40.2
Portland, Oreg.	220	64	29.1
Providence, R.I.	168	122	72.6
Reading, Pa.	72	47	65.3
Richmond, Va.	106	69	65.1
Rochester, N.Y.	255	72	28.2
St. Louis, Mo.	1,061	422	39.8
St. Paul, Minn.	970	41	4.2
San Francisco, Calif.	342	192	56.1
Scranton, Pa.	125	7	5.6
Seattle, Wash.	65	9	13.8
Toledo, Ohio	438	60	13.7
Trenton, N.J.	100	7	7.0
Washington, D.C.	235	163	69.4
Wilmington, Del.	78	33	42.3
Worcester, Mass.	145	145	100.0

Source: U.S. Department of the Interior, Census Office, *Report on the Social Statistics of Cities in the United States, Eleventh Census, 1890,* compiled by John S. Billings (Washington, D.C., 1895), pp. 58–62.

spread over ever-larger areas. City officials had to choose between cleaning all the streets sporadically or concentrating on the main thoroughfares and ignoring the rest. With either choice, the result was inadequate service. For example, as Boston began absorbing suburbs such as Brighton and West Roxbury, the street-cleaning force was insufficient to meet the new demands. The Boston Board of Health complained that the printed list of streets to be swept rapidly became obsolete "owing to the impossibility of covering the entire area laid out," and that "the work was largely done by general orders to work where the dirt was the greatest." During those times of the year when the main streets were relatively clean in the central city, the whole Boston street-cleaning crew was transferred to the suburbs for a quick cleaning. During the winter months, the force was used primarily to collect ashes and neglected street cleaning almost entirely.[54]

Increases in the street-cleaning budget were no assurance of clean streets. The cost per capita for street cleaning varied widely across the country, with little or no correlation with the cleanliness of streets. One survey indicated that in the 1880s, citizens of Buffalo, New York, paid an average of 5¢ per year for street cleaning while New York City residents paid 71¢ per year. Translated into cost of street cleaning per mile, people in Buffalo paid $34, and New York City residents paid a staggering $1,870.[55] Graft and corruption played a large part in this disparity. In an 1895 article in *Harper's Weekly*, F. W. Hewes charged the New York municipal government under Tammany Hall domination with mishandling of street-cleaning department funds. In Hewes's survey, other cities spent less than one-fifth what New York did, and the streets of New York were still deplorable. Similar charges had been raised earlier with the same conclusion: the street-cleaning department was rife with corruption and manned by political appointees.[56]

As city authorities began to recognize the inadequacies of collection and disposal methods, they turned to municipal or sanitary engineers for answers. At first, municipal engineers offered as a solution the adaptation of British and Continental disposal technologies, giving considerably less attention to collection practices. Like sanitarians and civic leaders, they considered refuse a health problem, but one which could be resolved through technical processes. As the field of sanitary engineering matured in the early twentieth century, faith in a technical solution to the refuse problem would give way to a more comprehensive approach.[57] In the meantime, American engineers marveled at the way many European countries had apparently solved their waste problems efficiently and "scientifically," especially through the use of fire as a "disinfectant." Burning waste at high temperatures seemed to be the perfect disposal method—no stench-ridden dumps, no pollution of streams and other watercourses, no unsanitary landfills. One doctor called cremation of garbage "a great sanitary device." Another physician, from Wheeling, West Virginia, stated that the health department of his city had experimented with incineration of wastes and concluded, "at last we have secured a means of entirely destroying these substances and their power to do evil."[58] In 1888, Dr. Kilvington told an audience attending the annual meeting of the APHA:

> Everywhere interest in the question of cremation is awakening and the present points to the future—a near future—in which every city, large or small upon the American continent will consider the crematory a necessary part of its municipal outfit; forward to a time when our cities will be redeemed from the curse of accumulating waste, when the rivers will be unpolluted by the sewage which now converts them into common sewers, when the cess-vault and the garbage-pit and the manure-heap and even the earth cemetery will be abandoned, when the age of filth-formation will be superseded by the era of filth-destruction, when fire will purify alike the refuse of the living and the remains of the dead—but also it is allotted to each one of us to help to bring in the coming of this sanitary consummation.[59]

The Europeans' technical expertise and leadership in the practice of cremation and other new sanitation techniques was predictable. London and other European cities had been grappling with the problem of waste long before the United States was founded, and, furthermore, they did not have land or water available for dumping. Out of necessity, less primitive methods had to be invented. The first systematic cremation of refuse at the municipal level was tested in Nottingham, England, in 1874. In Manchester, two years later, Alfred Fryer built an improved "destructor" (Americans used the terms "garbage furnace," "cremator," and "incinerator" to describe their systems), which became the model for much subsequent development of the technology. Fryer's effort marked the beginning of large-scale implementation of incinerating devices throughout England and the rest of Europe. The success of the British method of disposal led to the construction of cremators and incinerators in the United States. In 1885, Lieutenant H. J. Reilly of the United States Army built the first American garbage furnace on Governor's Island, New York. In 1886–1887, engineers installed the first municipal cremators in Wheeling, West Virginia; Allegheny, Pennsylvania; and Des Moines, Iowa.[60]

Municipal officials throughout the country quickly took note of these experiments, and orders for incinerators were brisk. The Engle Sanitary and Cremation Company of Des Moines was an early pioneer in the field. By 1894, it had installed incinerators in cities throughout the country, from Portland, Oregon, to Coney Island, New York, and from Milwaukee, Wisconsin, to Saint Augustine, Florida. The company also constructed an incinerator for exhibition at the Chicago World's Fair and built several systems for cities in Latin America.[61] Regulated disposal by fire was hailed as a technological panacea, though some engineers and scientists cautioned against expecting too much from the cremation method, or from any other single method of disposal.[62] In 1890, the Boston Health Department acknowledged that burning waste was the "best and safest" means of disposal but, because of the high cost of commercial cremators, recommended burning waste in home kitchens.[63] That, of course, was a highly impractical solution for most urbanites. Yet there was finality about burning waste that was attractive to many city dwellers, no matter what the procedure or method.

Unfortunately the first generation of incinerators based on the British model was not successfully adapted to conditions in the United States. Of the 180 furnaces erected between 1885 and 1908, 102 were abandoned or dismantled by 1909. Criticism was widespread about incomplete combustion of the waste and the consequent generation of noxious smoke. Some critics argued that British engineering expertise had not been taken into account in building crematories in America. Others complained about the excessive use of additional fuels to augment the burning, which sharply increased the cost of operation. Part of the problem, the argument went, was that American refuse contained a higher water content than English refuse, and thus required higher temperatures to burn. Other factors, not always noted by contemporaries, also contributed to the demise of the first-generation incinerators in the United States, including greater availability of cheap land for dumping, lower fuel costs which made long hauls of waste less expensive than in Europe, and the poor quality of the furnaces produced by American manufacturers.[64]

In addition to cremation, other technological breakthroughs from Europe were attracting attention as alternatives to primitive methods of disposal. In 1896, a company in Buffalo, New York, introduced the so-called

Vienna, or Merz, process for the extraction of oils and other by-products through the compression of city garbage. The "reduction process," as it became known, offered cities a method of disposal that provided recoverable and resalable materials from waste. Utilization of waste was an old idea, but reduction was a new development. By-products could be sold commercially as lubricants, perfume bases, or fertilizers. Cities that adopted this process, such as Saint Louis, Detroit, Buffalo, and Milwaukee, sought means of recovering some of the costs of disposal through the sale of the by-products. Thus, reduction offered another dimension to disposal methods. Some cities employed both reduction and incineration in efforts to convert what waste they could into profitable by-products and burn what was not commercially viable. But like incineration, reduction was an imperfect disposal method. Foul odors emanating from the plants and the high cost of construction and operation limited widespread use of this technology. In fact, reduction plants went out of service faster than first-generation incinerators.[65]

Thus in the 1880s, the refuse problem finally came to the attention of American city dwellers. Sanitarians and public health officers gave compelling health reasons for concern about it, civic groups offered aesthetic reasons for improvement in its collection and disposal, and engineers provided new technologies to attempt to eliminate it. Yet little real progress was made in controlling the refuse problem. Hope and expectation had yet to be translated into practice. Public awareness was crucial, but even more crucial was the willingness to commit municipal funds to provide adequate sanitary services for all sectors of the constantly expanding cities and workable systems of refuse management to contend with the ever-increasing volume and complex array of wastes. Furthermore, public awareness had to evolve into public responsibility; citizens had to become convinced that proper sanitation was a personal as well as a municipal obligation. With the appointment of Colonel George E. Waring Jr. as street-cleaning commissioner of New York City in 1895, came the first practical, comprehensive system of refuse management in the United States.

TWO THE "APOSTLE OF CLEANLINESS" AND THE ORIGINS OF REFUSE MANAGEMENT

On October 1, 1898, Spanish and American diplomats met in Paris to negotiate a treaty ending the four-month war between their countries. The brief but significant conflict toppled the anemic Spanish Empire in the Caribbean and the Pacific, leaving the Philippine Islands, Guam, Puerto Rico, and Cuba under American control. Fearing a yellow-fever epidemic in occupied Cuba, President William McKinley appointed the noted sanitary engineer Colonel George E. Waring Jr. as special commissioner of the United States government to investigate health conditions in Havana as a preliminary step toward the establishment of a comprehensive system of sanitation there. While he was in Havana, Waring contracted yellow fever and died on October 29, soon after returning to New York City.[1]

The reaction to Waring's death was emotional and heartfelt. At Cooper Union on November 22, a memorial service was held at which several prominent civic and political leaders praised the public servant. As the *New York Times* reported, "More than 5,000 men, women, and children assembled within the walls of Cooper Union last night to pay tribute to the memory of Col. George E. Waring Jr., Commissioner of Street Cleaning during the administration of ex-Mayor Strong. In eulogistic speeches the good deeds of the dead man were told and retold by those who had been his close friends during his life."[2] Among those who took the podium were Jacob A. Riis, the famous muckraking journalist; Felix Adler, a leader in humanistic religion and founder of the Ethical Culture Society; Seth Low, president of Columbia University, labor reformer, advocate of Negro rights, mayor of Brooklyn from 1881 to 1885, and the future mayor of New York City from 1901 to 1903; Carl Schurz, lawyer, newspaperman, reformer, political orator, and former Republican senator from Missouri; and William Strong, former mayor of New York City, under whom Waring had served as street-cleaning commissioner. The New York State Chamber of Commerce established a permanent memorial to Waring raised by public subscription. Interest on the $100,000 collected was paid to his widow and daughter during their lifetimes and then reverted to Columbia University to establish the Waring Memorial Fund for instruction in municipal affairs.[3]

Tributes in many of the popular periodicals and newspapers of the day extolled the contributions Waring had made to the promotion of sanitation. Albert Shaw, editor of *Review of Reviews,* an ardent Progressive and an admirer of Waring, called the fallen municipal engineer "the greatest apostle of cleanliness." In her eulogistic poem, which appeared in the *Century,* Helen Gray Cone described him as "Fever-Slayer, yet slain by the breath abhorred." The *Outlook* declared that Waring's work in New York City had been "epoch-making." Arturo Fernandis, a Cuban official, wrote to Mrs. Waring that her husband was "[a] notable figure among brilliant personalities who by his skill and knowledge was preeminent in the mighty American Union." One of the colonel's admirers, an embittered anti-imperialist reckoned that Waring was a victim of American expansionism. "Waring was sent to Cuba," he declared, "on an errand as foolish as most of the 'expansionist' policy. . . . If we knew the things which make for our peace and prosperity, we should regard the life of a man like Waring as of more value to the American people than the whole island of Cuba and all that it contains."[4]

Sudden death has a way of making martyrs out of public figures. Certainly the testimony of Waring's friends and admirers exaggerated the accomplishments of the man. Yet there was something about his career as sanitary engineer, agriculturalist, popular writer, and municipal official that made his death a public, as well as a personal, loss. Despite his widely publicized flamboyance, his posturing and theatrics, his drive for personal recognition and material wealth, Waring was a pioneer in the field of sanitary engineering, an important urban environmental reformer, and the father of modern refuse management. His brief stint as street-cleaning commissioner of New York City from 1895 to 1898 formed a bridge between the primitive collection and disposal practices of the nineteenth century and the increasingly sophisticated methods of the twentieth century.

Waring's career spanned the second half of the nineteenth century, a time of radical changes in the nation. In keeping with the times, as the United States was transforming itself from a rural agrarian society into an urban-industrial one, he shifted his career from farming to municipal engineering. He was, therefore, a transitional figure with one foot in the past and one in the future.

Born in Poundridge, New York, on July 4, 1833, Waring spent much of his boyhood in nearby Stamford, Connecticut. He had a typical middle-class upbringing; his father was a manufacturer of stoves and agricultural implements in Stamford. He was educated in both public and private schools in the 1840s, and after graduation from Bartlett's School, in Poughkeepsie, New York, he spent a year in the hardware business and about two years managing a rural gristmill. In 1853, Waring became a student of the renowned agricultural scientist James J. Mapes, who conducted many experiments in farming techniques and edited the *Working Farmer.* Through Mapes's guidance, young Waring launched on a career in scientific agriculture, which provided him with extensive practical training in several areas, including drainage engineering, and brought him into contact with influential people in rural New York and New England. He lectured throughout Maine and Vermont on scientific-farming techniques during the winters of 1853 through 1855. In 1855, he became manager of Horace Greeley's farm near Chappaqua, New York. He held that position for two years and then accepted a similar position on noted landscape architect Frederick Law Olmsted's farm on Staten Island in 1857.[5]

Waring published his first book, *The Elements of Agriculture,* in 1854. In it he lauded the virtues of scientific methods of agriculture, encouraging farmers to employ modern "book methods" to increase crop yields, preserve soil fertility, and utilize manure properly. The book stressed the idea that human beings could harness nature for their own uses: "[The practical farmer] knows nothing of the first principles of farming, and is successful by the 'indulgence' of nature, not because he understands her, and is able to make the most of her assistance." Waring's assertion that human beings could play a significant role in improving the physical environment grew out of his belief that scientific methods could help control malevolent nature.[6]

Waring's association with Frederick Law Olmsted also helped to shape his environmentalism and proved to be a significant turning point in his career as an engineer. According to Olmsted biographer Laura Wood Roper, by initiating the rural-park movement in the United States, Olmsted tried "to humanize the physical environment of the cities and to secure precious scenic regions for the use and enjoyment of all the people."[7] When he set about constructing Central Park in the late 1850s, he looked to Waring to help with the project. In August 1857, he appointed Waring as drainage engineer and charged him with responsibility for most of the agricultural work in the park. Waring received such high praise for his efforts that he obtained several other urban assignments and soon abandoned most of his farming activities. His friendship with Olmsted and their professional association not only influenced Waring's new career orientation but also made him more sensitive to the need for his engineering skills in improving living conditions in the city. Several years after Central Park was completed, when Waring was street-cleaning commissioner of New York City, he publicly expressed gratitude to Olmsted for his influence on his profession.[8]

The Civil War temporarily halted Waring's promising engineering career. In May 1861, he was commissioned a major in the Garibaldi Guards (the Thirty-ninth New York Volunteers), who acquired the nickname "Organ Grinders" because of their red blouses and Bersagliere hats. Waring served briefly with the Army of the Potomac at the First Battle of Bull Run and then was sent to Saint Louis to recruit troops for General John C. Fremont. He organized a battalion of cavalry, the Fremont Hussars, which was deployed against guerrillas striking in Missouri during the summer of 1861. In January 1862, the Hussars were consolidated with other units to form the Fourth Missouri Cavalry, at which time Waring was commissioned a colonel. He served with the Fourth Missouri until 1864, primarily in the southwestern part of the state.[9]

Waring's appreciation for military principles, decorum, and discipline never deserted him, nor did the title "Colonel," which he bore until his death. Like many other young men who survived the war, he looked back on his military experience with nostalgia tinged with bravado. Several years after the war he said:

> It is a pleasant thing to be a colonel of cavalry in active field service. There are circumstances of authority and responsibility that fan the latent spark of barbarism which, however dull, glows in all our breasts, and which generations of republican civilization have been power-

> less to quench. We may not have confessed it even to ourselves; but on looking back to the years of the war, we must recognize many things that patted our vanity greatly on the back—things so different from all the dull routine of equality and fraternity of home, that those four years seem to belong to a dream-land, over which the haze of the life before them and of the life after them draws a misty veil.[10]

As he stated in the *Outlook* in 1898, "The most complete and lasting happiness of which we are capable comes from a sense of duty done."[11] Certainly much of his pomposity and arrogance was an outgrowth nurtured by the military bearing he cultivated during the Civil War, but so was his faith in his organizational skills and administrative ability, which served him well in his later career.

After being mustered out of the service in 1865, Waring set out to build a successful business career. Some abysmal failures in oil and coal ventures sent him back to scientific agriculture. In 1867, he assumed the management of Ogden Farm near Newport, Rhode Island, a position he held for ten years. He plunged vigorously into his work and became particularly interested in husbandry and horticulture. In 1868, he organized the American Jersey Cattle Club and edited *Herd Book;* two years later, he introduced the trophy tomato to horticulturalists. One commentator observed, "[The tomato] was practically unknown as a table luxury a quarter of a century ago. Its usefulness began with the production of the variety called the trophy. And the trophy tomato was one of Colonel Waring's contributions to agriculture and to modern luxury of life."[12]

In the late 1860s and early 1870s, Waring's interest in agriculture faded once again, and he gave more attention to engineering problems such as drainage and sewerage. He gave up all his gardening operations in 1872, and by 1877 he had abandoned farming entirely. His book *Draining for Profit and Draining for Health* was published in 1867. A practical guide to the construction of drainage systems, it was the first of several books and articles that he wrote on the subject during the 1870s and 1880s. In the 1870s, he began accepting commissions to build drainage and sewerage systems on the East Coast. He constructed sewer systems for Ogdenburg, New York, in 1871; for Saratoga Springs, New York, in 1874; and for Lenox, Massachusetts, in 1875–1876. At Lenox, Waring built the first "separate system" in the United States.[13] The separate system, also called the "Waring system," channeled rainwater and raw sewage into different pipes for ease of disposal. Although it was a source of much controversy in the engineering community, the separate system elevated the colonel to the forefront of the nation's sewerage designers.[14]

Waring's design and construction of a separate system in Memphis, Tennessee, brought him widespread attention but also embroiled him in one of the greatest controversies of his professional career. For years, Memphis had suffered from inadequate sewerage and had been decimated by yellow-fever epidemics. In 1879, Waring was appointed to a special commission of the National Board of Health charged with examining conditions in the city. He learned that in two short years, yellow fever had claimed approximately five thousand victims, and that in the resulting panic, two-thirds of the city's 40,000 residents had fled. After considering several plans for a sewerage system for Memphis, the commission awarded the contract to Waring over the opposition of several civil engineers. The project was carried out in 1880. Many contemporaries lauded the new system, claiming that it was responsible for saving the city from ruin. Civic leaders from other urban areas, impressed by the favorable publicity that the Memphis system received, sought out the colonel to have him build separate systems for their cities.[15]

Waring and his separate system were not without their detractors. Some engineers preferred the combined system, which employed a single large pipe for carrying rainwater and sewage to its final destination. An investigation of European sewerage systems, conducted for the APHA in 1880 by the noted sanitary engineer Rudolph Hering, indicated that both systems were equally sanitary but that the combined system was cheaper for densely populated cities while the separate system was cheaper for less heavily populated cities. The debate raged on as engineers defended their favorite methods. Waring held the American rights to the separate system, and some charged him with harboring crass financial interest in promoting it rather than having a genuine commitment to it as the most sanitary disposal method. "The Memphis system," one critic stated, "was the most conspicuous [of the various separate systems], although a comparative failure, a fact which the people of the city naturally suppressed for business reasons for many years." Indeed, the system was innovative but not well constructed. It provided small sanitary pipes that clogged easily, had flush tanks that malfunctioned, lacked manholes for easy access to the underground pipes, and gave very limited attention to storm drainage.[16]

Criticism of Waring's system was justifiably harsh, and charges that he gained financial advantage from his engineering accomplishments are not without merit. His promotion of the "earth closet" is a good example. In the 1860s, Waring publicized the earth closet as a revolutionary innovation in household sanitation, even after the water closet had proved to be more efficient and practical for American homes. He not only tenaciously defended the device, but also tried to induce his friends, even Olmsted, to invest in it.[17]

Waring's commitment to his sanitary principles was not guided exclusively or even primarily by materialistic interests, however. His engineering accomplishments and his writings demonstrate his growing commitment to the positive implications of environmental sanitation. His service as a special government agent charged with compiling social statistics of cities for the tenth United States census (1880) gave him broad insight into urban problems throughout the country. His association with the ill-fated National Board of Health also gave him access to a cross section of public-health officers, engineers, and sanitarians, as well as adding a national perspective to his views on health and sanitation.[18] His environmentalism quickly moved beyond the small towns of New England and the mid-Atlantic states to the major urban centers of the nation. By the mid-1890s, his reputation as a national leader in sanitary reform was widely accepted; his list of publications on sewerage, drainage, and general sanitation multiplied rapidly. He also continued to ride the crest of popularity generated by several travel accounts of jaunts to Europe and by guides on horsemanship and bucolic narratives which he churned out in abundance.[19] By most measures, and despite periodic setbacks and professional rivalries with other engineers, Waring was a success.

The colonel had a unique opportunity to test his expertise in sanitation as street-cleaning commissioner of one of the world's great cities. In 1894, the Committee of Seventy—leaders of the civic reform movement in New York City—successfully engineered the defeat of Tammany Hall in the mayoral race and elected their candidate, William L. Strong, who had been a bank president, corporate director, and successful dry goods merchant. According to historian Justin Kaplan, "It was confidently expected that Strong would apply to the affairs of the city the business virtues by which he had amassed a personal fortune of about a million dollars. (It was at the counting table, the faith of the day ran, that philosopher-kings were to be trained.)"[20] Although politically naive and lacking the experience of a seasoned veteran of the wards, Strong was honest and attempted to move municipal government beyond the personal rule of Tammany Boss Richard Croker. Strong failed to achieve many structural changes in city government, but he was able to enlist a few potent reformers, who brought some acclaim to what otherwise was a lackluster administration. Police Commissioner Theodore Roosevelt and street-cleaning commissioner George Waring were the bright spots of the Strong administration (Gifford Pinchot, the noted conservationist, declined the post of park commissioner). Apparently Roosevelt was offered the street-cleaning post first, but declined in favor of the appointment as police commissioner, which proved to be a stepping-stone to the governorship of New York. With the endorsement of people such as Francis Kinnicutt, of the Street-Cleaning Aid Society, Waring was offered the post of street-cleaning commissioner.[21] At the time, Waring was assistant engineer for New Orleans. "I made it a condition of my acceptance," he later noted, "that I should be entirely exempt from interference and free of all political obligations. I would undertake to clean the streets if I could do it in my own unhampered way, and not otherwise."[22]

At first the New York City press mocked the appointment. Waring was often photographed astride his well-groomed steed, sporting his expertly waxed handlebar mustache, and dressed in riding togs, pith helmet, and riding boots. He gave the appearance of a parody of the military officer he had been during the war. Those who were aware of his professional background, however, looked with great favor on the appointment. *Engineering News* noted that the appointment "insures the services in this important department of a man of marked executive capacity and one aggressive in methods and well posted in the sanitary engineering of cities."[23] As streets became cleaner and garbage was more efficiently eliminated, the initial criticism turned to lavish praise. The former cavalry officer, drainage engineer, and agriculturalist ultimately found his most appreciative audience in New York City.

In his tenure as street-cleaning commissioner, Waring applied everything he had learned about sanitation. He implemented a wide variety of reforms, many of which were attempted piecemeal in the previous thirty or forty years. At the core of his views on sanitation was an adherence to Hippocrates' adage "Pure air, pure water, and a pure soil" and the belief that cleanliness was a gauge of civilization. "There is no surer index of the degree of civilization of a community," he said, "than the manner in which it treats its organic wastes."[24] Waring's environmentalism, however, rested upon the filth theory of disease, which he tenaciously supported even with the growing sentiment favoring the germ theory. Many of his writings emphasized various aspects of that flawed view of disease transmittal, especially the dangers of "sewer gas," which he held responsible for many of the health problems of the day.[25] "This much decried and insidious sewer-gas," he exclaimed in 1880, "is probably entitled to most of the blame it receives for its own direct action, and to as much more from the fact that it so often acts as a vehicle for the germs, or causative particles of specific diseases."[26] Never an original thinker on matters of health, and untrained in medicine, Waring accepted the prevailing wisdom of the anticontagionists. Since the germ theory and the implementation of bacteriological laboratories did not become widespread until after Waring's death, it is somewhat unfair to criticize him for not abandoning the filth theory. Had he lived beyond 1898, he might have become willing to discard anticontagionism. In fact, some of the colonel's later writings indicate that he had begun to recognize the role of bacteria in causing infection and disease.[27]

Despite his advocacy of an outdated theory of disease, Waring instinctively recognized many of the potential dangers of unsanitary surroundings. In *Sewerage and Land-Drainage* he wrote: "While we know, thus far, relatively little of the exact causes of disease, our knowledge at least points out certain perfectly well-established truths. One of them is that man cannot live in an atmosphere that is tainted by exhalations from putrefying organic matter, without danger of being made sick—sick unto death."[28] He may have attributed too much to sanitary reform as a cure-all for disease, but his environmentalism was built on a relatively broad base, which recognized that human beings deserved unpolluted surroundings for aesthetic as well as physical reasons, and that citizens should not shirk their responsibility for preserving a livable environment. He asserted, "It has hitherto been—and, in fact, it still is—the practice of the world to consider its wastes satisfactorily disposed of when they are hidden from sight. In spite of an almost universal outcry about sewer-gas, filth diseases, and infective germs, the great mass, even of those who join in the cry, pay little heed to defects in the conditions under which they are living so long as they are not reminded by their eyes or their noses that their offscourings are still lurking near them."[29]

In the long run, Waring would be better known for his deeds than for his words. His reformist credentials were built on action rather than on medical or scientific expertise or discovery. Toward that end, he emphasized the need to muster human resources to solve sanitation problems. In his early writings, he stressed the responsibility of the individual for improving household health conditions. He soon came to believe that programs of community action were essential for solving health problems beyond the home, and he advocated programs to educate not only the general public but the social and political leaders as well. He once observed, "Until we can convince the country physician that his most important obligation to his community lies in a supervision of the conditions under which it lives, it is hardly worthwhile to waste breath upon the average members of the community."[30] In Waring's mind, community action had to be guided from the top by a social, political, and technical elite trained to solve health problems and responsible for encouraging citizens to do their part. He was paternalistic, but not impractical, and clearly action driven. The simplest way to initiate reform, he believed, was to inspire those in positions of authority to effect those changes. Before taking control of the New York City Department of Street Cleaning, Waring had organized several "village improvement associations." The purpose of such an association was to improve the appearance of the village, which should be "a wholesome, cleanly, tidy, simple, modest collection of country homes, with all of its parts and appliances adapted to the pleasantest and most satisfactory living of its people."[31] These associations were the archetypes of his civic-involvement programs in New York City. Throughout his career, the marshaling of people to solve problems was at least as important to Waring as technological innovation—a view his fellow engineers rarely shared.

As street-cleaning commissioner, Waring did more than simply adapt village programs to the conditions of the big city. In 1881, in a long two-part article in *Scribner's Monthly* entitled "The Sanitary Condition of New York," the colonel demonstrated his understanding of the complexities of large cities. He analyzed the causes of New York City's sanitary problems and proposed remedies. He suggested that the city had some "remarkable natural advantages," such as its location near the sea, its natural drainage, and its prevailing winds, and he concluded that the "causes of unhealthfulness" were "removable." Always the pragmatist, he knew that "the city cannot be torn down, and its sewers and drains dug up, and the whole work begun 'de novo.'" Instead, Waring suggested, the city should make the best possible use of the existing drains, pave the streets adequately for proper drainage, implement an effective street-cleaning program, discharge waste more efficiently, conserve the water supply, and institute household sanitation programs. Although Waring tended to oversimplify the physical condition of the city, he nonetheless grasped the totality of its sanitation needs and offered realistic solutions to many of the problems. Indeed, many of the solutions he suggested in 1880 would become part of his reform program in 1895.[32]

From his first day on the job as street-cleaning commissioner, Waring practiced the art of the possible. He recognized that his immediate task was to gain control of a department that had been little more than a source of patronage for Tammany Hall. "[The department] was hardly an organization," he stated. "There was no spirit in it; few of its members felt secure in their positions; no sweeper who was not an unusually powerful political worker knew at what moment the politician who had got him his place would have him turned out to make room for another." Waring vowed to put "a man instead of a voter" behind each broom and to expel the political cronies who would not accept his leadership. Claiming not to have any specific grudge against those who supported Tammany, he retained almost half of the current employees. It was reported that a superintendent in charge of snow removal hired during the previous administration was summoned to the new commissioner's office. The colonel told the man that he had been accused of being a "rank Tammany man." "Whenever you want my resignation, it is at your service," the indignant man replied. "Don't be quite so fast," Waring said. "Let me hear your version of the case." The superintendent spoke up: "Do you know what a Tammany man is? It is a man who votes for his job. I have been a Tammany man, and a faithful one. I have worked for the organization; I have paid regular contributions to it. But I am a Waring man now." Detecting a smile on the colonel's face, he added: "Don't misunderstand me. If Tammany comes into power again, I shall be a Tammany man again." Impressed by the man's candor, Waring kept him on.[33]

As a major figure in Mayor Strong's reform government, Waring was a constant target for Tammany attacks. In April 1895, critics charged the colonel with a conflict of interest.[34] Apparently his engineering firm, Waring, Chapman, and Farquar, had connections with the Sanitary Security Company, which inspected and certified houses for compliance with municipal sanitation ordinances. Waring was cleared of that charge, however. On several other occasions, Tammany-connected individuals accused him of increasing the cost of administering the department without apparent increase in benefit to the community. Waring dealt with this and other accusations with little difficulty. In 1897, however, he brought charges of malicious libel against Boss Richard Croker. The source of the alleged libel was a campaign document printed in the *New York Morning Telegraph* accusing Waring of inefficiency, use of public property for private purposes, gross extravagance in public office, and speculation with municipal funds. The stormy relationship between Waring and the machine lasted until 1898, when Tammany successfully turned out Strong's reform administration and the colonel lost his job.[35]

Waring's reorganization of the Department of Street Cleaning was well received by all but those intimately tied to the machine. Given his penchant for military discipline and his faith in engineering expertise, it is not surprising that he selected graduates from technical schools or men with military backgrounds for supervisory positions in the newly structured department. "Knowing that organizations of men are good or bad according to the way in which they are handled," he said, "that 'a good colonel makes a good regiment,' I paid attention first to those at the top—to the colonels." In establishing his departmental leadership, he chose men from schools throughout the country. Several of his district supervisors and his master mechanic were young; the superintendent of final disposition was twenty-five years old.[36]

The most far-reaching personnel changes came in the lower echelons, among the street cleaners. Waring believed that human labor was superior to machines for keeping streets clean and that an efficient work force was essential for effective operation of the department. He was also aware that the street cleaners were the most immediate link between the department and the citizens. The public was certain to gauge the department's success by the work of the street crews. Using his natural talent for public relations, Waring sought not only to create an efficient force but also to give the street cleaners what would be called in later years "a change of image." Few people saw much nobility in street cleaning, especially since the workers were often drawn from some political roster or from the lowest economic station. Waring organized his more than two thousand workers into a "military" unit. They were issued uniforms, were required to attend morning roll calls, and were subjected to fines or dismissal for breaches of an elaborate set of rules ranging from absence without authorization to entering a saloon during working hours. In a speech delivered in 1897, Waring said, "One of the [department] rules that was made in reference to the men was, that no man should go into a liquor saloon in uniform or during working hours. (Applause) I was told recently by a gentleman who owns a block on East River, near one of our dumps, where 150 laborers go five or six times a day, that before I 'ruined' him he had four liquor saloons rented there at $1,000 each; but now three of them are closed and the other paid only $300 rent."[37]

Waring's most dramatic attempt at image molding for his street cleaners was dressing them in white uniforms. What appeared at the time a foolish stunt proved to be a masterstroke. According to municipal law, the street crews must be uniformed so they could be distinguished as city workers. Under the previous administration, they had worn overalls and brown jumpers and caps with the department's initials. Waring wanted something more eye-catching, something that would have a greater impact on the public. After considering various combinations, he recalled, "at last I suggested to my wife I would try white. My wife, who was a frank person, said that I was a fool!" Nevertheless, he tested the white uniforms in a single district and liked the results so well that he ordered all the sweepers to purchase white uniforms and caps (later cork helmets). As impractical as the uniforms seemed, the public quickly began identifying the workers with doctors, nurses, and others in the health professions. This was the image for which Waring strived.

His theatrics did not stop with the uniforms. Simultaneously, he instituted citywide parades to show off his legions of "White Wings," as they became known. On May 26, 1896, he staged the first of several annual employee parades of the street-cleaning department. Astride his mount, Waring led the columns of about fourteen hundred sweepers, six hundred drivers (the latter dressed in brown), and twenty-three bands as they marched down Fifth Avenue. Some sweepers carried their brooms, some pushed carts, and a few carried a banner proclaiming "Four Hundred and Twenty Miles of Streets Cleaned Every Day." The White Wings were first met with hisses and boos, but by the time the spectacle of two thousand workers and twenty-three bands passed the mayor in review, the applause of the crowd was uproarious. An editorial in *Garden and Forest* declared: "[The parade] was an inspiring sight when it was recalled through what a storm of distrust and abuse Colonel Waring had to make his way when he proposed to turn politics out of his department, and especially when he put his men in uniforms—and white uniforms at that." In this circus atmosphere, Waring brought his department to the attention of the public.[38]

Public approbation was not all that Waring sought in reorganizing and reorienting his work force. He also wanted to build an *espirit de corps* among his men that would result in more effective street cleaning. Of course, the uniforms, parades, and other military trappings could not by themselves produce the desired results. Waring, a pragmatist, knew that theatrics had limits. They would not get his men to perform their tasks, nor would rigid discipline without compensation. Consequently, he initiated some significant labor reforms in the department. He increased salaries to sixty dollars a month (two dollars a day)—almost double the pay of most unskilled laborers. He also instituted an eight-hour work day, which was shorter than the work days of most laborers. After studying the Belgian method of "arbitration and conciliation" and examining labor conditions among masons and bricklayers in the United States, he established a rather elaborate system of grievance committees and arbitration boards to hear complaints and pass judgment on disciplinary matters. Waring's reforms were not introduced out of altruism. He realized that his workers would respond better to material rewards than to moralizing. Additionally, as a man who could not tolerate opposition to his authority, he wanted his workers beholden to him and not to aggressive labor leaders or to Tammany Hall. Thus there were clear limits to his perception of labor reform. In the press, he often exaggerated the dedication and loyalty of his employees, failing to mention the inevitable strikes and nagging grievances that rack every large, diverse organization. Despite his sometimes autocratic manner (or paternalism), Waring's efforts markedly improved the image of the street cleaner in the eyes of the public and, most important, made the streets much cleaner.[39]

Street cleaning in New York, as in every other large city, required constant attention. The streets were crudely constructed, littering was acceptable public behavior, only the main thoroughfares or streets in fashionable neighborhoods received regular service, and horse manure was everywhere. Under Waring's guidance, the department undertook a systematic sweeping program that relied on the army of White Wings rather than technical innovations such as machine sweepers. Although a number of cities had come to the conclusion that sweeping machines were the best way to cut costs and perform the cleaning task effectively, Waring continued to place his faith in hand sweeping.[40] He also objected to the clouds of dust that the rotary sweepers circulated and the way the machines obstructed the streets at night, when most of the sweeping had to be done.

Waring's street-sweeping program required almost 60 percent of his work force and accounted for 40 percent of his department's budget. New York City had 433 miles of paved streets, on which 1,450 sweepers worked, which meant that each sweeper was theoretically responsible for about a third of a mile of street. In actual practice, however, each sweeper was responsible for an area of about seven miles. To organize the sweeping, the department set up a citywide district system with inspectors and supervisors in each district. Under this system, streets could be cleaned one to five times a day, depending on the need. Part of Waring's success was due to the vast resources available to him for the task, though he often initiated public debates with the city comptroller over appropriations and other budgetary matters. Nonetheless, his organizational skills in establishing his cleaning system offered a good example to other cities interested in initiating or upgrading street-cleaning systems.[41]

Snow clearing was an annoying street-cleaning problem for northern cities, but Waring's restructured department met the challenge. One ardent admirer declared, "Presently the snow came [in the winter of 1896]. Under former regimes the populace would have trampled through the slush for weary weeks. But long before the snow had ceased falling the streets were full of wagons and carts. . . . Military orders had been given for each [worker] to report at his particular station as soon as the snow should reach two inches in depth. Presto! The invading army took possession of the streets, and in a space of time that seemed almost incredible the obstruction had been removed. The streets were dried out by the first sun, and it was necessary to go to Central Park to discover that the sleighing was excellent."[42] The department applied the same degree of vigor to dealing with snow that it did to street sweeping. Waring often boasted that "[In] five consecutive *weeks* of 1895 more snow was removed, and for less money, than in all of the five *years* beginning with 1889." On one day Waring's crew removed 55,773 loads of snow. The following was Waring's uncharacteristic response to the success of the snow-removal program, "I have been told by the president of the United States Rubber Company that this snow removal, together with the abolition of mud from the streets at all seasons, has cost that company $100,000 per year by reason of the decreased demand for rubber boots and shoes. What this means to the poorer people of the city, as compared with their previous suffering, need not be said."[43]

H. L. Stidham, the department's snow inspector, claimed that for the crowded tenement dwellers snow clearing meant improved health. His reasoning was somewhat specious: "With the crowding of the immense tenement population into that human beehive, the East Side, there had been an actual bulging out from the houses to the now clean asphalt streets.

Whether it be winter or summer, the people must have this additional room opened up for them, and a delay in the removal of the almost knee-deep snow and befouled slush is at the cost of much sickness, and probably many lives, each winter."[44]

Street cleaning and snow removal were only two responsibilities of the department. It also had complete charge of the collection and disposal of refuse. As with street cleaning, Waring discovered that the methods employed before 1895 were primitive and that the service was erratic. Although he did not introduce many revolutionary programs in those areas, he did apply his organizational skills to coordinating the collection and disposal programs—a task no one had been willing to attempt until that time. He was extremely critical of random methods of waste collection and disposal: "The 'out-of-sight, out-of-mind' principle is an easy one to follow, but it is not an economical one, nor a decent one, nor a safe one."[45] Waring realized that refuse was diverse in type, requiring various methods to achieve success. His goal was to collect household waste quickly and efficiently, to recover whatever economic value the discarded material might have, and to dispose of the remainder in ways appropriate to its composition.

His most ambitious project was to devise a program for efficient collection of household and commercial wastes. "Source separation," which had been advocated for years but never attempted on a large scale, was attractive to him as an answer to New York's collection problem. The rationale for the system was that mixed refuse limited the options for disposal while separation of wastes at the source allowed the city to recover a portion of its costs of collection through the resale of some items and the reprocessing of others. Furthermore, Waring believed that the street crews could handle the wastes more easily if they were separated. His plan for primary separation required each householder and business establishment to keep garbage (organic waste), rubbish, and ashes in separate containers until the department collected them. In 1896, Mayor Strong assigned forty policemen to the street-cleaning department to explain the separation plan to every householder and businessman and assure compliance with the new ordinance. Waring's program also ended New Yorkers' long-standing practice of placing refuse containers on the sidewalks where they could easily be tipped over. All receptacles now would have to be kept within a stoop line. As might be expected, there was considerable public resistance to primary separation; citizens were becoming conditioned to accepting sanitation programs as a municipal rather than a personal responsibility—an ironic twist considering the effort that had been made to encourage that point of view. Under the Waring program, those who resisted complying with the rules were sometimes fined and even arrested. Despite the initial unpopularity of the plan, by 1898 it had proved to be fairly successful and was receiving much acclaim from city leaders, if not from an obdurate public.[46]

Because of their complexities, the utilization and disposal of refuse proved to be Waring's most difficult problems. While he achieved some success in these areas, he was not able to establish a completely efficient and economical system of final disposition of waste during his tenure as commissioner. He did, however, make substantial progress by experimenting with various methods, hoping to find a practical solution to the city's monumental disposal problems. He especially made progress in reducing the amount of waste New Yorkers dumped in the ocean. Until Waring's time, the city's dry waste was dumped ten miles beyond Sandy Hook, where the tide was supposed to carry it out to the open sea. Waring believed that ocean dumping was "theoretically a perfect disposal" but that it did not work in practice. The obvious sign of its failure was the abundant debris cluttering the beaches of Long Island and New Jersey—at least until the residents of several fashionable estates screamed for relief.[47]

Unlike some of his engineering colleagues and some city officials, Waring was not persuaded that any one method offered a ready solution. He was especially skeptical of cremation as a panacea, considering it "a costly and a wasteful process." He further described it as "an art which has reached a high degree of development, and which in its best form and under proper guidance may be accepted as good, from a sanitary point of view, and as being practically free from offense. At the same time, cremation means destruction and loss of matter which may be converted into a source of revenue."[48] If there was any consistent element in his plans, it was to recycle or utilize waste economically to recover some of the money the city invested in disposal. Municipal leaders usually responded favorably to his determination to recover expenditures and granted most of his requests for appropriations even when the amounts requested grew substantially higher.

The colonel advocated or attempted various programs of resource recovery or waste utilization that matched specific kinds of refuse with appropriate disposal methods. Beyond the obvious reasons for such an approach, there was an effort to increase municipal authority over services generally provided in the private sector. For example, the city's pushcart men had made a living by collecting discarded items that could be resold to junk stores. Waring argued, "public authorities might with advantage take control of the whole business of the collection of rubbish. This would probably be necessary to the securing of the great pecuniary return." Waring justified this plan by arguing that the "push-cart man who jangles his string of bells through the streets" carried on "a more or less illicit traffic with domestic servants." The city fathers would not only be enriching the public coffers but also increasing "the public safety."[49]

Waring attempted to usurp the work of "scow trimmers," who rummaged through the heaps of waste on the dumping scows searching for rags, shoes, carpets, paper, and anything else with a resale value. Until 1878, the city had paid the trimmers for their services, allowing them to keep what they salvaged. From 1878 to 1882, the subsidies were curtailed, and the scavengers were allowed to take what they wanted. Beginning in 1882, the city charged a flat rate for the privilege of trimming, realizing that the business had heated competition among rival scavenging crews. Most of the trimmers were Italian immigrants, organized into crews by local padrones. According to Waring, Italians were "a race with a genius for rag-and-bone picking and for subsisting on rejected trifles of food." Prejudice aside, the colonel saw an opportunity to bring the city into the arrangement in a more forthright manner and recommended that the city take over the operation completely. He said, "Dickens' 'Golden Dustman' and the accounts of the rag-pickers of Paris have made us familiar with the fact that there is an available value in the ordinary *rejectamenta* of human life. We learn by the work of the dock Italian of New York that to regain this value is a matter of minute detail; it calls for the recovery of unconsidered trifles from a mass of valueless wastes, and the conversion of these into a salable commodity."[50] In a short time, the city took most of the resulting profits from scow trimming and in January 1898, Waring established the first rubbish-sorting plant in the United States.[51] His schemes for resource recovery demonstrate the thoroughness of his commitment to sanitation as a municipal responsibility.

As with rubbish, he devised methods for the utilization of garbage, ashes, and street sweepings. Although he never became interested in incineration, he was enthusiastic about experiments in garbage reduction that coincided with his philosophy of utilization of waste. During his tenure as street-cleaning commissioner, New York City had a contract with the Sanitary Utilization Company, which extracted grease, other liquids, and dry residuum (for use as fertilizer) from the city's waste at its plant on

Barren Island. Waring also encouraged further experimentation to find more efficient and economical methods of reducing waste and utilizing the by-products. Eventually he hoped that this process would also be placed under municipal control.[52]

A long-term advocate of land reclamation for agricultural and other purposes, Waring saw the possibility of using ashes and street sweepings for landfill. Under his direction, a bulkhead was constructed around a shoal at Riker's Island in the East River for the purpose of beginning a fill. Pumping scows regularly steamed to the island to dump loads of ashes and street sweepings. The city also provided fill material at no cost to private owners of shore flats and participated in experiments to turn ashes and organic materials into fireproofing blocks.[53]

Waring was never content to make the street-cleaning department simply a servant of the people. Throughout his term of office, he used his position as commissioner to persuade citizens that sanitation was not only a municipal responsibility but also a community and an individual responsibility for which the department simply provided the leadership. His White Wings parades were meant to stimulate public interest in the work of the department, his program of primary separation demanded individual commitment to proper collection and disposal methods, and his public addresses and publications spoke of civic pride and civic duty. Beyond these efforts, Waring actively sought to involve specific segments of the population in his sanitation campaign. Waring was a master at utilizing human resources. As he said in October 1896, "I believed that if the people were once interested in [the refuse problem], and it were known that it was possible to disregard the ideas of the politicians, they would show such a desire for reform in this particular, that political influence would have no weight against them. This was the reason I did so many things that were considered 'injudicious,' 'dramatic,' and, perhaps even foolish. My plan was to force the department on to the attention of the people in every possible way, and I knew that the easiest way to do this was to introduce the personal element, and to make myself as Commissioner as conspicuous as I could."[54] To that end, Waring formed an advisory committee of civic leaders to help him analyze sanitation conditions and to offer possible solutions. He also regularly appeared before groups such as the Good Government Club, the City Improvement Society, the Ladies' Health Protective Association, the University Settlement, the College Settlement, and the Committee of Seventy. All the groups tried to help the department in its work.[55]

His most celebrated and innovative effort to encourage public involvement was the establishment of the Juvenile Street Cleaning League. Taking note of the participation of children in New York's Civic History Club and other patriotic organizations, Waring concluded, "it seemed possible to enlist their interest in the cleanliness of the city."[56] According to historian Daniel E. Burnstein, Waring, as a Progressive Era reformer, shared the conviction "that the environment, including surrounding street conditions, could greatly affect the moral trajectory of the child." Waring added, "Many reformers believed that children were the key to gaining greater community cooperation, that the young were the most important agents in changing the consciousness of large numbers of individuals regarding civic cleanliness." Organizing children in this way also was regarded as a way to offset the impact of street gangs and to instill civic pride and responsibility.[57]

Waring had several goals in mind in establishing the league. The children of the city could act as eyes, ears, and noses for the department in discovering unsanitary conditions and their perpetrators. The colonel assumed that the young league members might have greater success than adults in gaining the cooperation of litterers. Waring also wanted to educate the children in the ways of sanitation and civic pride: "[The children] are being taught that government does not mean merely a policeman to be run away from, but an influence which touches the life of the people at every point."[58] He hoped that the message would ultimately find its way to the parents. The popular, but bigoted, notion that citizens in working-class ethnic neighborhoods were the least sanitary and most prone to littering led the department to initiate the league on New York's East Side. As David Williard, a Department of Street Cleaning supervisor, observed, "To arouse a civic pride among New Yorkers is not distinctly within the province of the Department of Street-Cleaning. It is desirable, however, that an interest in the observation of the simple necessary rules of the Sanitary Code be awakened in the minds of at least the ignorant foreign population crowded into the East Side districts. To use for this end the influence of the children, who are recognized by their parents as superior to them in education and intelligence, is not a new idea, but one practically untried to any extent."[59] Waring argued, "If nothing is gained to the city except in a negative way, at least the neutrality of thousands of children has been purchased and the streets, are the cleaner from the fact that so many are kept from making them dirty."[60]

After a slow start, the Juvenile Street Cleaning League became a rousing success. In 1896, Waring directed Reuben S. Simons to seek the permission of the New York City Department of Education to deliver addresses on sanitation in the public schools and to try to organize the children into neighborhood leagues. Coaxing the boys and girls to join was no simple task; many of them were averse to volunteering for fear the department wanted to use them as spies. After intensive proselytizing, several groups were formed at settlements and in the public schools. In a very short time, there were forty-four leagues with 2,500 participants; by 1899, there were seventy-five leagues and 5,000 participants. The number of children in each club generally varied from twenty to fifty, with some branches having as many as 150 to 200 children. In 1913, Harlem housewife Sophie Loebinger formed a "Junior League" that roughly paralleled the work of Waring's juvenile street-cleaning clubs.[61]

Waring could not resist the temptation to organize the juvenile and the junior street-cleaning leagues along military lines. The members held weekly meetings, took a civic pledge, wore little white caps, and were issued badges. The leagues even established a ranking system—"helpers," "foremen," "superintendents"—depending on the services the children provided to the city. League members, sometimes five hundred strong, marched in parades as the White Wings had marched before them. At their meetings, the children sang songs with obvious messages, to the tunes of "Baby Mine," "As We Go Marching On," and so forth. Here is a sample stanza from "And We Will Keep Right On":

> There's a change within our city, great improvements in our day;
> The streets' untidy litter with the dirt has passed away.
> We children pick up papers, even while we are at play;
> And we will keep right on.[62]

The duties of the league members were rigidly outlined. Each week all the boys and girls were to record the number of people they had spoken to about sanitation, the number of bonfires they had extinguished, and the number of fruit peels they kicked into the gutter. The weekly reports were filled with good deeds done and diligence to duty. One youngster wrote:

> Col. Waring.
>
> Dear Sir:—While walking through Broome St., . . . I saw a man throughing [sic] a mattress on the street. I came over to him and asked him if he had no other place to put it but here. He told me that

> he does not no [sic] any other place. So I told him in a barrel, he then picked it up and thanked me for the inflammation [sic] I gave him. I also picked up 35 banana skins, 43 water mellion [sic] shells, 2 bottles, 3 cans and mattress from Norfolk St.
>
> Metropolitan League[63]

The Juvenile Street Cleaning League was one more example of Waring's efforts to bring the work of his department directly to the attention of the community at large. As Burnstein incisively concluded, "Waring's association of clean streets and moral tone signified the potent meaning ascribed to sanitation in the Progressive Era."[64] The idea of the juvenile leagues seemed farfetched at first, but they effectively publicized the need for civic involvement in resolving sanitary problems in particular and community-wide social problems in general. Although there is no way of quantifying the degree of success the leagues achieved, the response of the press and city officials from around the nation was very favorable. The participation of so many youngsters indicated that the colonel reached them in some basic way. Impressed by the results, other cities around the country, including Philadelphia, Brooklyn, Pittsburgh, Utica, and Denver, established their own juvenile leagues. In 1916, Philadelphia's leagues had ten thousand members. Unfortunately, the Juvenile Street Cleaning League of New York was disbanded in 1900, largely because of the lack of enthusiastic leadership that Waring had offered. It was eventually revived in 1909, however, by followers of Waring. In a few years, there were three hundred leagues throughout the city.[65]

Colonel Waring's whirlwind approach to reforming the street-cleaning department ended abruptly when Tammany Hall succeeded in turning out Mayor Strong in 1898. Much of what Waring had accomplished was undone or neglected (at least temporarily) under machine rule.[66] Yet Waring's comprehensive program, despite his untimely death, received extensive national attention and inspired the implementation of similar programs in many cities.[67] Waring had been able to produce a workable model for sanitary reform first and foremost because his message was clear: waste is a menace to health and to palatable living conditions and can be eradicated only through the coordination of municipal authorities and civic action. Waring's varied career and his ebullient personality gave him the necessary tools to promote his message. He was a skilled publicist and promoter who took to heart a famous punch line from an old joke: ". . . but first you have to get [their] attention." Waring exhibited an unbridled faith in elite rule, like many other Progressive Era reformers, but he also realized the importance of seeking a congenial public forum to promote his reforms. He was not solely a publicist and promoter, he was a man of action with an uncompromising belief in the art of the possible. He was not an ideologue, nor was he an environmentalist in the broadest sense of the term. He recognized the interrelationships among myriad health and sanitation problems, but he did not perceive the city in a broad, holistic sense. For good or for bad, he accepted the city on its own terms, never questioning its form or structure, the nature of its growth, or its economic activity. He was primarily concerned with the quality of human life and the need for proper sanitation. The thrust of Waring's sanitary programs are best described as "pragmatic environmentalism." Publicizing the department's activities and mustering all available human and material resources, Waring proved that change was possible without elaborate facilities and without blind adherence to a technological panacea. Waring was naive about the possible achievements of environmental sanitation in promoting health, but his instincts about acting upon obvious problems and his tenacity in promoting community involvement in sanitation paid rich dividends. Unfortunately, future generations of reformers would sometimes lose sight of the practical approach the colonel took in New York City.

Waring also came to be recognized as the leader in his field because his timing was perfect. He spoke the language of the progressive reformers, who sought to bring some order out of chaos created by the emergence of a new industrial society. Industrialization and urbanization challenged Americans' values, way of life, and state of mind. After the Panic of 1893, reformers began to realize that Americans must reconcile their old ways to the new, and accommodate themselves to the many changes a modern industrial-urban culture would bring. The fervent optimism at the heart of progressivism grew out of the belief that, given the proper physical, social, economic, and moral environment, Americans would overcome their problems.

Waring's nascent environmentalism, with its emphasis on improving the quality of urban life, coincided well with mainstream progressive thought. His goal to place public welfare above private gain mirrored progressive support for various civic improvements. Waring also was quick to declare his optimism about the prospects for change, and he exhibited a familiar faith in the inherent good of the people. In characteristic pose, he recoiled in moral outrage at the physical and social degradations of city life and envisioned himself as a noble crusader, much like a Jacob Riis or a Lincoln Steffens, against the forces of evil. Like other progressive reformers, however, Waring was paternalistic and an elitist whose faith in the people was tempered by the belief that leadership in reform must come from those best suited to the task, especially experts. This belief accounts for Waring's support of sanitation reform through municipal channels—at least those channels freed from machine dominance. In a real sense, therefore, Waring was a product of his time. He had harsh words for the old order that was still clinging to power, but he provided hope and enthusiasm for the wide-eyed reformers, who were confident that they could find solutions to the pressing problems of the day.

Refuse management was never the same after Colonel Waring. The "apostle of cleanliness" demonstrated that cities could move beyond the primitive and haphazard practices of the nineteenth century. A controversial figure to the end, he nonetheless pointed the way to modern refuse management and helped encourage an urban environmental consciousness which would produce some positive results in the twentieth century.

THREE

REFUSE AS AN ENGINEERING PROBLEM

Sanitary Engineers and Municipal Reform

In a 1906 issue of *Charities and the Commons* (later called the *Survey*), the editor proclaimed the rise of sanitary engineering as "a new social profession." This profession, he stated, "is neither that of physician, nor engineer, nor educator, but smacks of all three. It levies on autocratic powers, kin to those of ancient tyrants, but at the same time depends upon the sheerest democracy of information and co-operation to give its work effect."[1] Sanitary engineers were the twentieth-century heirs of Colonel Waring and his sanitary reforms. In one sense, they were highly trained (or experienced) specialists in the increasingly complex field of public works. They were the experts upon whom reform politicians depended to solve the pressing problems of advancing industrialization. In a larger sense, however, sanitary engineers were generalists when it came to broad questions concerning the maintenance of a viable physical environment. In the years before the appearance of professionally trained and academically educated ecologists, sanitary engineers were among the small minority of technocrats who possessed a comprehensive knowledge of the urban ecosystem.[2] It is no wonder that they came to dominate refuse reform in the early twentieth century. Their efforts to more scientifically measure the extent of the refuse problem, to devise modern collection and disposal methods and technologies, and to implement business efficiency in administering public works departments went a long way in giving credence to Colonel Waring's claim that refuse was more than a simple nuisance—that it was a serious environmental problem.

In the early twentieth century, sanitary engineers superseded health officers and sanitarians as the leaders of refuse reform in the United States. They accomplished this feat as much by default as by their training or environmental vision. In the wake of the bacteriological revolution of the late nineteenth century, many professionals in the health field came to regard environmental sanitation as an inconsequential means of combating disease, and virtually abandoned it. They placed their faith instead in the germ theory of disease transmittal, which led to the establishment of bacteriological laboratories and the widespread use of inoculation and immunization to eradicate communicable diseases. Dr. Fred B. Welch, commissioner of health of Janesville, Wisconsin, noted that the science of bacteriology and parasitology had "completely changed man's concept of environmental sanitation."[3] Doctors and health officials recognized the advantages of eradicating filth and cleaning physical surroundings, but they had been frustrated for years by their inability to prevent communicable diseases through sanitary measures alone. As medical historian John Duffy wrote, "The discovery of specific pathogenic organisms [especially through the work of Louis Pasteur, Robert Koch, and others] enabled public health workers to understand for the first time precisely what they were fighting."[4]

The transition from the widely accepted miasmatic, or filth, theory to the germ theory was not a simple one. Through the mid-1880s, anticontagionists offered strong resistance to the germ theory. Most people were unable to comprehend that something invisible and intangible could be the cause of disease. In 1878, even *Scientific American* chided the advocates of the new theory for accepting such a farfetched notion. Impressive reductions in the mortality rate between 1860 and 1880, attributable in some measure to good sanitation practices, further weakened the case for bacteriology. The death rate in most cities fell from twenty-five to forty persons per thousand in 1860 to sixteen to twenty-six per thousand in 1880. Rates were especially low in cities with sound sanitary practices. Nonetheless, the tenacity of the germ-theory advocates and the inconsistent results of environmental sanitation perpetuated the controversy. Not until the turn of the century did the contagionists successfully topple the miasmatic theory. The verifiable successes of immunization and inoculation and the advances credited to the bacteriological laboratories far exceeded the erratic record of "municipal housecleaning" and other sanitation practices.[5]

The contagionists' victory over the anticontagionists was only a qualified success. Adherents of the germ theory too easily accepted bacteriology as the sole means of preventing disease and too quickly dismissed environmental sanitation as a valuable practice in disease prevention. In the eyes of many in the health field, the emphasis on the environment as a root cause of disease had been misplaced, or at least exaggerated. The scientific base of environmental sanitation had been seriously flawed, but its goal—removing potential breeding cultures of disease from the range of human senses—had validity.[6]

The demise of the filth theory led to a critical appraisal of environmental sanitation as a function of health departments. Several public health officials questioned the necessity for health workers to supervise or direct the collection and disposal of waste. They recognized that refuse was in some respects a health problem, but few believed that its solution required the active involvement of municipal health departments. The views of Dr. Charles V. Chapin, superintendent of health of Providence, Rhode Island, and one of the pioneers in the American public health movement, typified the contagionists' beliefs:

> Though abandoning the time honored [filth] theory which was taught him, the writer has not abandoned the fight against filth. Filth is a nuisance, and is usually an evidence of some one's carelessness of his neighbor's comfort. The state or city should certainly protect its citizens against such nuisances. Good sewerage, well swept streets, prompt scavenging, public baths, clean tenements, are all parts, desirable and essential parts, of our civilization. They would be worth what they cost even if they had no relation to health; but the proper disposal of excreta and cleanliness of person doubtless do have much to do with the prevention of the spread of many communicable diseases. Much is to be gained by promoting cleanliness, but nothing by fostering false notions of the dangers of filth.[7]

Chapin's observations led him to conclude that the "filth nuisance" should not be a health-department responsibility. The health officer, he asserted, "should be free to devote more energy to those things which he alone can do. He should not waste his time arguing with the owners of pig-sties or compelling landlords to empty their cesspools."[8]

Chapin recognized the need to employ health experts with the highest possible levels of training and skill. In many ways, however, his opinions suggested a return to the nineteenth-century view of filth as simply a nuisance. While paying respect to the value of municipal cleanliness, he underrated the problem of waste as an environmental threat and underrated the importance of health-department supervision of refuse collection and disposal. His views were by no means the most extreme. Others in the health field were quick to abandon the old theories and were strident in their criticism of environmental sanitation.

At the 1912 meeting of the APHA in Washington, D.C., a very lively session focused on the relationship between public health and municipal waste. Dr. P. M. Hall, commissioner of health in Minneapolis, provoked an emotional debate when he contended that, although "unsanitary conditions of the home and unclean surroundings do not cause or originate infectious disease, . . . they do have a great influence upon the severity of the attack, and consequently have a direct bearing upon. . . mortality." He added that the origin of infectious disease was not "a closed book" and that the "teachings of centuries" should not be automatically abandoned but should be taught in conjunction with newer theories. "Why should we ignore the surroundings?" he concluded.[9] Hall's comments brought forth a flurry of rejoinders. M. N. Baker, an eminent sanitarian, flatly stated, "I do not think you can find any proof whatever that garbage collection and disposal has any material relation to health." Colonel J. R. Keane concurred: "There is no relation between garbage removal and public health." The sentiments of these men and others in attendance ran strongly against health-department responsibility for collection and disposal of waste. Dr. Gillett, of Colorado Springs, agreed that boards of health should not take up the question of removal of garbage, but, he asked, "[I]f they do not, who will?"[10]

Dr. Gillett's question was of central importance for those who continued to support the control of refuse collection and disposal by health departments or boards of health. It appears that those who adhered to the germ theory and were disenchanted with the results of environmental sanitation may not have given serious thought to who should or would be responsible for sanitation measures. Some of the advocates of health-department responsibility feared that the progress that had been made in providing sanitary service to the city would be lost as private contractors once again assumed major control and direction of municipal sanitation. At the 1917 meeting of the APHA, Edward D. Rich, state sanitary engineer of Lansing, Michigan, suggested that health-department supervision over collection and disposal of municipal waste would justify itself "if it did no more than to protect the municipalities from the unscrupulous vendor or untried or worthless devices."[11]

Regardless of the controversy over responsibility, most health departments did not control the collection and disposal of waste in the early twentieth century.[12] A survey of eighty-six cities, conducted by the American Child Health Association and published in 1925, showed that in forty-seven cities operating their own collection service, only nine (or 19 percent) gave the responsibility to departments of health (that figure represents only 10.5 percent of all cities surveyed).[13] In most cities where health departments did not oversee collection and disposal, they did have enforcement power over nuisances and, in some instances, supervisory power over the selection of contractors. Such compromises, however, often led to jurisdictional disputes among municipal departments and failed to resolve the question of whether health departments should retain full or at least partial control. The outcome was that health officers lost much of their influence over defining and controlling waste.

The engineer was the obvious choice to replace the health officer. The growing assumption that environmental sanitation was a task primarily requiring effective administrative and technical expertise, supplemented by some basic medical and scientific knowledge, pointed directly to engineers. Edwin T. Layton Jr. suggested that they were to be the "stewards of technology" in an increasingly mechanized and technically advanced world.[14] In the words of David F. Noble, "As he strove to create a professional identity for himself, the engineer commonly tried to present himself to the public as 'technology' itself, the great motive force of modern civilization."[15]

By 1900, engineering was the second-largest profession in the United States, following closely behind teaching. In that year, there were about 45,000 engineers in the United States. By 1930, their ranks had swelled to 230,000. Civil engineers were the first professional engineers. They emerged in the great canal- and railroad-building era of the early nineteenth century. By 1900, professional civil engineers had been joined by professionally trained mining, mechanical, electrical, and chemical engineers. As early as 1852, civil engineers had established their own professional organization, the American Society of Civil Engineers. Shortly thereafter, the American Institute of Mining and Metallurgical Engineers (1871), the American Society of Mechanical Engineers (1880), the American Institute of Electrical Engineers (1884), the American Institute of Chemical Engineers (1908), and other similar groups were formed.[16]

Inevitably, large numbers of engineers looked to the cities as their primary arenas of enterprise and opportunity. Rampant urban growth created massive physical problems that engineers were often best trained to address. By the late nineteenth century, engineers were playing major roles in American cities as consultants to city officials or as administrators and employees of various municipal departments, and were chief among the technocratic elite. Engineers acquired this role because they successfully promoted themselves as problem solvers through their expanding professional networks and organizations.[17]

Some municipal engineers became widely known for their accomplishments. Colonel Waring, who had built a lucrative business as an itinerant consultant before accepting the street-cleaning position in New York, was by no means the only engineer with a national reputation. Rudolph Hering was equally prominent but less flamboyant. His experience ranged from park surveys to bridge building, from sewer construction to water supplies. Among his best-known projects (all carried out in the late nineteenth century) were the extension of Fairmont Park in Philadelphia and the construction of sewerage systems for Washington, D.C.; Mexico City; and Santos, Brazil. He was also a member of the important Drainage and Water Supply Commission of Chicago between 1885 and 1887. Probably his most important contribution to sanitary engineering was his supervision of many investigations of water supplies, sewerage systems, and refuse collection and disposal practices. He traveled throughout the country and the world; his investigation of water supplies alone resulted in reports on more than 150 cities. He became an active or honorary member of almost every major organization in his field, and his expertise and public recognition led to his election as president of the APHA in 1913. In 1953, the APHA honored Hering by naming him the "Father of American Sanitary Engineering."[18]

Other important leaders in municipal engineering during the period were Ellis Sylvester Chesbrough, George Soper, Colonel William F. Morse, and William Mulholland. Chesbrough was the first city engineer of Boston (1851–1855), but he was best known as the engineer of Chicago's Sewerage Commission (1855–1879). Under his guidance, Chicago became the first city to implement a systematic sewerage plan, which included raising the level of the streets so that liquid waste flowed off into Lake Michigan. He also developed a system whereby the city water supply flowed through pipes under the lake and into the city.[19]

An expert in sanitary science and water purification, Soper was responsible for the sanitary rehabilitation of Galveston, Texas, after the disastrous hurricane and flood of 1900, which took six to ten thousand lives and destroyed $20 million in property. He spent most of his career in New York State attempting to eradicate communicable diseases. He gained a reputation as an autocrat—he was known to burn down infected houses and seize schools for makeshift hospitals—but he also was regarded as a dedicated and successful sanitary reformer.[20]

Colonel Morse, today probably the least-known sanitary engineer of the era, gained a significant reputation as a designer of modem incineration systems and a prolific writer on the subject of refuse collection and disposal. Like several of his colleagues, including Colonel Waring, Morse was sometimes accused of conflict of interest, especially with respect to his design and marketing of garbage crematories. Yet he made some of the most thorough examinations of collection and disposal techniques carried out at the time, which helped cities select sanitation systems more intelligently.[21]

Mulholland, like Chesbrough, was a major figure in the development of municipal water supplies. According to historian Michael Robinson, for fifty years Mulholland "led development of water supply systems that enabled [Los Angeles] to become a great municipality."[22] Born in Ireland in 1855, Mulholland emigrated to the United States in 1874. A self-taught engineer of the old school, he rose through the ranks of the Los Angeles City Water Company, a private firm that supplied the city with water. When the city assumed control of the system in 1902, Mulholland was retained as chief engineer, a position he held until his retirement in 1928. A major proponent of the famous Los Angeles Aqueduct, he was embroiled in the controversy over building the aqueduct from Owens Lake to Los Angeles, and exploiting inland Owens Valley for the sake of the burgeoning coastal metropolis. During the campaign for its construction, he was asked to predict the consequences of relying on the existing supplies. He bluntly answered, "Well if you don't get it now you'll never need it!"[23]

Legions of less well-known municipal engineers assumed leadership in established departments of streets, street cleaning, and public works, or were responsible for spearheading the establishment of "engineering departments" with a wide range of duties. Soon sanitary and other engineers were entrenched in several municipal governments throughout the country. According to historians Stanley Schultz and Clay McShane, "Labeling themselves neutral experts, engineers professed to work above the din of local politics. Usually they tried to isolate themselves from partisan wrangles, and often succeeded. In the creation of administrative bureaucracies, engineers apparently were the earliest municipal officials to achieve anything like job security."[24]

As the growing concentration of people in the central cities in the nineteenth century strained the meager city services, and as suburban residents clamored for the extension of these services to their communities, municipal authorities called on engineers to improve existing conditions and often deferred to them in making policy on these matters. The need for safe water supplies, adequate sewerage, well-ventilated housing, and efficient refuse collection and disposal required the engineers' technical expertise and the public health officers' knowledge of sanitation. A hybrid profession—sanitary engineering—emerged to try to meet the environmental challenge of the burgeoning industrial cities.

The origins of the new profession were to be found in Europe, not in the United States. During the 1870s, the emergence of sanitary engineers, primarily in England and Germany, coincided with the development of the biological sciences and the implementation of water filtration and sewage treatment in London and other major European cities.[25] In the United States, Hering, Waring, and others had led the way toward the professionalization of sanitary engineering. By 1890, a few professionally trained sanitary engineers were graduating from technical schools. By the early 1900s, leading private schools, such as the Massachusetts Institute of Technology, Carnegie Technical School, Harvard, Yale, Cornell, and Columbia, as well as major state universities of Ohio, Illinois, and Michigan, were offering courses and even whole curricula in sanitary engineering. Harvard and MIT jointly sponsored a Graduate School of Public Health, which trained students to hold administrative positions in the health field. Among those allowed to enroll were sanitary engineers.[26]

Like their European counterparts, American sanitary engineers received their first practical education in developing public water supplies and constructing citywide sewerage systems. In the post–Civil War years, most large cities were forced to abandon many of their local water sources—wells, cisterns, and local springs—for less-contaminated and larger-volume sources usually far from the city limits. By 1896, engineers were helping to provide more than three thousand new water sources, and by 1910, more than 70 percent of cities with populations over thirty thousand were maintaining their own waterworks.[27] The introduction of running water into urban residences and business establishments led to tremendous increases in water use. Per capita water consumption increased from about two to three gallons a day to between fifty and one hundred gallons a day. The convenience of running water also led to the widespread adoption of water closets; by 1880, about one-third of all urban households had them. This sanitary innovation added greatly to the growing consumption of water. The dramatic increases in water use placed excessive burdens on existing cesspools and privy vaults, since waste water had no place to go but into the soil or into the yards of adjacent houses. The health hazards implicit in this phenomenon and the inconvenience of overflowing water led city officials to support the construction of citywide wastewater systems.[28] After 1880, most major cities adopted sewerage systems to accompany or combine with their stormwater systems (in the form of underground sewers or sometimes open gutters). Naturally, sanitary engineers were asked to design and construct them.[29]

The efforts of the sanitary engineers produced impressive results. In 1860, there were only 136 municipal waterworks in the country; by 1880, there were 598. Increases in sewer lines of all kinds were no less extensive. The miles of sewers increased from 8,199 (in cities of more than ten thousand people) in 1890 to 24,972 (in cities of more than thirty thousand people) in 1909.[30] The central role of the sanitary engineers in the construction of abundant and effective water- and wastewater-carriage systems propelled the profession into the forefront of environmental sanitation in the United States. Sanitary engineers seemed to offer the best balance between technical and health expertise in dealing with the water problem.

Faith in technology fostered the belief that since the water-carriage problem had been solved by technical means, refuse could likewise be mastered through the skills of the sanitary engineer. An editorial in *Engineering News* commented, "Besides bad politics and general inefficiency in municipal administration, the chief hindrance to putting refuse collection and disposal on a satisfactory basis is the failure of the public and of non-technical city officials to recognize that the most difficult of the problems involved are engineering in character and will never be satisfactorily solved until they are entrusted to engineers."[31]

The public expected sanitary engineers to find immediate solutions to the refuse problem. Refuse, however, had always been more than an engineering problem. Despite growing adherence to the germ theory, refuse remained a health matter to be seriously considered. It presented even more confounding administrative problems than sewerage had raised. If engineers had relied solely on their technical expertise, their contribution to refuse reform would have been minimal. Instead, trained as environmental generalists as well as technical specialists, sanitary engineers advanced refuse management significantly, although they did not provide the quick solutions expected of them. They capably defined the range of issues associated with refuse collection and disposal through a more comprehensive understanding of the refuse problem than their predecessors had. Opportunities for thoughtful solutions, therefore, were greatly improved. Some contemporaries recognized the true nature of the sanitary engineers' skills. In 1901, John H. Emigh, city engineer of North Adams, Massachusetts, addressing a meeting of the American Society for Municipal Improvements (later the American Society of Municipal Engineers), commented on the engineer's role:

> It has been said that the problem of garbage disposal is becoming more and more a question of engineering. This is true. If I interpret aright engineering may be defined as a science that manipulates, applies and makes the best use of the laws of nature; and I am sure that every municipal engineer will agree with me that the problems connected with the manipulations of human nature are quite as intricate and difficult as any with which he has to deal. He cannot avoid having his acts modified by the opinions and sentiments of people competent or incompetent, right or wrong. Here, then, rests also his high duty and privilege, namely, that of leading his constituency into right conclusions, however tedious and torturous such processes may be. The intent of the mass of people and its tendency is to have conditions that are right and best; but it is deplorably unfortunate that so many mistakes are made, and faulty conditions prevail as the result of ignorant or selfish planning.[32]

Other contemporary evaluations of the sanitary engineer reflected an emphasis on the environmental generalist. In 1898, in his *Elements of Sanitary Engineering,* Mansfield Merriman suggested a broad definition of sanitary science: "Sanitary science embraces those principles and methods by which the health of the community is promoted and the spread of disease is prevented. Hygiene properly relates to the individual or to the family, but sanitary science has a wider scope and includes the village, the city, and the community at large."[33] Prevailing wisdom dictated that, to fulfill his assignment as the new protector of the community's health, the sanitary engineer should be broadly trained. William Paul Gerhard, a noted sanitary engineer, asserted: "The mere fact that a man is qualified in a single special branch—for instance, in house drainage or in the plumbing work of buildings—does not entitle him to be regarded as a sanitary engineer." Gerhard suggested that the sanitary engineer should be "a man with the broadest possible general culture. Only a person combining a liberal education with broad views can expect to attain a high position in modern life."[34] Gerhard thought that beyond basic engineering, the sanitary engineer should be acquainted with biology, chemistry, physics, medicine, architecture, law, and the social sciences.[35] Although this broad training suggests the ideal, educating practicing sanitary engineers and formalizing the curriculum at major technical schools, colleges, and universities brought some standards to the profession.

The contemporary view of the goals and responsibilities of sanitary engineers also indicates strong emphasis on the development of environmental generalists. According to A. Prescott Folwell, author of *Municipal Engineering Practice,* one of the most influential early guides to sanitary engineering, no one individual could have an expert's knowledge of all necessary subjects, but "a capable city engineer should have a good general knowledge of the most important; a pretty complete one of the most strictly municipal branches, such as paving and sewers; and sufficient common sense and moral courage to recognize when he needs the assistance of an expert in any line, and to secure it."[36]

Beyond the regular duties of sanitary engineers, some perceived a higher calling: sanitary engineers had to transcend their training and seek larger roles as community leaders in philanthropic and political capacities, especially as members of civic commissions, as municipal administrators, and even as officeholders. As Ellen Swallow Richards, instructor in sanitary chemistry at MIT, a pioneer in the field of ecology, and a leader of the home-economics movement, stated, "The sanitary engineer has a treble duty for the next few years of civil awakening. Having the knowledge, he must be a 'leader' in developing works and plants for state and municipal improvement, at the same time he is an 'expert' in their employ. But he must be more; as a health officer he must be a 'teacher' of the people to show them why all these things are to be. The slowness with which practicable betterments have been adopted among the rank and file is, partly at least, due to the separation of functions, of specialization, and partly to the exclusiveness of agents in the work."[37] Richards coined the phrase "public [or civic] engineer" to describe what she perceived to be the goal of the sanitary engineer in society.[38] Gerhard suggested an almost spiritual calling for the sanitary engineer: "Much of the sanitary engineer's work is necessarily of a missionary character, as the public must be educated to appreciate the benefits of sanitation."[39]

The views of Gerhard and Richards typify the elitist mentality that accompanies professional identity. There is always the danger that a professional will assume that he or she has a monopoly on the answers to society's ills; in this way, sanitary engineers came to view themselves as guardians of the urban physical environment. The benefit, however, of this growing professional identity for sanitary engineers was that it helped to formalize their training as both environmental generalists and technical specialists. As they grew in stature as leaders in sanitation reform, this dual role for sanitary engineers would ensure a relatively sophisticated view of the problem.

Sanitary engineers became a powerful force in municipal affairs not simply because of their apparent suitability for the task, but also because of the powerful influence they exerted nationally and internationally through their professional organizations. Involvement in various engineering societies and other professional group activities, plus access to a network of technical publications such as *Engineering News* and *Municipal Journal and Engineer*, provided appropriate media for the communication of new ideas and specifically for the reinforcement of the growing conviction that refuse was an engineering responsibility. One of the first and best-known groups dealing with the refuse problem was APHA's Committee on the Disposal of Garbage and Refuse. Many of the most respected sanitary engineers of the time, including Rudolph Hering, were members of this committee, which was created in 1887 to inquire into and make recommendations about the waste problem in the United States. In 1897, after a decade of research and the assimilation of thousands of pieces of data, the committee issued its seminal report, certainly the most thorough evaluation of American collection and disposal practices of its time. It incorporated information from approximately 150 cities, including evaluations of collection and disposal methods, costs, and comparisons with European methods. The 1897 report was the first step in developing a consensus on many of the refuse problems of the day and how to deal with them.[40]

The American Society for Municipal Improvements concentrated on the refuse problem. The ASMI was the first national organization to try to unite all municipal engineers into one group. In 1894, George H. Frost, publisher of *Engineering News;* M. J. Murphy, street commissioner of Saint Louis; and a few others brought together more than sixty city officials from sixteen cities at an organizational meeting in Buffalo, New York. In 1897, the ASMI resisted absorption by the newly formed League of American Municipalities, but in doing so lost almost all its mayor and councilman members to the new group. Thereafter, ASMI's membership became narrower, and the organization evolved into an engineering society that disseminated information about the newest municipal engineering techniques and encouraged professional exchanges and social interaction to forge bonds of cooperation among municipal engineers. The annual convention became the primary medium through which most of the ideas were transmitted, and in order to attract the widest representation, meeting places were often selected outside the manufacturing belt of the East, in such cities as Dallas, Texas; Birmingham, Alabama; and even Montreal and Toronto, Canada. Although the ASMI never achieved the numerical strength of other engineering societies, it had a steady growth throughout the early decades of the new century. In 1894, the membership totaled 53; by 1916, it was 552. Fewer than 20 cities were represented at the original Buffalo meeting; 266 cities were represented at the 1916 meeting in Newark, New Jersey.[41]

Other engineering groups devoted at least some attention to the waste problem. The American Society of Civil Engineers (ASCE) usually included sessions on refuse at its conventions and periodically established ad hoc committees to investigate garbage and related problems. From time to time, industry professionals, especially engineers, formed specialized groups not directly associated with major engineering organizations to meet a specific need or to provide a vehicle for the dissemination of information. For instance, in 1915 the Society for Street Cleaning and Refuse Disposal of the United States and Canada was established to "guide the thought and concentrate the effort to secure better conditions in street cleaning and refuse methods." Like other groups, it gathered statistics, held conventions, and distributed information about the newest techniques and equipment available to cities.[42]

Since the engineering profession had always had strong ties abroad, it is not surprising that many American sanitary engineers belonged to international organizations with shared interests in sanitation. In August

1900, the International Committee on Street Hygiene was formed as an outgrowth of the important International Congress for Hygiene and Demography. The committee consisted of seventeen members, with British engineer H. Alfred Roechling as chairman. The ubiquitous Hering was also a member. The committee met in Brussels in 1902 to outline its aims of the "furtherance of good sanitation" through street construction, repair, cleaning, and planting; public water closets; and proper disposal of waste. In 1904, at the International Congress of Engineers, three sessions were devoted to sanitary engineering, including one on the disposal of refuse. Most significantly, the International Association of Public Works Officials regularly conducted conferences on sanitary matters, including street cleaning.[43] Many sanitary engineers established global—although mostly trans-Atlantic—ties that helped broaden their perspectives and provided comparative models to judge domestic refuse-management programs.

Possessing broad environmental consciousness, administrative experience, and technical training, sanitary engineers were better prepared to confront the refuse problem than were the city bureaucrats and politicians who hired them or sought their counsel. Although some of their recommendations were criticized or ignored by municipal leaders in the politically charged atmosphere of city government, sanitary engineers offered realistic alternatives to the primitive methods still being widely practiced. They also provided the first national, and even international, perspective on refuse problems, which further helped discredit the shortsighted programs of the past.

Imbued with a penchant for the systematic and the orderly, sanitary engineers understood the importance of gathering and collating data about past and present collection and disposal practices as a necessary prelude to offering possible solutions. The first stage in the evaluation process was to take account of local conditions that affected the waste problem. Hering believed that local conditions had to be observed to "protect health, to avoid nuisance and to require an expenditure that is comfortably within the available means of the community."[44] To understand a community's problems in context, however, required more than evaluating local conditions, which is why the sanitary engineer's involvement in national and international organizations was helpful. From the late nineteenth century onward, engineering societies—often in conjunction with other groups interested in sanitation—applied their resources to conducting surveys and collecting data on collection and disposal practices in North America and abroad. As Hering stated, "The problem of refuse has . . . become more complex than formerly, and this complexity may not yet have reached its limit. In order that correct solutions for the best methods of disposal may be found, both from the standpoint of sanitation and economy, it is necessary to inquire into details far more than formerly, so as to have more definite facts and figures with which to solve the problem. The more accurate information now required is necessary for the varying special conditions existing in different communities. In short, we must have more special data and statistics before we can indicate the best methods for the disposal of a particular town's refuse."[45] The importance of the surveys often transcended their use as envisioned by Hering. Such surveys ultimately provided a national focus for the refuse problem, allowing investigators to observe trends and patterns not recognizable when the problem was examined from the narrow local perspective. Indeed, much of the information produced a consensus among engineers about the important aspects of the waste issue.

The surveys varied in quality and comprehensiveness. Some were based on extensive statistical samples, while others were less scientific. Despite these problems, a veritable reservoir of information became available to city governments. APHA's 1897 "Report of the Committee on Garbage" was a model upon which many subsequent investigations were based. Although the committee did not actually conduct its own local investigations or experiments, it did collect and collate statistical information, inspect disposal works, and evaluate the evidence and opinions of others. The committee also examined the contemporary literature on the subject. The report in many ways simply restated the obvious, suggesting that local conditions often dictated collection and disposal practices, but it presented the first panoramic view of the refuse problem in the United States. Unfortunately, because of a shortage of funds, the report was published without the accompanying statistical tabulations. It nonetheless provided encouragement to other groups hoping to refine the study and determine its applicability to their city or region. Between 1900 and 1917, the report inspired many additional studies and special municipal investigations. Cities such as Buffalo, Chicago, Louisville, and Washington, D.C., carried out extensive inquiries.[46] With the flurry of investigations in the early twentieth century, the APHA's garbage committee, realizing the problems inherent in interpreting disparate data from a wide array of sources, devised the "Standard Form for Statistics of Municipal Refuse" in 1913 to bring some order to the collection of data.[47]

Collection of data and investigations of city practices offered a sounder base for evaluating the sanitary needs of communities than the qualitative measures the nineteenth century had provided. Sanitary engineers were not satisfied simply to collect statistics, however. Implementing new programs or revising old ones was a necessary second step, which required effective administration. Advising municipalities on ways of placing their public works departments on a business footing, sanitary engineers sought uniformity in organizational methods. Several of them, such as the noted Samuel A. Greeley, looked to European practices for guidance: "One who has investigated refuse disposal work abroad, cannot but be impressed with the uniformity of the methods used there, more especially for the house treatment and collection of refuse, but also for the disposal. German cities are approaching a common standard in this work. Differences in the effectiveness of the service are in large part due to differences in the ability of the department chiefs in charge of the work."[48] Sanitary engineers vigorously debated the advisability of implementing specific European programs, but they tended to share Greeley's admiration for European organization.

Careful record keeping came to be regarded as one of the best ways to improve municipal refuse management. For example, determining the amount of moisture and combustible matter in garbage or considering the potential by-products in the wastes helped city officials determine the most efficient and effective methods of disposal. Cities often failed to take such factors into consideration before choosing new disposal methods. As a result, thousands of dollars were squandered on expensive equipment that did not produce the desired effects. According to *Engineering Record,* "It is not enough to make a few estimates in the winter and in the summer, but a year's records obtained at least once every month and preferably more often should be secured. Unless this is done, any system of collection and disposal that is adopted is based on guess-work, and every intelligent citizen has had enough opportunity to observe the results of guesswork in municipal affairs to be loathe to encourage anything of that sort."[49]

Sanitary engineers almost universally supported municipal control of sanitation functions. Although commitment to community responsibility in street cleaning and garbage removal was spotty during the 1880s and 1890s, the increasing emphasis on "home rule" in the larger cities and the subsequent expansion of municipal bureaucracies made municipal control of sanitation seem more practicable. It might seem that a large portion of the engineering community would disapprove of municipal control, especially since many sanitary engineers made their livings as itinerant consultants to city officials throughout the country. Some engineers, like Colonel Morse, channeled their engineering expertise into entrepreneurial ventures, such as manufacturing and marketing collection and disposal equipment. Yet the itinerant consultant and the engineering entrepreneur had vested interests in municipal control of sanitation. At the time, municipal control provided a greater degree of permanence than contracted services. Particularly in the large cities, new sanitation programs required a large investment and considerable effort to produce complex management systems. Furthermore, city-provided sanitation services ensured jobs for sanitary engineers, jobs that afforded them substantial responsibility, authority, and power. The sense of permanence and stability that municipal control implied offered the best atmosphere for the sanitary engineer to promote his career and the necessary reforms required to produce a well-organized program of sanitation. In their influential book *Collection and Disposal of Municipal Refuse,* Hering and Greeley stated bluntly, "The collection of public refuse is a public utility."[50]

Any support for the contract system among sanitary engineers was modified to include more rigorous municipal supervision of contracted work. Not everyone accepted consulting engineer Louis L. Tribus's matter-of-fact conclusion that "unsightliness and uncleanliness have gone hand in hand with private collection," but many tended to agree.[51] M. N. Baker condemned what was considered the greatest abuse of the contract system, the short-term (annual or biannual) contract. According to Baker, the contract was "one important cause for the unsatisfactory condition of garbage disposal in the United States." The short-term contract came under attack not primarily because it granted exclusive responsibility to the private sector but because it provided little long-range planning for the city. Baker repeated an oft-stated conclusion that contracts were granted for political reasons, and thus suspect. He also strongly emphasized that the duration of such contracts was too short to allow for the adoption of new disposal methods, which required extensive coordination and planning, not to mention substantial costs.[52] Unlike political reformers who spoke about the contractor's indifference and lack of commitment to civic betterment, the sanitary engineer tended to emphasize the technical and administrative problems associated with the contract system.

Surveys conducted from the 1890s to the outbreak of World War I indicate a substantial shift away from the contract system and toward increased municipal responsibility in street cleaning and garbage collection and disposal. By 1880, a significant number of American cities (approximately 70 percent) had made important strides in the municipal control of street cleaning, and a survey of 150 cities taken in 1914 indicated that about 90 percent of the cities had assumed responsibility for street cleaning.[53] In collection and disposal, which had traditionally lagged far behind street cleaning, surveys from the 1890s onward indicated a steady rise in municipal responsibility. By World War I, at least 50 percent of American cities had some form of municipal collection system, as compared with only 24 percent in 1880.[54] Whether the sanitary engineers influenced this trend is difficult to determine, but there is little question that municipal control of sanitation provided them with a solid base of operations in municipal government.

Once sanitary engineers began exploring the magnitude and complexities of the refuse problem, they came to realize, more than any other group had before them, that tinkering with the existing programs would be insufficient. Consequently, they advocated centralization of refuse oversight through municipal departments operating under the newest managerial techniques. Their administrative and organizational procedures borrowed heavily from private industry as well as from successful water- and wastewater-carriage programs. While these techniques had validity for refuse management, solutions to the problems of collection and disposal required further examination and study. As William T. Sedgwick, a leader in water-pollution control, suggested, "Doubtless one reason why the refuse problem is still so vexing in most communities is because the collections, of necessity, must be intermittent, and yet the system must function smoothly and without interruption."[55] Sedgwick, of course, revealed only a small portion of the issue, since no single method of collection and disposal could be universally applied.

Although sanitary engineers were unable to find the ultimate technological methods of collection and disposal, they did make strides in evaluating the relative strengths and weaknesses of primitive methods and the potential applicability of new techniques. The luxury of hindsight allowed them to understand many of the limitations of nineteenth-century methods. They almost universally condemned, or at least criticized, such practices as land and sea dumping, open burning, and filling with untreated wastes. In an informal discussion at the annual meeting of the American Society of Civil Engineers in 1903, John McGaw Woodbury, the street-cleaning commissioner of New York, asserted, "It is of the utmost importance that dumping at sea be stopped, not simply because it makes the beaches unsanitary and unsightly, but because it is a waste of valuable material."[56]

Woodbury and other critics of primitive methods were not in a position to take into account ecological considerations, but the ever-increasing quantities of refuse convinced sanitary engineers that more comprehensive criteria had to be devised in selecting methods of waste disposal. As pragmatists, sanitary engineers placed emphasis on the need to examine local conditions thoroughly before suggesting appropriate collection and disposal methods for a given community. In planning new systems, they tried to consider not only the types and quantities of wastes but also the quality of local transportation facilities, the composition of the agency in charge of the work, and physical characteristics of the city that might determine the type and location of the disposal system. Although they were no more sensitive to questions of race, ethnicity, and class than other bureaucrats at the time, engineers also began to consider political and social factors that might indicate the receptivity of the local government and citizenry to changes in sanitation practices. Sanitary engineers' reliance on a technological solution to the refuse problem, therefore, was increasingly tempered out of necessity by considerations that were relatively broad in scope and not directly related to refuse as an engineering problem. In a sense, sanitary engineers, as implementers of new systems, began adapting to the complexity of the issue, relying more on their roles as environmental generalists than on their positions as technical specialists.[57]

Because of their low regard for primitive methods of collection and disposal, sanitary engineers focused their attention on newer methods that they hoped would fulfill the sanitary and financial requirements of the various cities. Incineration and reduction were most often discussed as the logical alternatives to the older methods. Each method had its advocates and detractors, but both came under greater scrutiny than they had when first introduced in the late nineteenth century. The naive view that either method offered the ultimate answer was gone. The maturity of the sanitary-engineering profession, and the knowledge gleaned from hasty implementation of poorly tested equipment, brought a more measured response to the unqualified advocacy of a single method.

The success of the English destructors abroad had caused the APHA's refuse committee to give its provisional endorsement to incineration in 1897 and had also led to the rapid implementation of the method in several cities.[58] As further study of the British systems' application to American needs indicated, however, some municipalities had acted rashly in building untested or inappropriate equipment. Colonel Morse had argued that several failures occurred because of insufficient professional analysis of incineration principles, faulty incinerator designs, overconfidence in the capabilities of the apparatuses, and unskillful management of the crematories.[59] With further study, it became clear that British destructors could not be employed without careful adaption to American needs, both financial and technical.[60]

The reduction process, which appeared in the United States at about the same time as incineration, went through a similar evolution: impulsive implementation, severe criticism, and reevaluation. The 1897 report had indicated a great deal of interest in the method because of its promise of returning revenue to the city.[61] Experiments by Colonel Waring and others had generated considerable interest, but after a period of operation in some cities, several undesirable side effects caused increased criticism of the method. The foul odors emanating from the plant raised loud protests. Memphis mayor J. J. Williams complained: "The air for miles and miles around [the reduction plants] is so contaminated that the courts and lawmakers have been appealed to, and have, as a rule, given relief to the sufferers by abating the foul, diseasebreeding [sic] business. . . . if for no other cause, the laws should prohibit these establishments, because it is degrading and inhuman for human beings to spend their days in such an occupation as assorting the filth of our cities."[62] In 1916, ASMI's Committee on Refuse Disposal and Street Cleaning recommended a compromise. Reduction was fine for large cities where the revenue derived from it might warrant its use, but for small cities, incineration appeared to be more sanitary and less costly.[63]

The days of haphazard collection and disposal practices appeared to be ending with the rise of sanitary engineers. The careful accumulation of data, the design and evaluation of new equipment, and the organizational structure of public works departments were the kinds of improvements admired in an increasingly complex—and confounding—age. In the chaos of the newly emerging urban-industrial society, dependence on experts seemed imperative. Not only technical expertise but managerial skills and efficiency were especially prized talents. As Samuel Haber suggested, many Americans were having a love affair with efficiency: " . . . the progressive era gave rise to an efficiency craze—a secular Great Awakening, an outpouring of ideas and emotions in which a gospel of efficiency was preached without embarrassment to businessmen, workers, doctors, housewives, and teachers, and yes, preached even to preachers."[64]

Yet the impact of sanitary engineers on refuse management relied at least as much on their ability to perceive the totality of the urban environment as on their skill in tinkering with its parts. Had there been greater cooperation between health departments and public works departments during the period, that perspective might have been broader yet. Nonetheless, sanitary reform came a long way under the guidance of the sanitary engineers. They effectively refined Colonel Waring's rudimentary environmentalism and formalized it through their young profession.

Operating from within the institutional structure of municipal government, sanitary engineers had little occasion (or inclination) to define their responsibilities in terms of popularizing refuse reform or promoting civic involvement in finding solutions. They often failed to grasp that to the nonspecialist and to the layman, refuse was primarily an aesthetic problem—a nuisance. Responding to sanitary reform with emotional zeal, early in the twentieth century voluntary citizens' groups—especially women's organizations—added an aesthetic perspective and a social consciousness to the sanitary engineers' technical and organizational reforms. Somewhere between the time of Waring's death and World War I, refuse reform found expression in two distinctive, though not totally divergent, groups: one technical and organizational, the other aesthetic and civic-oriented.

FOUR

REFUSE AS AN AESTHETIC PROBLEM

Voluntary Citizens' Organizations and Sanitation

In his widely circulated technical tract *Garbage Crematories in America* (1906), William Mayo Venable commented optimistically on the growing interest in sanitary reform in the early twentieth century, "The reason why the problem of refuse disposal is receiving an ever-increasing amount of attention from engineers, municipal authorities, and from the American public does not lie in the newness of the problem, but rather in an intellectual awakening of the people. The same spirit that leads men to realize the corruption of politics and business, and to attempt to remedy those conditions by adopting new methods of administration and new laws, also leads to a realization of the primitiveness of the methods of waste disposal still employed by many communities, and to a consequent desire for improvement."[1] The sporadic protests against the garbage nuisance in the nineteenth century had made way for better-organized and more comprehensive reform efforts in the twentieth century. Yet, as sanitary reform broadened in scope and appeal, it also splintered into two distinctive, though not totally independent, factions. Sanitary engineers, a technically elite group that functioned within the municipal government, dominated the first. The second group, primarily composed of citizens' organizations, operated in the public realm. Seeking the same end, each faction approached the refuse problem from a different perspective. Sanitary engineers attempted to identify the problems associated with waste collection and disposal and offered administrative and technical solutions. Although they were in a good position to influence policymaking within the city bureaucracy, sanitary engineers' professional allegiances and obligations tended to limit their effectiveness as popularizers of reforms at the grassroots level. Citizens' groups and civic organizations, on the other hand, had to rely on some form of public support or public participation to give their protests momentum. These two strains of reform were not at odds; they appealed to different audiences, different constituencies.

The significance of the civic dimension of refuse reform did not lie principally in efforts to devise new collection and disposal programs or in the writing of new ordinances. The institutionalization and bureaucratization of refuse management placed responsibility for the former directly on the public works and engineering departments. Writing new ordinances, although important, simply resulted in updating existing laws or attempting to make them more enforceable. The major impact of civic reform was educational: it publicized and popularized a new environmental ethic of cleanliness and efficiency, and intended to promote greater civic involvement.

The time was right for public action. Especially after the Panic of 1893, a municipal-reform impulse permeated all facets of urban life. The depression of the 1890s emphasized the cost that unregulated economic growth exacted from the country, especially from major urban centers dependent on industrial production. Would-be reformers, reflecting on the benefits versus the costs of rapid economic and physical growth, argued that the time had come to consider the priorities of the new industrial society. In 1896, Thomas C. Devlin wrote:

> This magical growth of cities has been the pride of the people. The metropolis of each state is a sort of Mecca for its people. The cities have become a gigantic power in the political and social life of the nation. It is the intellectual force of the American cities which shapes legislation through state and national conventions. Within them are all the allurements and excitements of modern life. . . . This movement is necessarily attended by many evils, and much has been said and written to counteract the crowding of cities. Such efforts are useless. The present population of cities is permanent. The evils which have been incidental to their rapid growth must be eliminated. Living in cities must be desirable.[2]

"Living in cities must be desirable" captures the essence of urban reform in the late nineteenth and early twentieth centuries, especially the civic phase. Urbanites began responding to the need for what Roy Lubove called "the creation of a socially integrated and physically beautiful city."[3] The determination to "humanize" the urban environment was not original to reformers of the post-1893 period, nor were their proposed solutions. Reforms that had received only sporadic local attention in the 1870s and 1880s became national issues in the late 1890s. Only a year after the Panic of 1893, the United States was experiencing the most intense public interest in municipal affairs in its history. In early 1894, there were fewer than 50 urban reform groups; by December, the total had risen to well over 180. Every large city, and many smaller ones, had at least one reform organization by 1896. Enthusiasm for the groups' activities broadened as well, especially among academics, city officials, and journalists. In 1909, there were more than one hundred periodicals dealing primarily with urban affairs.[4]

The momentum generated by the municipal-reform groups touched many nagging problems that had befuddled and annoyed city dwellers in the past and brought attention to other problems that had been ignored or had gone undetected. Civic groups confronted issues in areas ranging from education to entertainment, from parks and public buildings to water purification, from smoke abatement to transportation. For the moment at least, the mood for reform had no bounds. It was the anarchy of the reformist mood, the boundlessness of zeal, and the depth and breadth of enthusiasm, as much as the substance of the problems themselves, that provided a public forum for many civic issues.[5] Municipal governments would, of course, be responsible for the urban revival. As Devlin wrote:

> To protect the lives and property of the people, to care for the morals, health, education, and protection of the community, to foster its interests, to manage its finances judiciously, to demand and secure the best service in every work under its supervision, are all within the purview of city government. Shall these be done well or shall the evils that have followed in the wake of great national progress finally engulf the whole? There is more national patriotism and local pride than ever before, but less civic spirit. This too is being aroused, and the intelligence and energy which have accomplished so much in the development of industries and the building of cities will also find good forms of government.[6]

The proliferation of municipal-reform organizations had several causes: growing civic pride, intercity rivalries, new interest in community-wide projects and building programs, humanitarian concerns, and physical disasters and epidemics. The realization of common goals drew citizens closer together. Voluntary citizens' organizations and taxpayers' associations were the first groups to become interested in civic betterment (several of them originated even before the reform upsurge). Some professed a general interest in municipal reform, taking such names as the City Improvement Society, the City Reform Club, and the City Government Club. Others had more specific interests and goals. The Anti-Spoils League of New York, for instance, sought to abolish the patronage practices that had undermined public-service jobs. Another group, the Society for the Prevention of Crime, attempted to make the city a safer place to live.[7]

Municipal officials and bureaucrats also acquired group identities at this time. The professionalization of service departments and the reorganization of city governments through charter revisions or through the adoption of new forms such as the Commission Plan and later the City Manager Plan provided the impetus for these associations.[8] Some formed alliances with citizens' organizations and vied for leadership of reform movements. Among the first organizations of municipal officials, both of which had specialized interests, were the Municipal Finance Officers Association and the National Association of Port Authorities. Most significant, however, was the rise of groups composed of executive officials, such as the Conference of Mayors and the Conference of City Managers. Like the various engineering societies, the executive officials of these organizations sought out their peers to form state and national associations whose interests transcended local problems.[9]

The National Municipal League (NML), an outgrowth of the first National Conference for Good City Government (held in 1894), was the best known of the municipal associations. Herbert Welsh of the Municipal League of Philadelphia organized the NML and attracted representatives from New York City, Brooklyn, Chicago, Boston, Baltimore, Minneapolis, Milwaukee, Albany, Buffalo, Columbus, and, of course, Philadelphia. The league began in May 1894 with 16 affiliated local groups. By the next year, the ranks had swelled to 180 branches, and 80 or 90 more were added in 1896. The core of the membership came from the mid-Atlantic states, especially New York and New Jersey.[10] A rival organization founded in 1897, the League of American Municipalities acquired strength from small and medium-sized cities instead of from the larger metropolises, where the National Municipal League dominated. Despite the emergence of these major associations, state organizations continued to flourish, largely because they concentrated on manageable issues relevant to the regions they served. In 1901, there were only eleven state organizations, but by 1915, their number had increased to twenty-nine.[11] In the congenial reform atmosphere of the 1890s, environmental problems in general, and sanitation problems in particular, gained wide interest. Water, smoke, noise, and refuse pollution became important subjects of community protest, since they affected a large cross section of the population. The progressive spirit that dominated the reform milieu of the day was geared to view social problems in an environmental context—human beings, it was believed, would seek "the good" if the physical and social environments were made less hostile. As sanitary engineers borrowed the gospel of efficiency from progressivism, citizens' organizations and their allies borrowed its civic-mindedness, its emphasis on aesthetics, and its moralistic tone. Waste, by polluting the physical surroundings, threatened health and promoted squalor. Primitive collection and disposal practices were signs of backwardness and barbarity; civilized societies were well kept and sanitary. One could hardly expect citizens to seek moral and material progress in a despoiled habitat polluted by litter and disease-breeding refuse. In the broadest sense, filth bred chaos, while cleanliness promoted order. Civic pride and responsibility, therefore, were necessary attributes of urbanites committed to municipal improvement.[12]

Historian Melvin Holli has argued that most urban reforms of the late 1890s were "basically refinements of older concepts."[13] That is certainly true of sanitation reform. Civic groups continued to emphasize, much in the vein of the protests of the Ladies' Health Protective Association in the 1880s, that waste was a health problem. The rhetoric of progressivism, however, changed the tone of sanitation reform, and national publicity secured for it a larger and more diverse audience. This publicity for sanitation reform was achieved largely through its association with the City Beautiful Movement, which swept the nation in the 1890s. At the Columbian Exposition in Chicago in 1893, which marked the four hundredth anniversary of the discovery of America, citizens were treated to the spectacle of the White City: "In contrast to the sprawling, ugly industrial cities that were becoming common at that time, the classical buildings rising from blue lagoons, with their white plaster-of-Paris facades gleaming in the sun, seemed to be a vision of a lost utopia. Observers called it the White City, and it epitomized cleanliness, grandeur, beauty, and order. The lush green lawns and the frequent display of statuary combined with the architecture to impart a classical flavor to the entire exposition."[14] The visual impressions of the exposition provided dramatic impetus for an aesthetic revival in American cities. David Goldfield and Blaine Brownwell remarked, "The city no longer need be perceived as an ugly, ungainly giant, but rather as a beautiful, graceful physical creation of human beings. The entire city could be downtown."[15] Almost overnight many projects for city beautification were begun, and major redevelopment plans were designed for Washington, D.C., and Chicago.[16]

The link between City Beautiful, with its grandiose objectives, and sanitary reform, with its more modest goals, might seem tenuous at first glance. The connection was natural and complementary, however. In a perceptive article, historian Jon A. Peterson effectively demonstrated the importance of the civic improvement idea as a pillar of City Beautiful. Unlike those who claimed that City Beautiful was primarily an effort to promote Classic and Renaissance architecture and monumental planning, Peterson argued that the movement "had other meanings and origins and that their recovery enables us to recognize the phenomenon as a complex cultural movement involving more than the building arts and urban design." Three distinct concepts helped to launch City Beautiful: municipal (or decorative) art, outdoor art (city park development and landscape architecture), and civic improvement. Only after 1901 did City Beautiful appear to shift toward urban planning in the most traditional sense.[17]

The civic improvement legacy, according to Peterson, began as a "laymen's cause" primarily in small and medium-sized cities, especially through village improvement societies. These groups were formed as early as the mid-nineteenth century. Civic improvement gained national attention through the National League of Improvement Associations (NLIA), established in 1900. Because of the close relationship between the goals of civic improvement—order, cleanliness, moral uplift—and the roles of women as mothers and housewives, women reformers dominated many of the local groups. The NLIA, renamed the American League for Civic Improvement in 1901, defined its goals as: promoting outdoor art; public beauty; and town, village, and neighborhood improvement. It emphasized civic improvement over the narrower village-improvement concept of the past. Through this maneuver, as Peterson noted, "the League aligned itself with the reform ethos of the era." For many people, City Beautiful was "the aesthetic expression of turn-of-the-century urban reform."[18]

The emphasis on the visual was City Beautiful's major contribution to sanitary reform. Although there was an indirect connection between sanitary improvement and the architectural and landscape modes of City Beautiful, there was a direct relationship between the cleansing of the physical surroundings and civic improvement. For example, the Civic League of Saint Louis sponsored programs such as a "Keep Our City Clean" campaign, and helped secure the appointment of a female sanitary inspector.[19] Sensory considerations—sights and smells—had often played a role in the pre-1893 sanitation protests, but City Beautiful institutionalized reform for aesthetics' sake on a national scale.

By promoting sanitary reform chiefly as an aesthetic and civic issue, citizens' organizations tended to underrate other important considerations. This meant emphasizing health and comfort, moral uplift, and city beautification while downplaying scientific management and efficient administration of public works. The layperson's approach to sanitary reform was broad enough to encompass a wide range of issues that the populace could identify as improving the urban quality of life. Reformers equated City Beautiful with city cleanliness. Rev. Caroline Bartlett Crane, a well-known health expert, noted: "We gladly hear much to-day of the movement for civic art; but it is well to remember that civic art without civic cleanliness is a diamond ring on dirty hands. The adornments of a dirty city do but emphasize its dirtiness, while cleanliness has not only a virtue but a beauty of its own."[20] Some reformers carried the argument a bit too far, however. The Friday Conversational Club of Monongahela, Pennsylvania, published the following:

> DID YOU EVER STOP TO THINK THAT:
> A clean town means a sanitary and healthful town.
> A clean town means a more beautiful town.
> A clean town means an increase in the value of our property.
> A clean town brings business to our merchants.
> A clean town induces a better class of people to locate here.[21]

Fusing visual appeal of City Beautiful with the more traditional health argument broadened the impact of sanitary reform in the public realm without fundamentally changing its purpose or goals. In an atmosphere of civic-mindedness, neighborhood protests with only local significance became citywide crusades with state and even national ramifications. Periodicals and published proceedings of civic groups were filled with stories describing the refuse problems of various cities and towns and the need for immediate solutions.[22] Demands for municipal management of collection and disposal were greater than ever before. Most reformers accepted the idea that city taxes should provide decent living and working conditions, not simply protect lives and property. This view substantially expanded the perceived role of municipal government in citizens' lives.

The private contract, as might be expected, became the most obvious symbol of corruption and inefficiency. Reformers claimed that the contract system of collection and disposal was an unworkable compromise between personal responsibility and municipal authority. In a speech delivered at the annual convention of the League of American Municipalities in 1903, one staunch advocate of municipal responsibility stated a widely accepted view of urban reform: "When the proposition is finally accepted that municipal governments are incapable of doing every class of public work cheaper and better than it can be done by letting the work to private contractors, then the failure of municipal government is conceded, and the inability of the people to govern themselves finally established."[23]

The fear of private monopoly, not uncommon among reformers of all kinds in the period, lent support to the anti-contract sentiment. Many environmental reformers in particular feared that privately controlled water supplies or energy supplies (particularly coal) threatened a vulnerable urban society. Public ownership, whatever weaknesses it had, was a better gamble than contract systems.[24] A private monopoly in garbage collection hardly posed the same potential danger as a privately controlled water supply, but the principle was the same: public works were meant for the benefit of all the citizens; they should not be the means of private gain. For civic reformers, the question of municipal responsibility transcended questions of efficiency; municipal responsibility was a safeguard against damage to the public welfare.

In the early twentieth century, reformers increasingly demanded the termination or alteration of contracted collection and disposal systems. In 1898, the Philadelphia Municipal Association sent a message to the city council strenuously opposing the awarding of garbage contracts to the same combination of bidders who always won them. Charging favoritism and declaring that the bidding process was "a cunning scheme for robbing the city," the association complained that there was an effort afoot to keep large contracts of that kind out of competitive bidding. In Saint Louis in 1912, members of the Civic League distributed circulars to civic groups throughout the city to promote the move to municipal collection of ashes and rubbish. They also lobbied for appropriations to implement a plan for municipal disposal that the city council had passed two years earlier. Efforts like these stirred up public dissatisfaction with the contract system and the handling of refuse in several cities.[25]

Despite their unwavering support of municipal control of public works, civic groups did not assume that all the problems of the past could be rectified by a new ordinance. They hoped that greater municipal commitment to such services as garbage removal and street cleaning would mean a greater sensitivity of government to the needs and demands of the people as a whole. Believing that they were spokesmen for the public at large, civic groups hoped that their activities would give them permanent access to the sources of political and administrative power.

These groups tried various methods of entering the decision-making process. Usually they were concerned with bringing the refuse problem to public attention. They wrote, printed, and disseminated thousands of pamphlets and handouts and also initiated or supported citywide investigations. The Citizen's Research Council of Michigan sponsored an extensive investigation of street-cleaning practices in Detroit. Relying on experts inside and outside the city to prepare the study, the council selected Raymond W. Parlin, deputy commissioner of street cleaning in New York City, and H. S. Morse, of the Detroit Bureau of Government Research, as chief investigators. Completed in 1917, the report was intended to "indicate conditions and problems, and suggest solutions which have proven successful in communities of similar size and conditions." The investigation team made a number of recommendations and suggestions, including the complete reorganization of the Detroit Street-Cleaning and Sanitation Department. The report also contained recommendations to extend service to areas not previously included, update the methods used to clean the streets, and offer greater incentives to sanitation workers.[26]

Significant investigations sometimes grew out of smaller or more specific projects, as in the City Club of Chicago's 1911 conference on the city's mosquito problem. The club's Committee on Public Health sponsored a local conference to consider the mosquito nuisance as it related to public health problems, and members established a special committee to continue the investigation. Later, the committee's functions were expanded to include a wide range of problems related to household pests, which inevitably led to greater attention to municipal cleaning practices.[27] By initiating investigations, civic groups often provided momentum for citywide action.

Some organizations chose to employ more direct action, with varying degrees of success. The Citizen's Health Committee of San Francisco, like many similar groups throughout the country, had applied public pressure to force local scavengers to improve their collection practices. In 1909, the committee members took some of the responsibility into their own hands and distributed three thousand garbage cans among the poor to keep the refuse from attracting rats. They also urged those who could afford it to purchase cans with tight lids. The garbage-can drive ultimately led to an ordinance in San Francisco making the use of receptacles compulsory.[28] In another important campaign, the Civic League of Saint Louis relentlessly sought a comprehensive plan for refuse collection and disposal in the city. In 1906, the Public Sanitation Committee of the league published a report including recommendations for source separation, increased numbers of collections, improved transportation of wastes, recycling programs, and the construction of a reduction plant and a rubbish destructor. The committee had been investigating practices in Saint Louis since 1903 and had rejected several piecemeal solutions offered by the municipal authorities. In 1908, the league announced that officials had accepted a comprehensive plan based on the committee's recommendations. The plan called for a cooperative venture between municipal government and private enterprise: the city would collect garbage, rubbish, and ashes, and the Standard Reduction and Chemical Company would be given a contract to turn the garbage into salable by-products at a relatively low cost. The second part of the plan was implemented in 1909, but the first part was snarled in red tape. Nonetheless, the league kept up the pressure and continued to lobby for its program.[29]

Civic organizations worked for improvements in street cleaning as well. Municipal control of street cleaning, which was almost universal by the turn of the century, had not proved to be a solution in many cities. Complaints of patronage, squandering of city funds, administrative bungling, poor cleaning practices, and citizens' neglect were typical. Rounds of indirect and direct civic-group action—investigations, demands for updated ordinances, publicity campaigns, and efforts to secure modern cleaning methods—often followed the complaints. A national street-cleaning conference held during the period indicated nationwide interest in the service.[30]

With more intensity than that which they had attacked the garbage problem, civic groups appealed for citizen action in the battle against dirty streets. After all, the streets were a truly community utility; if the streets—

especially the major thoroughfares—were full of litter and garbage, everyone was affected. Aesthetic and even moral outrage pervaded the demands for clean streets. As Frederick C. Wilkes concluded in an article on Pittsburgh's street problem: "Is there not presented in this matter a topic well worthy of the most serious consideration from every business man, and every citizen having in his heart the least degree of interest in his home and the welfare of this community? It is well established as an immutable law, that environment is what makes the character of men. Cleanliness of man makes a good man. Cleanliness of a city makes a good city. It elevates its moral atmosphere, and tends to encourage all that is good in humanity."[31] Many civic reformers concluded that citizens must not only assume the responsibility for paying someone to clean their streets but also take a personal role in assuring by good habits that the streets would remain clean.

The activities of the citizens' groups in their pursuit of sanitary reform point to their dual approach to achieving cleaner cities. They tried to influence city officials by supporting municipally directed programs, and also they sought to educate the public of the need for a personal commitment to civic cleanliness. The influence that citizens' groups enjoyed with city officials was inconsistent at best; more often they had to rely on public education and public arousal to achieve their goals. People had to be convinced that sanitary reform should have priority over other issues competing for municipal dollars. The public voted officials into office—and out again if they did not live up to expectations. Therefore, it is not surprising that citizens' organizations spent a great deal of time appealing to the officials' constituencies through publicity programs and programs of direct action.

It is also not surprising that women came to dominate the civic phase of sanitary reform. They were the most convincing speakers for such a cause, especially because they were perceived as having the greatest personal interest in it. In their roles as homemakers and mothers, women seemed to be the logical group to promote a clean and healthful environment. Writing about the period from 1890 to 1920, Lois W. Banner suggested that never before or since have "so many women belonged to so many women's organizations; not until the 1960s was feminism so vigorous." By World War I, the primary focus of the women's movement was suffrage, but before that time, most organized women were "social feminists" who displayed great sympathy for the disadvantaged, devoted considerable time to civic improvement, and hoped to obtain social justice for their sex. Social feminism was an outgrowth of several important factors. Like their male counterparts, women were caught up in the reformist mood of the day, confronting the excesses of rapid industrialization. Because of growing educational and professional opportunities, as well as expanded rights under the law, women took the opportunity to speak out on issues affecting themselves and the citizenry at large. Furthermore, women came to see in their own plight circumstances that they had in common with other groups, such as children, the disadvantaged, and the poor, all of whom were subject to the whims of society.[32]

By the mid-1890s, women's organizations and clubs had become bastions of many social reforms, such as child welfare, education, housing, health, and city beautification. Women played a vital role in all phases of environmental reform—noise and smoke abatement, sewage reform, pure-water campaigns, and refuse reform. For example, Julia Barnett Rice, a New York physician, became the driving force behind the Society for the Suppression of Unnecessary Noise, the largest and best-known antinoise organization of the period.[33] In 1892, the Ladies' Health Association of Pittsburgh helped gain passage of a major antismoke ordinance; similar campaigns were waged in Saint Louis; Cincinnati; Chicago; Baltimore; Salt Lake City, Utah; Youngstown, Ohio; and elsewhere. In Chicago, Mrs. Charles Sergel was elected president of the city's Anti-Smoke League.[34] The efforts of women's groups in sanitation reform also were well known through such organizations as the Ladies' Health Protective Association of New York.[35]

By the early twentieth century, the association of women with city-improvement projects was so common that the term "municipal housekeeping" became synonymous with sanitation reform. Mildred Chadsey, commissioner of housing and sanitation of Cleveland, Ohio, defined "municipal housekeeping" succinctly: "Housekeeping is the art of making the home clean, healthy, comfortable and attractive. Municipal housekeeping is the science of making the city clean, healthy, comfortable and attractive."[36] Rev. Caroline Bartlett Crane used the term to impress upon women their responsibility in civic cleanliness: "Municipal housekeeping! That is a word I want, to make women feel their share of responsibility for the cleanliness of their city. We say, when on a journey, 'Grand Rapids,' or 'Kalamazoo,' or 'Coldwater' is my home. And is not one's city in truth the extension of one's home?"[37]

The identification of municipal housekeeping with sanitation reform fit well with the strong aesthetic, moral, and civic cast of the women's organizations (there were, of course, other women's reform efforts during the period which did not fit this image). At the same time, however, municipal housekeeping imposed limits on the work of these groups, because it lacked the precision and comprehensiveness needed for a well-developed program of sanitary reform. For all its clarity of image, municipal housekeeping was in some respects a layman's perception of a complex environmental problem. Middle- and upper-class women sometimes promoted it to influence the poor and the working class to keep their homes clean and to participate in related community activities. As historian Susan Strasser suggested, "As public issues, poverty and trash were intertwined at the turn of the century: refuse was cast as an issue of poverty, not one of abundance as it is now."[38]

Municipal housekeeping also circumscribed the role of women as sanitary reformers. Since it so directly paralleled their domestic responsibilities, municipal housekeeping seemed a logical avenue of reform for women. In an article written in 1897, Edith Parker Thomson bemoaned the limitations of the reform model as it applied to women, especially in the public health field. She was astonished to discover how little the public knew about the significant role of women in public health during the late nineteenth century. She concluded: ". . . every attempt on the part of women to benefit the public is necessarily somewhat indirect. They do not hold the ballot, nor sit in legislative halls. Their only course is either to arouse public opinion or to present their cause to the public officials. Their part in public reforms is chiefly suggestive or cooperative. They can seldom of themselves carry anything to completion, as a Board of Health or a Street-Cleaning Department can do. Hence their work is frequently unmentioned in the public record."[39]

Although women's impact on sanitary reform would be much greater and better publicized in the twentieth century, Thomson pointed to important limits imposed on social feminists in the public health and sanitation fields. One reason for these limits was that many women reformers came from one economic stratum in the United States, the middle class, and many of the women were housewives. As Suellen M. Hoy noted, "Many of [the women reformers] were middle-aged, had children in school, and often hired servants to clean their homes."[40] The women involved in sanitary reform were strongly imprinted with their "proper" role in society. More recent scholarship, however, is suggesting that deep interest in social and environmental reform was not restricted to the middle and upper classes, and increasingly involved women of color and women of ethnic backgrounds.[41] Banner suggested that for women "of every class, ethnic group, and region, [life] was governed by the simple statistic that the overwhelming majority would marry and become housewives and mothers." In the twentieth century, approximately 90 percent of all American women had been married at some point in their lives. In 1900, only 5 percent of all married women in the country worked outside the home; ten years later, only 11 percent. "Thus for almost all women," Banner concluded, "marriage was a natural goal of life."[42] It is no wonder that municipal housekeeping garnered such immediate support from both women and men; it did not undermine the status quo. Men who reeled in horror at the mention of women's suffrage or more radical reform, such as Margaret Sanger's birth-control movement, saw little in sanitary reform that threatened the sanctity of the family or undermined the domestic role of women. Noted sanitarian Samuel A. Greeley was typical of male reformers. He spoke with high praise of the activities of women in city-cleaning projects and accepted the premise that women were particularly well suited to the work: "It is unfortunately true that cleanliness and efficiency in the house treatment of refuse are sometimes defeated by careless and infrequent collection service or improperly conducted disposal works. This condition must be very trying to the careful housekeeper, and is, perhaps, one reason why women have taken an active interest in promoting the efficiency of the larger phases of the work."[43]

Despite the narrowness of municipal housekeeping as a reform goal and the limitations women experienced at the time, the efforts of women and women's groups to improve the quality of urban life were valuable and helped raise public consciousness of the need for sanitary improvements. Municipal housekeeping took many forms, from the superficial to the substantial. Women's organizations, especially those in smaller cities, launched antilittering campaigns. The Woman's Club of Dayton, Ohio, distributed leaflets to every household decrying the practice of throwing wastepaper into the streets. The Civic Club of Huntington, Tennessee, and the Woman's Club of Green Cove Springs, Florida, provided trash cans for their towns. The Green Cove Springs Club even decorated the cans and emblazoned amusing rhymes on them:

> My name is Empty Barrel,
> I'm hungry for a meal.
> Pray fill me full, kind stranger,
> With trash and orange peel.

According to one observer, the "jolly cans so appealed to the public that refuse upon the streets became an unheard-of thing, and Green Cove Springs was known as the Parlor City of the South."[44] Other women's groups attacked the tin can problem, initiated civic rallies, and lobbied for ordinances prohibiting spitting in the streets or indiscriminately dumping rubbish.[45]

In some of the largest cities, where women's organizations often acquired substantial influence in municipal affairs, improvements in cleanliness were quite noticeable. In Boston, the Woman's Municipal League organized a traveling exhibit to educate the citizens about sanitation. The exhibit consisted of models contrasting dirty and clean meat markets, dairies, and tenements and provided information about the prevention and cure of tuberculosis. In cooperation with a Boston settlement house, the league paid the salary of a woman inspector for the Board of Health. The Indianapolis Sanitary Association also publicized good sanitation practices. The first women's group in the city's history to venture into municipal affairs, the Sanitary Association, held neighborhood meetings to instruct citizens on health matters, lobbied for improved garbage collection, and surveyed street-cleaning practices. In Louisville, the Women's Civic Association was most effective in its publicity campaigns. It organized an exhibit, published pamphlets, and even produced a motion picture through the cooperation of the Board of Trade, the Rotary Club, the Men's Federation, and the Jefferson County Medical Association. The film *The Invisible Peril* depicted "the travels of a discarded hat, and the disease which can be spread thru the open can, open wagon, open dump system of waste disposal." More than twenty thousand people saw the film. Through direct and effective action, the Waste Committee of the Woman's City Club of Chicago pressured the city to secure a scientific report and plan for garbage collection and disposal. The committee was also responsible for the decision by the Chicago City Council to establish a City Waste Commission, on which two club members served.[46]

Probably the best known of the activist women's groups was the Woman's Municipal League of New York City, which played as important a role in the early twentieth century as the Ladies' Health Protective Association had played in the late nineteenth century. The Woman's Municipal League appealed to "the intelligent women of New York for their co-operation in the present struggle for a morally and physically clean city."[47] An adversary of Tammany Hall, the league advocated improved methods of collection and disposal, including night disposal of wastes, and fought to increase benefits and improve working conditions for city sanitation workers. It also sponsored the Waring Medal, which was awarded to workers with the best performance records.[48] The league was most effective as a civic watchdog, monitoring the activities of the Department of Street Cleaning and evaluating its work. Lest anyone believe that the league played a merely ceremonial role in New York City's sanitary affairs, it should be noted that in 1907, it forced the resignation of MacDonough Craven, the street-cleaning commissioner. After investigating the department that year, the league charged Craven with incompetence, and after a strike by sanitation drivers aggravated the deteriorating condition of the department, Craven was allowed to resign.[49]

The flurry of projects initiated by women's groups throughout the nation produced several leaders who became synonymous with sanitation reform, including the highly respected Rev. Caroline Bartlett Crane of Kalamazoo, Michigan. After graduating from Carthage College in Kenosha, Wisconsin, in 1879, she taught school and worked briefly as a newspaperwoman. Her primary interest was theology, and in 1886 she was ordained in the Unitarian church. In October 1889, she was installed as pastor of the First Unitarian Church of Kalamazoo. Seven years later, she married Dr. Augustus Warren Crane. Rev. Crane had a strong inclination for the social, as well as the spiritual, improvement of her flock, and she plunged into projects ranging from education to police and fire protection. By the turn of the century, she was deeply engrossed in sanitation problems and had become a leading proponent of municipal housekeeping. In 1904, she founded the Women's Civic Improvement League of Kalamazoo, and by 1907, she had achieved national attention for her work in street-cleaning reform. She was constantly sought after as a lecturer, conducted many sanitary inspections throughout the country, and drew wide attention to sanitation problems with her incisive reports.[50]

Mary McDowell, of Chicago, also gained repute as a sanitary reformer. Like Crane, she had strong religious ties. McDowell taught classes in religion for young people and became a devoted follower of Frances Willard, founder of the Women's Christian Temperance Union. Committed to social service, McDowell later joined Jane Addams at Chicago's Hull-House in 1890. In 1894, she was asked to direct a new settlement, Packingtown, a project of the University of Chicago. Packingtown's desolate location, primitive facilities, and physical blight caused McDowell to gain an appreciation for proper sanitation measures. From her experience there, she acquired an interest in citywide sanitation reform and sought to attract popular support. She persuaded several Chicago women's groups to establish waste committees, and she chaired the City Waste Committee of the Women's City Club. Through her leadership, the women's groups were instrumental in the 1913 establishment of the Chicago City Waste Commission, which became influential in refuse reform. The "Garbage Lady," as McDowell became known, was recognized as a major force in the improvement of Chicago's massive waste disposal problems and a national figure in environmental reform.[51]

Interestingly, some of history's best-known women in other fields of endeavor also contributed greatly to sanitary reform. Jane Addams, the founder of Hull-House and a pioneer in the development of social welfare as a profession, was also a sanitary reformer. Among her many activities, Addams regularly pressured the Chicago City Council to improve collection service. A colleague of McDowell's in the Women's City Club, Addams lectured on sanitary reform and was named garbage inspector for her ward.[52] Ellen Swallow Richards also wrote and lectured on the importance of municipal housekeeping.[53] Through the leadership of women's organizations and the emergence of nationally recognized women reformers, by World War I, sanitary reform (indeed, much of environmental reform in general) was being identified with social feminism.

Social feminists spearheaded health education among urban children and enlisted droves of youngsters in sanitation activities. Several times the women drew on the efforts of Colonel Waring and his Juvenile Street Cleaning League. On a broader plane, the rise of progressive education during these years marked the start of civic involvement among children. As historian Lawrence A. Cremin noted: "Progressive education began as part of a vast humanitarian effort to apply the promise of American life—the ideal of government by, of, and for the people—to the puzzling new urban-industrial civilization that came into being during the latter half of the nineteenth century. The word 'progressive' provides the clue to what it really was: the education phase of American Progressivism writ large."[54] Cremin went on to suggest that Progressives sought, among other things, to use the schools to improve the lives of individuals, which meant broadening the function of the schools to incorporate concern for the quality of life and health, civic-mindedness, and vocational training. This approach was complementary to the environmental view of Progressives, which emphasized that, to improve human beings, the moral and physical environment must first be improved. Progressive education, however, was not elitist; instead, it attempted to appeal to children of all economic classes and ethnic backgrounds. In part, at least, progressive education was seen as an important tool for assimilating the children of the foreign-born into American society. Of course, the assumption was that assimilation was a worthy goal and that American values and culture were superior to those of immigrants from southern and eastern Europe and elsewhere.

An important thrust of progressive education was to provide children with training that was practical and immediately useful.[55] What emerged was a strong commitment to "civic education," that is, teaching the structure and functions of government and its legal system and instilling citizenship. In 1904, Delos F. Wilcox, an expert on municipal government and the publisher of Detroit's *Civil News,* declared what he believed to be the value of civic education in the urban setting:

> In cities human nature comes to the parting of the ways; allowed to drift along the lines of least resistance, it develops intense selfishness of the future, and those other characteristics of degeneration found in highly civilized society; but, properly trained, human nature in cities develops a wider social consciousness, a heartier spirit of co-operation, a more refined appreciation of the arts of life, a keener sense of responsibility to the future, and all those other characteristics of progress that are the hope of evolution and the justification of social effort. It is the character of civic education that will determine in the long run whether or not democracy can succeed in cities.[56]

Wilcox and many others perceived civic education as a goal for all citizens, but for practical reasons focused their attention on the young. "It is one of the paradoxes of reform," Wilcox said, "that no absolute social salvation can be brought about unless the children can be reached, while the only possible way to reach the children is through the grown people."[57] Responsibility for training the young, therefore, had to fall to the "enlightened" reformers who understood society's needs. The schools were the most expedient institution for reaching the greatest number of children. Within the schools, the most direct way to incorporate civic studies was through curricular additions or modifications. Manual training became a popular way to introduce practical skills to children; in the realm of civic education, the civics class became the most widely accepted method of inculcating citizenship. Civics classes appeared as early as grammar school and were carried over into high school. Student-government programs were established as extracurricular activities to provide practical experience in government organization and decision making.[58]

Outside the schools, but working in close concert with them, were citizenship organizations for children, usually managed by local civic organizations and women's groups. For instance, the Newport, Rhode Island, Civic League established a League of Good Citizenship with ten divisions in the public schools. In Jeanette, Pennsylvania, junior civic leagues planted trees and beautified stretches along railroad tracks. In Cleveland, Ohio, Hiram House Social Settlement set up "Progress City," a play city operated by the local children which included a city council, a court, a newspaper, and all the other trappings of a real municipality—without the graft and corruption, crime, violence, and unsanitary conditions of life in the real city.[59]

Inculcating civic pride and civic responsibility naturally led to attempts to give children a broad appreciation of their surroundings, including city cleanliness and beautification. The leap from the broader goals of civic education to a concern about sanitation needs was not great; indeed it was an integral part of the process. A citizen was just as responsible for obeying a garbage ordinance as for obeying any other law. City cleanliness was as much a mark of civilized society as was an honest and efficient government—or so it was argued. An observer noted: "To 'chuck' our fruit skins, wrappers, cigar stumps, sputum, everything, in fact, that no longer has value to us, is as natural as life when we are in public places, while in our homes those of us who are 'civilized' and 'cultured' would never think of such a thing. How deep is a culture that will keep rubbish and garbage off the floors of the private home and cast it at random upon the floors of the community home? Not very deep, nor very consistent, nor very valuable, we all know when we come to think of it."[60]

Civic pride through city cleanliness was an important ingredient of the various civic education programs for children. It was so important that it became the basis for many children's organizations, usually initiated and sustained by women reformers. Juvenile sanitation organizations were, therefore, a very significant element in the sanitation-reform movement.

Colonel Waring's Juvenile Street Cleaning League was an archetype. What had been a unique experiment in 1896 became widespread a decade or so later. In the summer of 1889, there were seventy-five leagues in operation with some five thousand participants. After Waring's resignation, the league lost its vitality and was disbanded in 1900. In 1909, it was reestablished with fifty leagues under the leadership of Reuben S. Simons, who had helped organize the original league. The new league operated through the schools—the result of the impact of civic education—largely under the guidance of women teachers. The activities of the new league were essentially the same as the first: the children inspected streets and alleys, reported litterers, held meetings, and, of course, conducted parades. The 1910 outing included more than fifteen thousand children from thirty-one leagues; the next year, more than twenty-five thousand children took part from the seventy-one leagues in operation.[61]

At the zenith of the league's popularity, other juvenile groups sprouted up throughout the country. In Philadelphia, Mrs. Edith W. Pierce, a nurse and the city's first woman street-cleaning inspector, organized a Junior Sanitation League with over ten thousand members. Every child wore a button bearing the seal of the city and emblazoned with the motto "For Clean Philadelphia Streets." The children were also furnished with instructions for volunteer inspectors. Pierce gained the cooperation of the city's teachers for her project, and occasionally the teachers required the children to write compositions about the league and their ideas about what a good citizen was. Leagues were organized in several other cities and towns throughout the country. The organizer of the Binghamton, New York, Junior Civic League, Mrs. Johnson, echoed the sentiments of others involved in civic education when she wrote: "We must picture to the children how each one of us has pride in his home, and wants it to be as perfect as possible, and how that feeling extends to its surroundings, to the lawns and streets in the vicinity. Then in a broader way we think of the city as our home." In Kenosha, Wisconsin, the Department of Education and the Department of Health organized Junior Health Leagues to produce "health missionaries." In San Diego, California, the city health officer organized a group of junior deputy sanitary inspectors to assist in a fly-prevention campaign. Other cities experimented with variations on these themes.[62]

Enlisting children was a logical step for sanitary reformers who sought to generate public support for their programs. A whole generation of urban Americans could be educated through the public schools to "reprogram" their habits of personal cleanliness and to become sensitive to their physical surroundings. Many reformers also believed that children, especially those from immigrant families, could transmit their experiences and knowledge to their parents and affect their behavior as well (substantial bigotry was attached to the notion that immigrants needed this education more than native-born Americans did). Beyond their educational value, the juvenile leagues were also outstanding avenues of publicity. The uniqueness of the organizations and the large number of participants assured that sanitation problems would obtain widespread public notice. The novelty would wear off sooner or later, but the civic programs continued.[63]

In raising public consciousness about sanitation, no program, not even the junior leagues, captured as much attention or enjoyed as much publicity as city cleanup campaigns. The earliest such campaigns were generally held for one or two days in the spring to encourage citizens to dispose of rubbish accumulated during the winter months. Drawing on the concept of municipal housekeeping, reformers promoted cleanup campaigns as citywide spring cleanings. Eventually, the campaigns became community projects, lasting a week or longer and emphasizing good sanitation practices, fire prevention, fly and mosquito extermination, home and neighborhood beautification, and many other programs. In their more elaborate forms, the cleanup campaigns relied on the cooperation of all segments of the community and brought together all groups concerned with sanitary reform, including civic and women's organizations, children's groups, social clubs, municipal authorities, and the press. Cleanup campaigns reached the proportions of a movement in the 1910s, spreading so rapidly that almost every city and town conducted at least one spring campaign. Some projects instituted during the cleanups were maintained year-round.[64]

The cleanup campaigns took many forms and varied in degree of success. Some had very limited goals. For instance, the Civics Committee of the State Federation of Pennsylvania Women sponsored a "Municipal Housecleaning Day."[65] In many other cities, the cleanup drives were designed to promote overall civic-mindedness. This was particularly true in smaller cities and towns, where the sense of community was strong. In Kirksville, Missouri, the Civic League and the Elks Lodge joined forces to produce a bulletin called "For a Cleaner and More Beautiful Kirksville," that was placed in every home in the city. Several businesses set up displays in their windows demonstrating the proper tools for cleaning and posted placards with catchy inscriptions. School children were asked to sign cards certifying that their yards at home had been thoroughly cleaned. Pastors of various churches preached sermons about the virtues of cleanliness. The league distributed flowers and vines to citizens who wanted to hide unsightly fences or other structures. For two years, the league also had direct responsibility for cleaning the streets.[66]

Some of the campaigns produced substantial and long-lasting results, especially the initiation of new and more effective sanitation ordinances. Those responsible for the campaign in Richmond, Virginia, realized that "the habit of littering streets, alleys and yards appeared to have been so thoroughly grounded in Richmond that the educational effect of the cleanup campaigns was not sufficient to eradicate it." As a result, the Society for the Betterment of Housing and Living Conditions recommended to the city authorities a more comprehensive ordinance to improve collection and disposal methods. The new law went into effect in February 1914. In Sherman, Texas, the women of the civic league launched a cleanup campaign that resulted in an ordinance requiring the city to sponsor four annual four-day cleanup periods. The first cleanup campaign was received with such enthusiasm that it lasted ten days. The Women's Civic League of Baltimore received high commendation from Chief Engineer August Emrich of the fire department for its cleanup "crusade" in 1912. According to Emrich, fire losses in that year were the lowest in thirty-four years, and he credited this achievement to the cleanup campaign's removal of rubbish and other inflammable materials from homes and business establishments.[67]

Civic organizations made great efforts to target children in cleanup campaigns or integrate children into the cleanup efforts. According to *American City,* "No other movement for civic betterment has made better use of the ability and energy of children than have the clean-up campaigns." In 1915, children played important roles in some five thousand local campaigns under the auspices of the National "Clean Up and Paint Up" Bureau—the only national organization of its kind. Thousands of children marched in parades promoting the cleanup activities, distributed pamphlets, organized cleanup brigades and wrote essays extolling the virtues of cleanliness. They also memorized appropriate pledges:

> I will not throw any paper into the streets, because I want our streets to be clean.
>
> I will take my own drinking cup to school with me.
>
> I will not bite anyone else's apple or chew anyone else's gum, because I do not want anyone else to bite my apple or chew my gum.[68]

Cleanup campaigns thus became crash courses in hygiene, sanitation, and civic education.[69]

Philadelphia's weeklong cleanup campaign of 1913 was the biggest of them all. It incorporated all the elements of successful drives throughout the country but on a larger scale. Ultimately, the annual campaign became such an integral part of the life of the city that the municipal government assumed leadership of it and made it a major civic event akin to a festival. In 1913, the city distributed 3,400 personal letters, 750,000 gummed labels, 260,000 bulletins, 20,000 colored display posters, 750 streamers, 1,000,000 cardboard folders, 300,000 badges, 300,000 blotters, 350,000 circulars, and various other promotional materials. All segments of the city participated. In 1914, during the second annual cleanup week, participants collected 140,000 cubic yards of rubbish. Special teams of cleaners scoured five hundred vacant lots, municipal buildings were thoroughly cleaned, and crews splashed paint on everything. The "Second Annual Clean-Up Week for a Spick and Span Philadelphia" was a huge success.[70]

The cleanup campaign epitomized the public phase of sanitary reform. It preached the City Beautiful civic improvement idea, and it promoted good public health practices. It also brought together the various groups that had been most responsible for the popularization of sanitary reform: civic organizations, women's groups, children, sympathetic journalists, and city officials. Its greatest contribution was in mustering citizen participation in cleaning the surroundings and in raising public consciousness about environmental problems. As a publicity device, it had dramatic appeal.

The cleanup campaign had serious shortcomings, however, and drew some criticism. When Charles Zueblin argued that "annual clean-up days are unnecessary in well-regulated cities," he addressed the flaw in the idea, namely, that cleanup campaigns in and of themselves were cosmetic activities, short-term substitutes for effectively enforced ordinances and efficiently conducted collection and disposal practices.[71] They gave citizens a sense of accomplishment, but they identified sanitary problems which could not be resolved in a burst of activity one week a year.

The cleanup campaigns also pointed out the constraints on the civic phase of sanitary reform. At their best, civic organizations promoted cleanliness and helped improve collection and disposal methods through publicity and direct pressure on city officials. Though often effective in raising public awareness about sanitation problems, the civic groups could not bring about policy changes. Sanitary reform was not simply a question of implementation. Embedded in the reformist effort were other inherent weaknesses. Foremost among them was that technical reformers (sanitary engineers) and civic reformers (citizens' groups) never joined in a single, broadly based movement. Each group possessed powerful reform tools, and each had links to important constituencies, but the gap between them was never bridged. The male-dominated, elitist, efficiency-oriented technical reformers, operated independently of the primarily female-dominated, aesthetics- and health-oriented civic groups. The gaps were not bridged among the myriad groups interested in various phases of environmental reform either. Lacking a sufficiently broad environmental perspective, antismoke groups, noise-abatement groups, and sanitary-reform groups pressed their specific causes independently, unable or unwilling to coordinate their efforts. Groups concerning themselves with the broader issues of environmental quality and the collective effects of pollution had not yet materialized.[72]

To say that sanitary reform had its limitations, however, is not to demean the efforts of those who grappled with the waste problem or to apply today's standards to their achievements. The early-day sanitary reformers, both technical and civic, were ahead of their time in seeking workable solutions to perplexing problems. In their refusal to accept untended waste as an inevitable by-product of industrialization and urbanization, they pointed the way for the reformers of the future.

FIVE

STREET-CLEANING PRACTICES IN THE EARLY TWENTIETH CENTURY

Those committed to ending the refuse problem—from engineers to journalists, from municipal authorities to civic leaders—looked optimistically toward a time when cities would be uniformly clean and free of pollution and disease. The flurry of reform activity and the profusion of rhetoric heightened the anticipation. As with any attempt at change there was a gap between expectations and achievements. Public works and sanitation departments made strides in street cleaning and in the collection and disposal of solid wastes, but vestiges of the out of sight, out of mind mentality persisted. Partial solutions and incomplete victories were the best that reformers could reasonably expect, especially since the improvements they demanded required major institutional, as well as attitudinal, adjustments by American city dwellers.

At the turn of the century, several informed contemporaries, while recognizing the improvements achieved since the 1890s, also realized that quick solutions to refuse problems were too much to expect. In 1905, a writer commented in the *Municipal Journal and Engineer*, "The whole question of economical and efficient methods of municipal waste disposal is just at this time in a changing, unsettled state; the tendency is consequently to 'go slow' and not to commit the city to the adoption of schemes and plans that have not been thoroughly tested by practical uses or that cannot be recommended by the endorsement of those competent to report upon the particular conditions governing each individual case."[1] To "go slow" was not simply to be deliberate, as the quotation implies. City officials hesitated to give priority to refuse management over other important city services and programs, such as police and fire protection, municipal building projects, park construction, and transportation. Reformers faced the unenviable task of trying to convince municipal leaders of the need to invest great amounts of time and money in the improvement of refuse management because of its intrinsic importance. Although government leaders may have been sympathetic to the reformers' case, they tended to react to immediate, as opposed to long-range, sanitary needs. Their lack of foresight, neglect of planning, as well as budget limitations were largely responsible for the delays in implementing effective programs.

Textbooks on sanitation written in this period echoed the reformers' concerns about the state of management. In evaluating past and present methods of street cleaning and refuse collection and disposal, these books both bemoaned the primitiveness of nineteenth-century practices and criticized highly publicized newer methods, such as incineration and reduction. In 1906, William Mayo Venable, in his book on garbage crematories, leveled a devastating broadside at the methods of refuse disposal practiced in American cities, "the crudity of the methods of refuse disposal in most of our American cities is almost incredible to an intelligent person when his attention is first directed to observe such matters, the disposal of garbage being in many cities less intelligently managed than among savages, and the disposal of litter, tin cans, waste paper, etc., a class of waste with which savages do not have to deal, being conducted in so slovenly a manner as to excite disgust in any person who realized the facts."[2] The textbooks also criticized the intransigence or disinterest of municipal governments that failed to take an active part in resolving refuse problems. According to M. N. Baker: "In the removal and disposal of city wastes we are far behind our attainments in providing municipal supplies. Only a small portion of the communities which enjoy public water supplies have the benefit of sewerage systems, and most of the latter discharge their sewage into the nearest body of water without regard to consequences. Garbage collection is strangely neglected in the majority of cities and towns, and its final disposal is one of the greatest blots upon American municipal administration."[3] Of the two major problem areas, street cleaning and refuse collection and disposal, street-cleaning practices made the greater overall strides. Civic reformers brought the problem to the attention of the public and city officials, while technical reformers instituted efficiency programs in public works and sanitation departments. Most important was the impact of technology. Improved paving made city streets more essential as transportation arteries and generated a greater demand for clean, unobstructed streets. Smooth pavement could accommodate a vast array of mechanical sweeping and flushing devices that significantly altered cleaning practices in both large and small cities.

The demand for more and better-paved streets, which had become widespread in the late nineteenth century, continued unabated in the early twentieth century. Although major federal support for a municipal highway system did not emerge until 1916, most major cities and many of the smaller ones were interlaced with elaborate street systems by that time, at least in central-city areas. Dr. Carol Aronovici, a leading sanitarian, was among those who recognized the important role of streets in the lives of urbanites:

> Until recently a street, from the standpoint of municipal government, was considered a thoroughfare, or a means of reaching various parts of the community without regard to the surrounding property, be that of a business or residential character. A closer observation, however, makes it clear that the street is essentially the means of approaching a home and of serving its conveniences. It is the hallway which connects the school and the church, the factory and the office with the home. From the standpoint of the tenement dweller, the street is the nursery and playground of the young, the social center and meeting place of the adult, the free market place for the transaction of business, and the display and distribution of the food supply. Not infrequently during hot weather the street is the common bedroom of the dweller in the congested, ill-ventilated and over-heated tenement house district. With such broad functions it is clear that the construction and care of streets implies more than the requirements of accessibility, easy grade and safety.[4]

The construction of elaborate street systems, however, produced a significant change in the role of streets in urban life. Use of the streets as marketplaces, social gathering places, and even extensions of homes was fading by the turn of the century. According to historian Clay McShane, "By 1900 many, possibly most, urban residents saw streets as arteries for transportation since house yards and porches in the new streetcar suburbs assumed the traditional social and recreational functions of streets." He added, "urban bureaucrats, who by 1900 controlled paving in most cities, had completely lost sight of the traditional functions of streets." Municipal engineers planned pavement to cope with increased traffic or to provide easier means of removing wastes from the cities. These changes predated the widespread use of the automobile. Asphalt, the paving material usually associated with the rise of private motorized transportation, had been used to pave nearly one-third of city streets by 1909, the year Henry Ford introduced the Model T. The impact of the automobile on urban street construction and transportation patterns was not significant until about 1914.[5]

In de-emphasizing the neighborhood functions of streets and accentuating their transportation functions, municipal officials acknowledged that extensive paving programs were vital to the economic interests of the city. As a result, street cleaning took on new significance. A clean, unobstructed street not only was healthful and aesthetically pleasing but also aided the flow of traffic. By World War I, most cities of more than thirty thousand residents had adopted a street-cleaning system, and, whereas 70 percent of the cities surveyed in the 1880 census had assumed responsibility for street cleaning, by around 1917, about 90 to 95 percent of American cities had municipally operated programs.[6]

The contract system of street cleaning was being abandoned in almost every major city. San Francisco, Indianapolis, and Washington, D.C., terminated the contract system in 1903, 1905, and 1911, respectively. Philadelphia, one of the last major cities to maintain the contract system, finally ended it in 1921.[7] Critics of street cleaning by contract charged that it was less efficient than municipally run programs and rarely provided adequate service for the money. The most serious problem was enforcement. Speaking before the Sanitary Engineering Section of the American Public Health Association in 1913, J. W. Paxton stated, "Street cleaning is probably the most difficult work to inspect of any which could be let under contract. No street is ever perfectly clean, only relatively so, depending on the accumulation of dirt and the amount of efficient cleaning. It is impracticable to specify that the contractor shall furnish clean streets or even clean streets at certain periods, but each detail of the required operation of street cleaning must be specified, inspected to see that each detail is carried out, and the city then must be satisfied with [the] results."[8]

Many sanitation reformers and some city officials believed that the termination of street-cleaning contracts and the establishment of municipally operated programs would assure improved service. This assumption was not always borne out. Municipally operated systems had to contend with serious problems in establishing effective street-cleaning programs.[9]

For instance, few cities cleaned all the streets uniformly. Because city officials associated streets with the vitality of the local economy, major thoroughfares in central business districts had the highest priorities. Streets in outlying areas and in working-class and immigrant neighborhoods had much lower priorities and received less frequent service—and no service at all if the streets were unpaved.[10] Moreover, street cleaning remained vulnerable to political influence, graft, and corruption. Even municipally operated street cleaning provided many opportunities for conferring political favors, exacting votes from grateful employees, and pilfering appropriations. Some officials lived by the axiom of the colorful Tammany ward boss and lay philosopher George Washington Plunkitt: "I seen my opportunities and I took 'em."[11] Thus municipal street-cleaning systems were not without their shortcomings in an era when under the best of circumstances services could not keep pace with the growing cities. George A. Soper, author of *Modern Methods of Street Cleaning* (1909), summed it up well when he wrote that street cleaning had "not yet emerged from the state of a nondescript kind of emergency undertaking to the position of an effective art."[12]

Nonetheless, a municipally operated street-cleaning program was generally regarded as the best means of ensuring effective service. In an attempt to improve the administrative structure of street-cleaning programs, several cities adopted Colonel Waring's organizational reforms as well as the new cost- and record-keeping methods that reformers were introducing into city management. Some officials cribbed directly from the master himself. Street Commissioner William A. Larkins of Baltimore gained a reputation as a leading sanitarian by reorganizing his staff, upgrading the equipment, and improving service citywide. His men were dressed in "spotless white and brown uniforms," and he initiated an annual parade of workers. He even had the street-cleaning carts painted in "resplendent colors."[13] The Philadelphia Bureau of Highways and Street Cleaning also adopted some of the trappings of the colonel's program: uniformed cleaners, street-cleaning parades, source separation, public-education programs, and cleanup campaigns.[14] Street Commissioner J. F. Fetherston of New York City, seeking to resurrect Waring's spirit, initiated some significant reforms in the Department of Street Cleaning, which had fallen into disarray under Tammany Hall domination. Chief among them was the establishment of a "model district," in which street cleaning was conducted "along the most improved and modern lines by the use of *the* highest developments in street cleaning and refuse removal apparatus."[15]

The reformers' enthusiasm for scientific management and efficiency programs manifested itself in the application of cost- and record-keeping techniques to city services, including street-cleaning and public works departments. In a paper read before the Sanitary Engineering Section of the American Public Health Association in 1913, consulting engineer S. Whinery stated, "Perhaps the most promising and fruitful advance that has been made in recent years is the greater attention given to the study of the details of street cleaning and the keeping and analysis of more complete and itemized accounts. The older accounts and reports of street-cleaning departments were so barren in details that they were generally of little or no value to either the student or the practical man."[16] Whinery believed that efficiency and economy in street-cleaning departments, rather than new machinery or new cleaning methods, were largely responsible for the noticeable improvement in street appearance. Although he acknowledged the importance of technical advances in street-cleaning devices in the first decade of the new century, he perceived "no radical or particularly notable advancement" in machinery or methods. "The flood of patents issued for street cleaning devices continues unabated," he argued, "but these patents relate largely to details and disclose little of value for practical use. This is somewhat surprising in this land of inventors."[17]

The results produced by the introduction of cost- and record-keeping methods in some of the major cities seemed to bear out Whinery's claims. The best example was Washington, D.C. In 1911, Congress passed a law terminating the street-cleaning contract for the capital city and turned over responsibility to the commissioners of the District of Columbia. One of the direct results of this change was the establishment of a modern cost-keeping system. After two years in operation, the new system had made some important gains, especially in helping determine labor efficiency and operating costs. The improvements under the "Washington system" were modest, but the initiators of the program had implemented it with deliberation, hoping to improve it over time. Instead of hiring an efficiency expert to develop a system in a few weeks or months, they charged a permanent employee with the task of developing and monitoring the system. As a spokesman for the new program noted, "The Washington system, as a whole, will probably fit no other department in Washington nor the Street Cleaning Department of any other city. Cost keeping is necessarily a process of first, evolution, and second, elimination. This department has gone through the first and the question now is can we do without this detail or how may we obtain it with less effort?"[18] Advancement in cost- and record-keeping and departmental reorganization, important as they were, did not attract the widespread attention that mechanization and motorization of street-cleaning equipment achieved. While these technological changes did not resolve all the problems associated with street cleaning, they did promote significant alterations in methods and in the variety of tasks which street crews could perform. New street designs and new paving materials led many American and European cities to employ machine sweepers. Between 1899 and 1905, about 85 percent of American cities with populations of more than 25,000 were using mechanical sweeping devices as supplements to or replacements for hand sweeping.[19] Soon, more than 55 percent of the smaller cities and towns had followed the larger cities' lead.[20]

The advantages of machine sweeping were many. The advent of smooth pavement made machine sweeping practicable, and advocates of the mechanical devices claimed that they were well suited for the new surfaces, offering efficient and economical service for every city in the nation and eliminating dust, the bane of urban dwellers. As Whinery noted:

> It is coming to be generally recognized that from both the sanitary and business point of view the most objectionable part of street dirt is the fine dust produced by the drying out and pulverization of the animal excreta and other matter that finds its way to the surface of the streets. The fresh, raw and usually damp excreta and rubbish are objectionable mainly to the sight, but when dried and ground dust floats in the air when disturbed, and disease germs contained in it are breathed into the nose, mouth and lungs of those exposed to it, where it may develop specific diseases. This dust, carried by the winds, enters residences and business houses to the injury of delicate goods or furnishings, and by reagitation may thence be carried into the human system. Any system of street cleaning that does not provide for the prevention or allaying of street dust cannot therefore be regarded as satisfactory.[21]

Advocates of machine sweeping also claimed that the army of laborers once needed to ensure street cleanliness could be eliminated or employed in other tasks. A reduction in the labor force would cut costs, as well as eliminate many worker-employer conflicts (this was an especially attractive argument for antiunion management).[22]

The fascination of Americans with machinery was nowhere more evident than in the promotion of mechanical cleaning devices. Private companies sprang up across the country, hawking all manner of devices to make the job of the city sanitarian easier, as well as to turn tidy profits for themselves. General Motors Truck Company promoted the Modern Flusher for Progressive Municipalities. The Sanitary (Automatic) Street Flushing Machine, manufactured by a Saint Louis company, "flushes—not merely wets—but scrubs, washes, scours, cleans, *rids the street of all health menacing unsanitary conditions"* with "no complicated mechanism. No skilled labor required." The Matchless Street Cleaner Company of New York sold a "simple, common sense machine" that did away with "all unnecessary work" and would clean 50 percent more surface than any other machine. The Austin-Western Company, Ltd., claimed, "All Unite in Pronouncing [their sprinklers] the best." General Vehicle Company exhorted its patrons to "Tow Your Street Cleaning Apparatus with Electricity."[23]

There were machines for every purpose, every street material, every climate, and every city size. There were mechanical brooms for general sweeping, squeegee cleaners that sprinkled and scraped the pavements, vacuum cleaners that sucked dirt into large canisters or bags, machines that swept and disinfected the streets simultaneously, and flushers that washed dirt into gutters and sewers.[24] R. N. Stevens invented an ingenious machine which passed over the street and subjected it to high temperatures, presumably to bake or burn up germs and dirt.[25] Needless to say, many of the mechanical devices were variations on a few basic types, and their claims of effectiveness and efficiency were proved only by trial and error—sometimes expensive trial and substantial error. Cities, however, kept several of the companies in business by continuing to seek new solutions to their street-cleaning problems. The practice of flushing streets gained many adherents. It promised not only to clean the streets of dust and disease-breeding filth but also to cool the pavement in the summer months. The machines were used on practically all surfaces—asphalt, brick, concrete, granite block, macadam, cobblestone, and even wood. The number of flushings a week varied with the seeming importance of the streets. Streets were flushed daily in New Orleans; in Los Angeles, the major business streets were flushed three times a week; other streets might be flushed once every two weeks. Denver streets were flushed daily in the summer, but sweepers were used in the winter months. Most of the cities using the flushing method continued hand sweeping or mechanical sweeping as a supplement.[26]

Although mechanical sweepers and flushers promised to revolutionize street cleaning, they did not always produce expected results. Several city officials and municipal engineers questioned their effectiveness and challenged the claim that machine sweeping was cheaper than hand sweeping. For example, in June 1907, Mayor George B. McClellan of New York appointed a commission to recommend a more effective system of street cleaning and waste disposal. The commission, whose members included the notable Rudolph Hering, conducted a thorough investigation. Its findings were critical of machine cleaning and sprinkling to settle dust. According to the report, sprinkling paved streets did not clean them but merely converted the dust into mud. The commission added that if the streets were properly swept, sprinkling would be unnecessary, and sprinkling costs—borne by private citizens—need not be incurred. The commission ranked flushing, hand sweeping, and machine sweeping in descending order of effectiveness. In terms of cost, flushing was the most expensive method. Hose flushing, with fire hoses attached to hydrants, cost about the same as machine sweeping, but machine flushing was more than twice as expensive as machine sweeping. Flushing wasted huge quantities of water needed for other purposes. With access to water becoming more difficult for major cities, this fact was a prime consideration in the decision whether or not to use the flushing technique. The commission concluded that, taking into account the strengths and weaknesses of the various methods, flushing combined with hand sweeping was the best solution. This method could properly clean the streets at a reasonable cost and with less water than would be required in a system totally dependent on flushing.[27]

Some cities resisted the temptation to commit themselves to machine-sweeping systems, due to reservations about effectiveness and unwillingness to invest huge sums in the equipment. Some city officials were reluctant to reduce the rolls of public employees, for obvious political reasons.[28] Many cities, nonetheless, caught up in the promise of a quick solution, adopted machine sweeping as the sole method of street cleaning or as a supplement to other methods. The promoters' claims of greater efficiency, lower costs, and the elimination of a cumbersome labor force often overrode the apprehension of critics.

Another technological advance that significantly altered street-cleaning practices in the United States was motorization of transportation and cleaning equipment. Economy and efficiency were most often cited as the primary reasons for the change, and street-cleaning departments were merely keeping pace with the transportation trends of the day.[29] Street-cleaning departments also came to realize that the trend toward motorization of private as well as public vehicles would make their jobs easier. The limitations of animal labor—a short productive life, feeding and grooming requirements, and manure—could be overcome with efficient and (seemingly) pollution-free motorized vehicles. Interest groups associated with the rising automobile industry, such as the Automobile Chamber of Commerce, painted glowing pictures of the potential benefits of the internal-combustion engine. H. W. Perry, secretary of the chamber, outlined four areas in which motor power could be beneficial in street cleaning: the substitution of motor vehicles for horse-drawn vehicles would reduce the cost of street cleaning, lessen the wear on street pavements, and help decrease the death rate of the population; the use of motor trucks by street-cleaning departments would reduce the costs of cleaning and of waste disposal; the elimination of the horse would lessen wear and tear on street pavements; and the replacement of macadam pavement with permanent pavement to provide easier travel for motor vehicles would ultimately lower street-cleaning and maintenance costs. Perry calculated that Chicago could save one million dollars on street cleaning alone if businesses and private citizens would substitute trucks and automobiles for the eighty thousand horses used for transportation in the city.[30] Street cleaners around the world heralded the arrival of the truck and the automobile. Few foresaw the long-term environmental impact of the automobile age or had any viable statistics about the effects of motor vehicles on the streets.

Although mechanization and motorization held out great hopes for improvements in street cleaning, these technological achievements could not change citizens' behavior. Sweepers and flushers may have helped improve the gathering of street debris, but they could not curtail littering. As Joseph J. Norton, supervisor of street-cleaning and oiling services in Boston, noted, "The greatest deterrent to clean streets in any community, particularly a cosmopolitan one, is the constant, unrestrained and promiscuous throwing of rubbish and waste materials into the streets."[31] No matter how efficient a street-cleaning crew was, it could not keep up with the litter. Edward D. Very, a New York City sanitary engineer, calculated that cleaning up litter took about 20 percent of a sweeper's time in Manhattan and about 10 percent of a sweeper's time in the other boroughs.[32]

Almost all cities had antilittering ordinances, but they were largely ignored, unenforced, or unenforceable. The Citizens' Research Council of Michigan contended that, although Detroit police and judges had adequate power to stop the littering of streets and alleys, the ordinances prohibiting littering and street obstructions continued to be "violated to an unusual extent."[33] In an address at the First New England Conference on Street Cleaning in 1910, one speaker argued that enforcement of antilitter ordinances was often lax because the police, who had jurisdiction over the ordinances, were independent of the street-cleaning forces and had very little sympathy with their objectives or appreciation of the importance of their work.[34]

Some cities intensified their efforts to bring littering under control. Reformers joined in the cause with cleanup campaigns and other publicity efforts. Some municipal authorities experimented with various new programs. In Washington, D.C., the police were instructed to intensify their efforts to enforce littering ordinances. For a while, arrests of violators increased, and the fine for littering was raised from one dollar to five dollars. The mayor of New Haven, Connecticut, admonished the police and the public works crews to help reduce litter. He offered prizes to schoolchildren for the best essays on the importance of maintaining a clean urban environment. Boston and other cities tried to curb littering by installing more trash barrels in the public parks and on street corners while doubling their efforts to educate the public about the evils of littering. No ready solutions were forthcoming, however.[35]

Any evaluation of the improvement in street cleaning in American cities has to take into consideration disparities in service. Local conditions—political, economic, geographic, and climatic—largely determined the quality of street cleaning. Often the question was not whether effective programs that utilized available organizational and technological tools could be initiated but whether city governments were willing to finance them. Under the restrictions imposed by local circumstances, it was very difficult for municipalities to determine what constituted adequate funding for street cleaning.

Expenditures varied widely in the early twentieth century. For example, according to a report issued in 1906, New York City spent more than $5.5 million on street cleaning in that year, while Philadelphia, the city with the second-highest expenditure, spent approximately $688,000.[36] Comparisons of cities' total annual expenditures were not very useful. Comparative and relative costs more effectively indicated disparities in service among cities. A report published in 1915 in the *Municipal Journal* indicated that, in thirty-one major cities surveyed, the average annual cost of street cleaning per thousand square yards was $0.355; however, costs ranged from $0.14 to $1.53. The annual report of the Baltimore Department of Street Cleaning for 1900 showed the disparity among four major cities. The per capita cost in Baltimore was $0.73, as compared with $1.76 in Boston, $1.36 in New York, and $0.96 in Philadelphia. One important conclusion drawn from several investigations was that per capita costs were greater in cities with populations over three hundred thousand. The financial burden of maintaining the streets was much heavier for residents of larger cities.[37]

Labor costs accounted for much of the expense of street cleaning. In 1909, more than 22,000 people were employed in street cleaning in cities with populations of 30,000 or more. Table 4 shows a breakdown of number of employees by size of city population. Table 5 shows the average number of workers employed in several cities in 1913.

Although the use of machine sweeping was increasing in the early twentieth century, human labor continued to be the mainstay of most large street-cleaning departments. Because of the magnitude of the street-cleaning task, the numbers of employees remained high. Good labor relations were vital to the efficiency and effectiveness of street-cleaning programs. Although Colonel Waring had encouraged superintendents not to take their work forces for granted, when machines began to do some of

TABLE 4

Total Number of Street-Cleaning Employees in American Cities, 1907 and 1909 (by population)

Size of Cities (by Population)	*1907*	*1909*
Over 300,000	14,176	13,963
100,000 to 300,000	3,702	4,052
50,000 to 100,000	2,500	3,072
30,000 to 50,000	1,460	1,473
Totals	21,838	22,560

Source: U.S. Department of Commerce and Labor, Bureau of the Census, *Statistics of Cities Having a Population of over 30,000: 1907*, (Washington, D.C., 1910), p. 474; U.S. Department of Commerce, Bureau of the Census, *General Statistics of Cities: 1909* (Washington, D.C., 1913), p. 133.

TABLE 5
Average Number of Street-Cleaning Workers in Selected American Cities, 1913

City	*Average Number of Employees*	*City*	*Average Number of Employees*
Albany, N.Y.	60	New York City, N.Y.	
Baltimore, Md.	310	Richmond Borough	96
Boston, Mass.	590	Queens Borough	196
Chicago, Ill.	1,220	Bronx Borough	930
Cincinnati, Ohio	175	Manhattan Borough	2,800
Cleveland, Ohio	450	Omaha, Nebr.	42
Columbus, Ohio	123	Philadelphia, Pa.	1,115
Detroit, Mich.	300	Providence, R.I.	190
Grand Rapids, Mich.	75	Portland, Oreg.	225
Indianapolis, Ind.	275	Rochester, N.Y.	125
Kansas City, Mo.	225	San Francisco, Calif.	170
Milwaukee, Wis.	250	St. Joseph, Mo.	24
Minneapolis, Minn.	425	Toledo, Ohio	45
New Orleans, La.	340	Washington, D.C.	400

Source: Chicago, Civil Service Commission, *Reports on the Bureau of Streets, Department of Public Works, City of Chicago* (21 April and 15 October, 1913), p. 65.

the jobs formerly carried out by human labor, many cities continued to recruit their employees from easily exploitable groups—transients, the chronically unemployed, and newly arrived immigrants. Even in New York City, where the colonel's program had emphasized improved working conditions and labor relations, street cleaners had lost most of their gains. Salaries in 1917 were only $2.50 a day, an increase of only about $0.02 a year since 1895, when Waring was appointed street-cleaning commissioner. Street cleaning continued to be hazardous. In 1917, according to Dr. S. I. Rainforth, chief physician of the New York Department of Street Cleaning, eight out of ten White Wings were physically disabled from job-related causes—a total of 5,484 cases of disability. Accidents, pneumonia, sunstroke, intestinal disorders, neuralgia, and other impairments related to weather conditions and poor sanitation were pervasive. Although New York City provided free medical care for its employees and required annual physical examinations, the hazards of the job, combined with low pay and other grievances, caused instability in street-cleaning departments. Efforts to improve working conditions, increase pay, provide pensions, and establish outlets for workers' grievances were slow in coming. Strikes and

TABLE 6
Frequency of Street Cleaning in Cities, 1909

Size of Cities (by Population)	*Percentage of Paved Streets Cleaned*	*Average Number of Cleanings per Week*
Over 300,000	49.5	4.8
100,000 to 300,000	36.8	5.0
50,000 to 100,000	42.7	5.0
30,000 to 50,000	28.6	4.4
Average	43.5	4.9

Source: U.S. Department of Commerce, Bureau of the Census, *General Statistics of Cities: 1909,* (Washington, D.C., 1913), pp. 124–32.

other demonstrations of discontent, such as frequent absence from work and poor performance, disrupted department operations.[38]

The obstacles to good street-cleaning service were legion: unfavorable local conditions, lack of adequate funds, lukewarm government support, and serious organizational and technical problems. Nonetheless, significant improvements were made in the early twentieth century. Technological innovations in street paving, the advent of motorized vehicles, and improvements in sweeping machines changed the methods of street cleaning. Cost- and record-keeping techniques offered means of monitoring expenditures and evaluating the effectiveness of programs. Most important was the establishment nationwide of permanent municipal street-cleaning programs. Regularity of service was another important measure of the effectiveness of the programs. Table 6, compiled from a statistical survey made in 1909, indicates how regular the service had become. Statistics on frequency of service or levels of expenditures do not, of course, tell the whole story. The quality of the service provided was a subjective matter, not easily expressed in statistics. By the measures available, however, it appeared that street cleaning as a municipal function had come of age.

Rudolph Hering (1847–1923), sometimes called the "Father of American Sanitary Engineering."
Courtesy of the Public Works Historical Society (APWA).

George E. Waring Jr. (1833–1898), former commissioner of the Department of Street Cleaning, New York City, helped initiate modern refuse management.
Courtesy of *McClure's Magazine.*

Mary McDowell (1854–1936), Chicago's "Garbage Lady," was a leading sanitary reformer who promoted "municipal housekeeping."
Courtesy of the Public Works Historical Society (APWA).

Caroline Bartlett Crane (1858–1935), a leading proponent of "municipal housekeeping," achieved national attention in Kalamazoo, Michigan, for her work in street-cleaning reform.
Courtesy of the Public Works Historical Society (APWA).

Pig in a basement on Mott Street, New York City. Pigs and turkeys, in particular, were widely accepted as useful scavengers in preindustrial America.

Courtesy of the Public Housing Administration, U.S. Archives and the American Public Works Association.

A horse-drawn garbage cart (1909) could easily be unloaded at a local dump because it could be tipped.

Courtesy of the Public Works Historical Society (APWA).

Convicts from Blackwell's Island unloading ash from garbage scows at Riker's Island. The ash was used for landfill at the site of New York's future prison.
Courtesy of *Scribner's Magazine*.

Sorting refuse by hand in nineteenth-century New York City.
Courtesy of *Scribner's Magazine.*

Hand-sorting refuse in New York City provided an opportunity to recover some of the cost of collection and disposal.
Courtesy of *Scribner's Magazine.*

A Studebaker flushing wagon (1890s) used to clean paved streets and to keep down dust on unpaved streets.
Courtesy of the Public Works Historical Society (APWA).

The Salvation Army (early twentieth century), like other charitable organizations, helped recycle used clothes and other goods through its collection drives.

Courtesy of the American Public Works Association.

An early 1900s gasoline-powered dump truck used by the Kearny, New Jersey, Street Department.

Courtesy of the American Public Works Association.

An early transfer station. A dump truck is placed on a hydraulic lift and unloaded into another vehicle.

Courtesy of the American Public Works Association.

(Above)
Street cleaner on rounds in Chicago (mid-twentieth century).
Courtesy of the American Public Works Association.

(Left)
New York City street cleaner in 1936, departing from the all-white uniforms of Waring's White Wings.
Courtesy of the American Public Works Association.

Collecting curbside refuse by hand was performed in cities throughout the U.S. well into the twentieth century when mechanized collection became more popular. In several communities, this hard work often was done by people of color or recent immigrants.

Courtesy of the American Public Works Association.

Philadelphia White Wings in assembly promoting antilitter campaign (early twentieth century).

Courtesy of the American Public Works Association.

Street cleaner in Newark, New Jersey, emptying sweepings in sidewalk bin.
Courtesy of the American Public Works Association.

A dragline working the Fresno Sanitary Landfill in 1940. The Fresno landfill was the archetype for this important disposal technology throughout the nation after World War II.
Courtesy of the James Martin Collection.

Jean Vincenz at the Fresno Sanitary Landfill (1981). The former commissioner of public works for the city of Fresno was the first to use the "trench" or "cut and cover" method of landfilling in the United States.

Courtesy of the James Martin Collection.

The three-wheel sweeper (circa 1948) sprayed water to eliminate dust. It could maneuver effectively along crowded streets.

Courtesy of the Public Works Historical Society (APWA).

(Left) Warning notice used by the New York City Department of Sanitation to enforce the sanitary code (1950s).

Courtesy of the American Public Works Association.

(Opposite, top, left) Signs used in New York City to notify citizens about littering and its consequences (1950s). The top three metal signs were normally posted on light poles; the "No Dumping" sign was posted at vacant lots; and the motor vehicle sign was distributed to gas stations, garages, and parking lots.

Courtesy of the American Public Works Association.

(Opposite, top, right) A wastepaper receptacle in Washington, D.C. (1959).

Courtesy of the American Public Works Association.

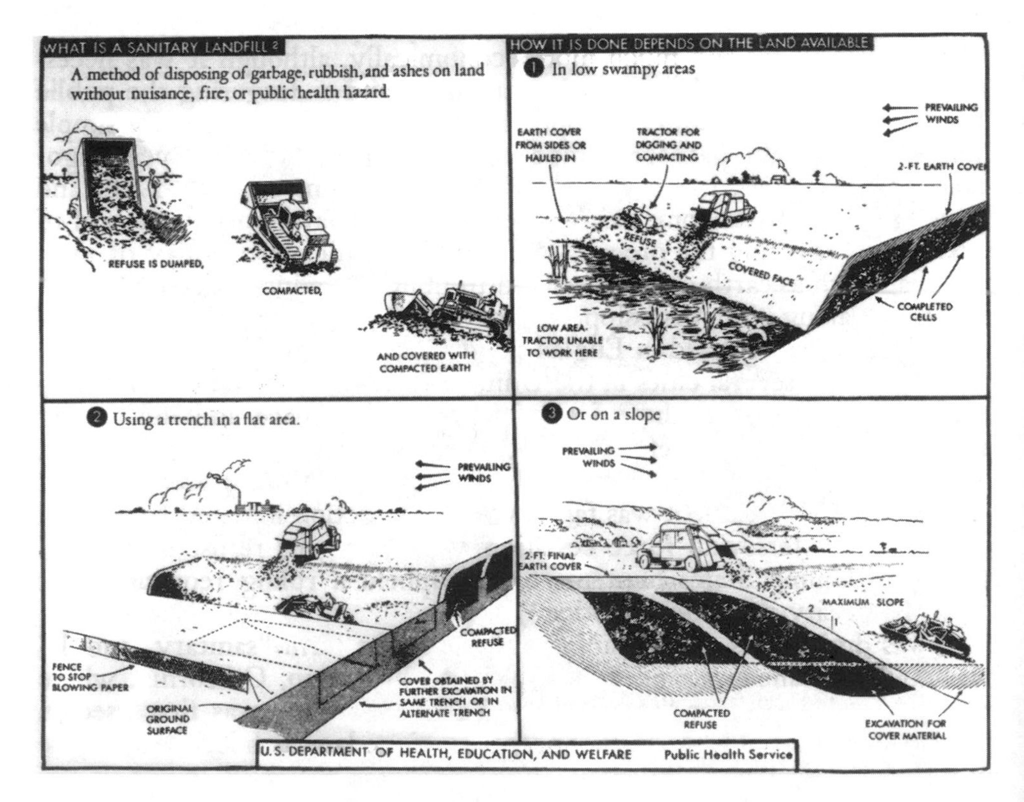

Various methods of landfilling in the 1960s.

Courtesy of the American Public Works Association.

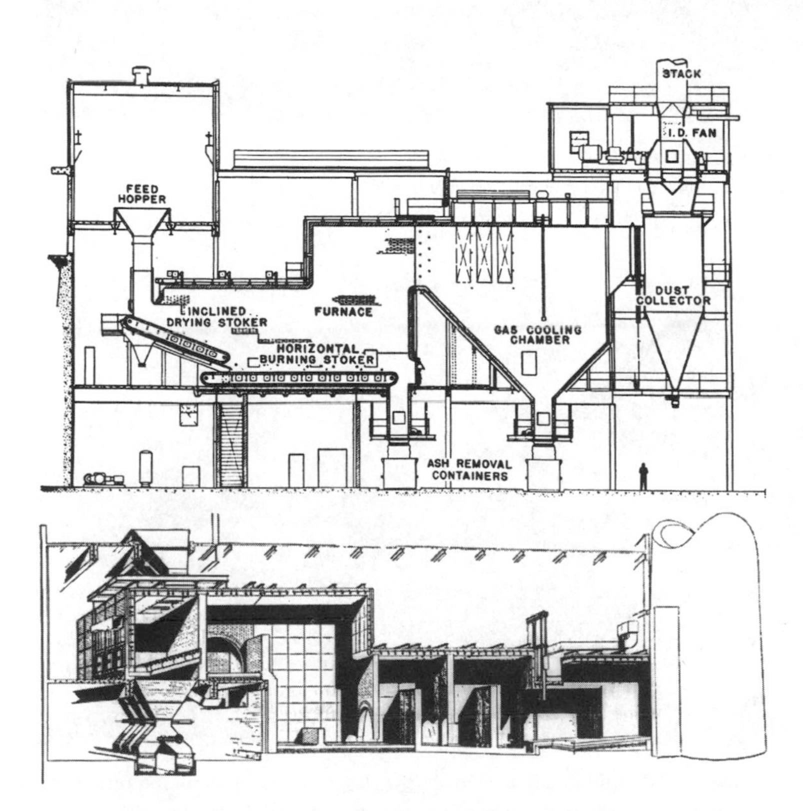

Cross-section drawings of a furnace at the Calumet Incinerator in Chicago (top) and the Northwest plan in Philadelphia (bottom), 1960s.

Courtesy of the American Public Works Association.

A Caterpillar Traxcavator used at the landfill operations in Fort Madison, Iowa.
Courtesy of the American Public Works Association.

Rock Island standard bulkhead flatcar loaded with 77 tons of compacted refuse in Saint Paul, Minnesota (1972), most likely to be shipped to a distant reclamation facility.
Courtesy of the American Public Works Association.

A modern front-end loader compactor truck used in commercial solid waste collection.
Courtesy of the Public Works Historical Society (APWA).

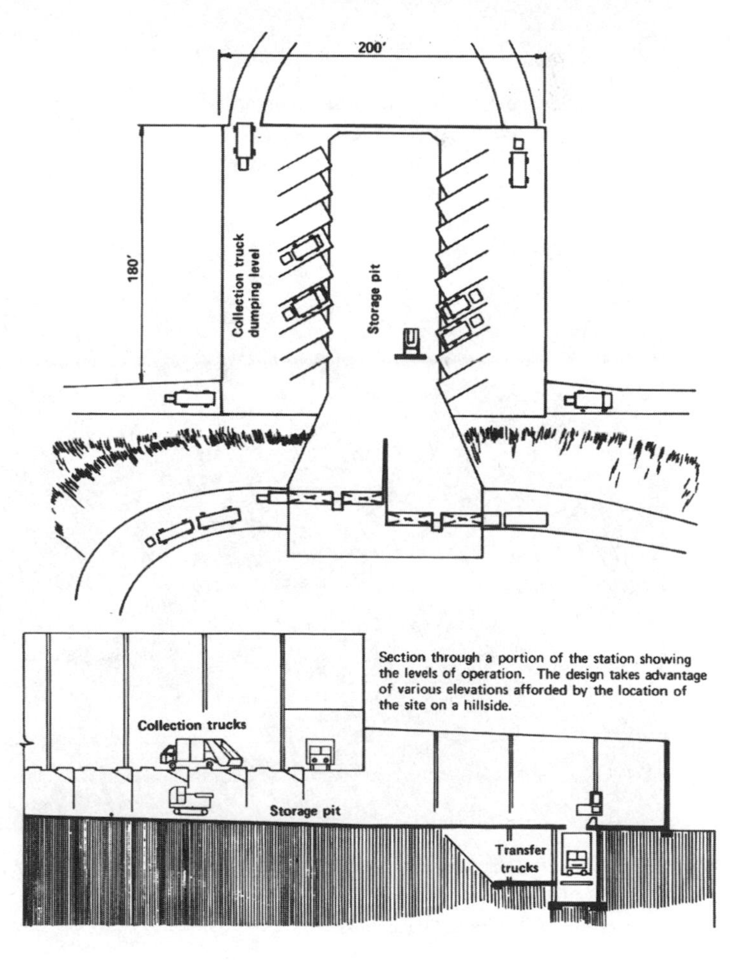

Drawing of a 2,000-ton-per-day transfer station in San Francisco (1971).
Courtesy of the American Public Works Association.

A side-loader truck with an automated arm that can reach over cars to pick up garbage bags. Bellaire, Texas, (early 1970s).

Courtesy of the American Public Works Association.

When Racine, Wisconsin, began recycling newspapers in 1974, it was one of the first cities to do so.

Courtesy of the American Public Works Association.

SIX

COLLECTION AND DISPOSAL PRACTICES IN THE EARLY TWENTIETH CENTURY

In 1908, refuse-disposal expert William F. Morse hypothesized that, as society became increasingly urbanized and as cities grew in size and number, methods for dealing with waste remained unchanged. Primitive methods sufficient for individual or family needs were simply applied to the new circumstances with little thought to the difference in context.[1] In an elementary way, Morse's perception was accurate—at least until the late nineteenth century. Colonel Waring made some progress after 1895 by convincing city officials that refuse was more than a personal inconvenience. As with street cleaning, increased public awareness of the physical environment and the activities of reformers had positive effects. Unlike street cleaning, refuse collection and disposal were not easily improved by technological innovations. The complexity of the tasks was only beginning to be addressed by World War I.

The naïveté of late-nineteenth-century reformers, who saw in Waring's programs quick solutions to the waste problem, was tempered by trial-and-error experimentation in the first two decades of the twentieth century. Those willing to accept partial victories as signs of progress expressed guarded optimism about attempts to improve collection and disposal practices. Rudolph Hering and Samuel Greeley contended that impressive results had been achieved in "collating data of experience" about the quantity and quality of refuse. Some observers noted with approval the in-

TABLE 7

Responsibility for Garbage Collection in Cities, Selected Years

	Year of Survey, Percentage and Number of Cities									
Form of Responsibility	*1880*	*1899*	*1901*	*1902*	*1913*	*1915*	*1915*	*1918*	*1918**	*1924*
City	24.0%	32%	39%	33.5%	57.0%	55.0%	50%	46%	54%	63%
	(48)	(12)	(11)	(54)	(16)	(31)	(84)	(48)	(7)	(60)
Contract	19.0%	46%	43%	30.0%	39.5%	36.0%	39%	34%	38%	25%
	(38)	(17)	(12)	(48)	(11)	(20)	(66)	(36)	(5)	(24)
Private	30.0%	3%	11%	25.5%	3.5%	3.5%	1%			10%
	(59)	(1)	(3)	(41)	(1)	(2)	(2)			(10)
Combined or Other	1.5%	6%				5.5%	6%	20%	8%	
	(3)	(2)				(3)	(9)	(21)	(1)	
No data or no system	0.5%	13%	7%	11.0%			4%			2%
	(1)	(5)	(2)	(18)			(7)			(2)
Total Cities	149	37	28	161	28	56	168	105	13	96

*Cities with 247,000 or more population.

Source: U.S. Department of the Interior, Census Office, *Report on the Social Statistics of Cities, Tenth Census, 1880* (Washington, D.C., 1886); "Garbage Collection and Disposal," *Engineering News* 42 (September 28, 1899): 214; "Methods of Garbage Disposal," *Municipal Journal and Engineer* 11 (September 1901): 123–124; William F. Morse, "The Disposal of Municipal Waste," *Municipal Journal and Engineer* 20 (February 7, 1906): 114; Rudolph Hering and Samuel A. Greeley, *Collection and Disposal of Municipal Refuse* (New York: McGraw-Hill Book Co., 1921), p. 106; *American Society for Municipal Improvements, Proceedings of the Twenty-second Annual Convention,* (1915), pp. 10–11; "Refuse Collection and Disposal," *Municipal Journal and Engineer* 39 (November 11, 1915): 725–27; William Parr Capes and Jeanne Daniels Carpenter, *Municipal Housecleaning* (New York: E. P. Dutton and Co., 1918), table vi; Pittsburgh, Commission on Garbage and Rubbish Collection and Disposal, *Report on Methods of Garbage and Rubbish Collection and Disposal in American Cities* (1918), pp. 14–15; "Garbage Collection and Disposal: A Compilation from Questionnaires Returned by 101 City Manager Cities in the U.S. and Canada," *City Manager Magazine* (July 1924): 12–14.

creasing number of municipal investigations being conducted to evaluate local needs and to uncover the shortcomings of refuse-management programs already in operation. *Engineering News* reported: "It is gratifying to observe the general attention now being paid to the disposal of refuse and garbage in American cities. While it cannot be said that any large amount of definite progress is to be recorded, the subject is now receiving the attention of the public, which should long ago have been paid to it."[2]

Other commentators and experts, exasperated by the snail's pace at which municipal officials responded to the refuse problem, could not hide their pessimism. M. N. Baker asserted that in no branch of municipal service had "so little progress been made in the United States as in the disposal of garbage."[3] Some individuals criticized the widespread public indifference and apathy, while others questioned the sophistication of available methods. Luther E. Lovejoy, secretary of the Detroit Housing Commission, fatalistically concluded, "The accumulation of garbage and rubbish is one of the penalties human society inevitably pays for the luxury of civilization."[4]

Despite the cynicism of some experts, statistics indicate a relatively rapid increase in municipal responsibility for collection and disposal of refuse. Of the cities surveyed in the 1880 census, 24 percent had municipally operated garbage-collection systems, and 19 percent contracted with private firms for the service. This meant that only 43 percent of the cities provided for some sort of collection. By the turn of the century, however, more than 65 percent of the cities surveyed had municipally sponsored collection, and in the period from 1910 to 1920, it rose to 89 percent or even higher. In 1915, for example, 50 percent had municipally operated collection systems, and 39 percent utilized some form of contract work (see table 7).

The growing trend toward municipal ownership of utilities and control of services led several cities to abandon the contract system because traditional arrangements were proving inadequate to meet the needs of the growing and more complex industrial cities. Contract terms varied greatly, but most were of short duration, running about three to five years. Renegotiation was a continual process and a mixed blessing. Frequent renewals gave cities the opportunity to reevaluate the provisions of the contracts and review contractors' performance, but they were also very time consuming, entangled in bureaucratic red tape, and usually disadvantageous to the cities. Contractors, uncertain of their long-term relationships with the cities, were unwilling to devise cleaning systems that required large capital outlays, such as for constructing expensive incinerators, reduction plants, or other permanent structures. Attempting to hold down costs, they often employed poorly trained workers, who received minimal benefits.[5]

Short-term contracts were abused in other ways. For example, city officials in Philadelphia took bids for the ensuing year in late November or early December. Because the period between the reception of the bids and the beginning of actual work was so short, a company lacking extensive resources and a well-developed program could not compete with the existing contractor. Thus, one contractor could maintain a long-standing monopoly built upon the yearly bidding system and the short-term contract. In Philadelphia and other cities, charges of political favoritism and bribery were common.[6] As with street cleaning, however, those critical of the contract system too readily assumed that municipal responsibility would assure better service and the elimination of corruption.[7]

Determining responsibility was only the first of many crucial decisions that city officials faced in establishing practical refuse-management systems. In the 1890s, Colonel Waring offered primary separation as a solution. However, investigations of collection practices throughout the country indicated that no single method was practicable in all cities. The choice had to depend on the special conditions of the city, the method of disposal in use, and, with the separation system, the availability and dependability of markets for the by-products. Simply to copy methods used in other cities or to accept uncritically the claims of those with financial interests in collection equipment was an irresponsible way to establish a proper system.[8]

In practice, cities had a choice between primary separation and combined-refuse collection. Advocates of primary, or source, separation echoed the colonel's arguments and were most persuasive in cities in which the reduction process was a major means of disposal. Their main argument was that primary separation promoted cleanliness.[9] Critics of the primary-separation method argued that combined collection was much easier for the householder, less complicated for the collection teams, and cheaper for the city. They insisted that the combined system was actually cleaner because objectionable garbage was mixed with other kinds of refuse. Cities relying primarily on incineration were more likely to employ the combined- or single-collection method. Indeed, combined collection was on the rise in Europe at that time because of the widespread adoption of incinerators.[10]

American cities rarely employed a single means of disposal. Primary separation might be used in some neighborhoods or under certain conditions but not throughout the city. Surveys conducted between 1902 and 1924 show that 59 to 83 percent of the cities practiced some form of separation, segregating only garbage, rubbish, or ashes. In less than half the cities with separation programs were all wastes separated along the lines of the Waring system.[11]

No matter which method was chosen, collection was a difficult municipal problem because it was the phase of refuse management that directly affected the greatest number of people. It was also the costliest phase. As table 8 indicates, the cost of collection was two to eight times as expensive as disposal. One reason for the high cost during this period was the increase in frequency of service. According to a study conducted at MIT in 1902, 79 percent of the cities surveyed (127 of 161) collected garbage on a regular basis.[12] Several factors determined the frequency of collection: the amount and nature of the wastes, population density, the physical layout of the city, climatic and seasonal variations, the financial resources of

TABLE 8

Comparative Cost of Collection and Disposal in Selected American Cities

			Cost per Ton	
City	*Year*	*Material*	*Collection*	*Disposal*
Albany, N.Y.	Estimated	Mixed refuse	$1.81	$0.41
Richmond Borough, New York, N.Y.	1911	Mixed refuse	1.64	0.54*
Seattle, Wash.	Estimated	Mixed refuse	1.30	0.62
Boston, Mass.	1910	All refuse	1.41	0.43
Buffalo, N.Y.	1907	Garbage	2.19	
Chicago, Ill.	1911	Garbage	3.42	0.38
Cleveland, Ohio	1911	Garbage	2.83	1.04†
Columbus, Ohio	1911	Garbage	1.88	0.77†
Milwaukee, Wis.	1910	Garbage	2.85	0.90
Minneapolis, Minn.	1910	Garbage and ashes	1.32	0.92‡
Buffalo, N.Y.	1907	Rubbish	4.90	0.04§
Averages			$2.42	$0.60

*Estimated from cost per cubic yard.
†Profit.
‡Cost of garbage disposal only.
§Cost after deducting profit from rubbish sorting plant.

Source: Rudolph Hering and Samuel A. Greeley, *Collection and Disposal of Municipal Refuse,* (New York: McGraw-Hill Book Co., 1921), p. 106.

the municipal government, and the form of transportation used. Collection and disposal vehicles were rapidly becoming motorized in the early twentieth century. Baltimore was the first city to use motorized garbage trucks. Some cities, including Chicago, New Orleans, and Cleveland, experimented with railway transport of refuse. Not until the 1920s, however, did horse-drawn vehicles relinquish their role in collection and disposal. In fact, horses were preferred to motorized vehicles for certain tasks, especially household pickups.[13]

Not surprisingly, collections were most frequent in the business districts, followed by those at central-city residences. Collections were least frequent in outlying areas. Almost half the cities surveyed between 1909 and 1918 collected garbage from businesses six times a week (the other half collected one to four times a week). In residential areas, it was most common to collect garbage at least two or three times a week; in more remote areas, one to three times a week. Of course, suburbs occupied by people in the upper socioeconomic classes received preferential treatment. Some cities, for example, diverted some of their collection teams from inner-city routes to upper- and middle-class suburbs.[14]

The efficiency of refuse collection largely depended on the volume of waste that the city had to remove. Amounts and composition varied widely from city to city, but some trends were obvious. By world standards—even in this period—the United States produced exceptionally large quantities of waste. Although consumption of goods varied widely according to economic status, the amounts of refuse were staggering overall. Between 1888 and 1913, the annual per capita weight of mixed refuse for fourteen American cities was 860 pounds (many other surveys indicated even higher figures for American cities through 1917). Comparisons with several European cities revealed that eight English cities generated 450 pounds per capita and that seventy-seven German cities produced 319 pounds per capita. This meant that the English cities surveyed produced only 52 percent of the solid waste produced by their American counterparts and the German cities produced only 37 percent.[15]

Examining the production of refuse by type also revealed some interesting trends. Between 1903 and 1918, the per capita production of garbage in American cities ranged from 100 to 300 pounds; rubbish, 25 to 125 pounds; and ashes, 300 to 1,500 pounds. Total per capita refuse ranged from one-half to one ton a year.[16] Furthermore, some kinds of waste, especially garbage, were increasing at an alarming rate. Between 1903 and 1907, Pittsburgh's garbage increased from 47,000 to 82,498 tons, or 43 percent. Other cities experiencing substantial increases in the same period included Milwaukee, 30,441 to 40,012 tons (24 percent); Cincinnati, 21,600 to 31,255 tons (31 percent); Washington, D.C., 33,664 to 44,309 tons (24 percent); and Newark, 15,152 to 21,018 tons (28 percent). Population growth, greater consumption of goods, and more efficient collection accounted for most of these increases. Obviously, not all cities experienced such rapid increases as those listed above. In Cleveland and Saint Louis, for instance, the volume of garbage actually declined over the same period. Yet few public works departments assumed that the volume of waste would diminish or that their workloads would decrease over time.[17]

Population size was an important variable in the impact of the refuse problem on American cities. Table 9 shows the average and range of annual tonnage of refuse for cities according to population in 1907. Besides having larger populations than their smaller counterparts, major cities supported arrays of business, hotels, restaurants, and services that contributed greatly to the tonnage figures.

Still other factors influenced the quantities of wastes in the various cities, such as the affluence of the citizens and the relative density of residential dwellings.[18] Although statistics are not particularly reliable for these variables, a few contemporary surveys produced some intriguing findings. A survey conducted in Chicago in 1912, for example, attempted to correlate nationality or ethnic background with the production of refuse. The survey demonstrated that "Americans" (presumably native-born white Americans of the pre-"New Immigrant" era) produced substantially more waste than their "foreign" (Italian, Polish, Bohemian, German, and Russian) counterparts. The average per capita annual production by Americans was 120.7 pounds of garbage and 630.7 pounds of ashes and rubbish, a total of 751.4 pounds of refuse. The foreign group annually produced 90.5 pounds of garbage and 582.3 pounds of ashes and rubbish per capita,

TABLE 9

Tons of Refuse Collected by City Size, 1907

Population	*Number of Cities*	*Average Tonnage*	*Range of Tonnage*
Over 300,000	14	521,009	3,042,308–75,000
100,000 to 300,000	26	59,405	301,211–4,868
50,000 to 100,000	34	25,449	67,501–1,047
30,000 to 50,000	49	9,030	22,893–400

Source: U.S. Department of Commerce and Labor, Bureau of the Census, *Statistics of Cities Having a Population of over 30,000: 1907* (Washington, D.C., 1910), pp. 452–57.

TABLE 10

Disposal of Garbage in Cities, Selected Years

Method	*1899*	*1901*	*1902*	*1902*	*1903*	*1913*	*1913*	*1915*	*1916*	*1918*	*1918**	*1924*
Dumping on land, used as fill, or buried	27% (10)	23% (6)	32.0% (7)	24% (39)	24% (44)	3.5% (1)	18.5% (25)	11% (21)	29% (10)	27.5% (29)		17% (16)
Farm use (fertilizer, animal feed)	16% (6)	19% (5)	13.5% (3)	21% (33)	32% (59)	14.5% (4)	28.0% (37)	8% (15)	14% (5)	20.0% (21)	15.5% (2)	38% (37)
Dumping in water	8% (3)	23% (6)	9.0% (2)	6% (10)	8% (14)	3.5% (1)	7.5% (10)	2% (4)		1.0% (1)		
Burning or incineration	16% (6)	8% (2)	32.0% (7)	22% (36)	20% (36)	25.0% (7)	26.5% (35)	30% (56)	43% (15)	13.5% (14)	15.5% (2)	29% (27)
Reduction	22% (8)	4% (1)	13.5% (3)	9% (15)	10% (19)	43.0% (12)	19.5% (26)	10% (19)	14% (5)	20.0% (21)	69.0% (9)	2% (2)
Combination of methods				12% (19)		3.5% (1)		7% (13)		17.0% (18)		1% (1)
Other methods												9%† (9)
No systematic method	11% (4)				6% (11)							
No data		23% (6)		6% (9)		7% (2)		32% (61)		1.0% (1)		4% (4)
Total cities	37	26	22	161	146‡	28	133	189	35	106	13	96

*Cities with 247,000 or more population.

†Sanitary landfill.

‡Responses totaled 183, but only 146 cities were surveyed.

Source: "Garbage Collection and Disposal," *Engineering News* 42 (September 28, 1899): 214; "Methods of Garbage Disposal," *Municipal Journal and Engineer* 11 (September 1901): 123–124; "Methods of Garbage Disposal," *Municipal Journal and Engineer* 13 (July 1902): 28; B. E. Briggs, "Cost of Collection and Disposal of Garbage," in *Proceedings of the Association for Municipal Improvements, Twelfth Annual Convention (1905)*, p. 149; Rudolph Hering and Samuel A. Greeley, *Collection and Disposal of Municipal Refuse* (New York: McGraw-Hill Book Co., 1921), p. 106; "Disposal of Municipal Refuse," *Municipal Journal and Engineer* 35 (November 6, 1913): 627; "Refuse Collection and Disposal," *Municipal Journal and Engineer* 39 (November 11, 1915): 728–30; "Methods of Garbage Collection," *Municipal Journal and Engineer* 41 (December 7, 1916): 701–702; William Parr Capes and Jeanne Daniels Carpenter, *Municipal Housecleaning* (New York: E. P. Dutton and Co., 1918), table vi; Pittsburgh, Commission on Garbage and Rubbish Collection and Disposal, *Report on Methods of Garbage and Rubbish Collection in American Cities* (1918); "Garbage Collection and Disposal," *City Manager Magazine* (July 1924): 12–13.

a total of 672.8 pounds. These figures are not representative of the entire nation, but they give credence to the theory that there is an important correlation between affluence and refuse production (if it can be assumed that the Americans were generally more affluent than the foreigners in Chicago).[19] It is curious, therefore, that those grappling with the refuse problem tended to place excessive responsibility on the immigrant populations for generation of wastes.

The huge quantities of waste pointed to the need for not only efficient collection but effective disposal as well. City officials faced substantially different problems in dealing with disposal as opposed to collection. For one thing, collection practices directly affected most of the urban population in the same way, while disposal problems impacted urbanites in dissimilar ways. City dwellers living close to open dumps, landfills, loading docks, reduction plants, and incinerators suffered the greatest inconveniences, annoyances, and dangers from disposal methods, while a large portion of the population encountered little direct contact with them. Thus city officials were less inclined to worry about whether disposal methods were offensive to the senses—especially if disposal sites were located away from population centers or in the neighborhoods of racial and ethnic minorities, the poor or working classes—and more inclined to select methods that were expedient and inexpensive or that offered the promise of much greater efficiency. Popular protests tended to be of much less consequence in the selection of a disposal method than they were in the choice of a collection method. It is no wonder that urbanites criticized the primitiveness of many disposal methods. Cities, however, were painfully slow in abandoning them. Breaking old habits was difficult, especially if lakes or streams were nearby or vacant lots in "less desirable" neighborhoods were available for dumping. Also the cost of conversion to newer methods frustrated many cities, especially smaller ones.

Table 10 shows the array of methods (many of them primitive) employed by cities to dispose of garbage in the early twentieth century. Table 11 shows the methods used to dispose of rubbish or combustible wastes between 1899 and 1913. Table 12 indicates the methods used to dispose of ashes in 1902. Dumping on land continued to be the primary method of disposal of rubbish and ashes, but no single method of garbage disposal dominated, despite the ballyhoo over incineration and reduction. As primitive methods faced greater scrutiny and more careful evaluation, city officials and sanitary engineers came to believe that they could be modified for or restricted to certain kinds of disposal needs. More cities began tak-

TABLE 11

Method of Disposal of Rubbish or Combustible Wastes in Cities, Selected Years

	Year of Survey, Percentage and Number of Cities		
Method	*1899*	*1902*	*1913*
Dumped on land	70% (26)	46.5% (75)	61% (17)
Dumped in water	3% (1)	2.5% (4)	3% (1)
Burning or incineration	16% (6)	29.5% (47)	7% (2)
Sanitary fill			7% (2)
Combination of methods		1.5% (2)	11% (3)
No systematic method	11% (4)	0.5% (1)	
No data		19.5% (31)	11% (3)
Total cities	37	161	28

Source: "Garbage Collection and Disposal," *Engineering News* 42 (September 28, 1899), 214; C. E. A. Winslow and P. Hansen, "Some Statistics of Garbage Disposal for the Larger American Cities in 1902," in American Public Health Association, *Public Health: Papers and Reports* 29 (October 1903): 141–53; Rudolph Hering and Samuel A. Greeley, *Collection and Disposal of Municipal Refuse* (New York: McGraw-Hill Book Co., 1921), p. 106.

ing into account what they expected to achieve by using a specific method, rather than simply accepting one because it was easy or rejecting it out of hand. This was the real revolution in collection and disposal practices in the early twentieth century, not technological innovation. A number of municipal leaders were abandoning the out of sight, out of mind attitude of the nineteenth century for a more thoughtful consideration of the means and ends of refuse management.

Dumping waste into water was the most universally condemned practice of all the primitive methods. As indicated in table 10, this method was declining by 1902. Several reasons—not necessarily environmental considerations—led to the end of water dumping. In the early 1900s, New York City temporarily curtailed dumping refuse at sea because too much of it floated back to the shoreline and also because officials in the street cleaning department believed that the waste could be put to better use as landfill.[20] A relatively common complaint about sea and lake dumping was that for all but the largest cities, it was too costly to tow wastes to deep water. This was especially true because a large portion of the material would not sink and washed up on adjacent beaches. Dumping into rivers

TABLE 12

Method of Disposal of Ashes in Cities, 1902

Method	*Percentage of Cities*	*Number of Cities*
Dumped on land (including fill)	79.0	127
Dumped in water	3.0	5
Burning or incineration	1.5	3
Combination of methods	0.5	1
No systematic method	0.5	1
No data	15.0	24
Total cities		161

Source: C. E. A. Winslow and P. Hansen, "Some Statistics of Garbage Disposal for the Larger American Cities in 1902," in American Public Health Association, *Public Health: Papers and Reports* 29 (October 1903): 141–53.

or streams also had serious legal ramifications. Downstream cities began filing lawsuits against upstream cities that used the rivers for dumping. In time, the federal government began trying to stop the disposal of refuse in interstate waterways to end these confrontations.[21] Evaluations of water dumping eventually moved beyond the following rationalization that appeared in a New Orleans Board of Health report for 1898–1899:

> To dump the garbage of a large city into a running stream from which is also derived the water supply of the city, might seem, at first glance, a rather crude and imperfect, as well as unsanitary, method of getting rid of the city's waste; but when it is remembered that the Mississippi River is at this point a half mile wide, from fifty to one hundred feet deep, with an average current of three miles per hour, as much as one million five hundred thousand cubic feet of water passing a given point during every second at the stage of high water, we may readily imagine how little influence a boat-load or two of garbage per day can have upon such an immense body of water in constant motion.[22]

However, in the late nineteenth and early twentieth centuries, dilution was considered to be the cheapest and easiest form of sewage disposal—if not refuse disposal—practiced in the United States. In its early uses especially, dilution was haphazard, but proponents—particularly among engineers—argued that it was a theoretically effective technique because of the ability of running water to purify itself.[23]

Like dumping into water, dumping on land came under increasing criticism in the early twentieth century. Discarding garbage on land, including landfills and burial, represented up to 32 percent of the disposal methods used. Depositing ashes or combustible materials on land, considered less objectionable than dumping organic matter, remained as high as 70 to 80 percent. The method was common because of its convenience. Many cities were resigned to dumping until they could afford to end the practice. Answers to a questionnaire mailed in 1913 to ninety cities with populations of 30,000 or more illustrate this point. Of the sixty-eight cities that replied, fifty-nine maintained dumps, and thirteen relied exclusively on dumping for the disposal of garbage. In answering the question "Is the public dump a proper means of garbage disposal?" however, sixty-four of the sixty-eight answered no.[24]

Criticism did not necessarily result in the abandonment of the practice, though the intensity of the protests might lead a casual observer to assume so. Boards of health and public works departments of several cities were continually hounded with complaints about open dumping. In a 1917 report, Cleveland's Committee on Housing and Sanitation stated that the dumps "caused the most complaint in the collection and disposal system for refuse." The city owned twenty-five dumps, and private concerns maintained many more. According to the report, they constituted "breeding places for rats and cockroaches and the homes in the vicinity are infested with them. Paper and other light articles blow over the surrounding territory and are a great source of annoyance. Fires break out on the dumps frequently not only endangering the adjoining property, but the smoke and smudge are very offensive especially when the fires often smolder for months."[25] C. E. Terry, a physician of Jacksonville, Florida, told an audience at the APHA convention in 1912: "In its simplicity and carelessness, as a means of waste disposal, the dump probably dates back to the discarding of the first apple core in the Garden of Eden, and its subsequent train of evils is ample testimony of the Eternal Wrath elicited by that act."[26]

As land and water dumping drew increased criticism, other primitive methods, especially filling and burial, began attracting the renewed interest of engineers and sanitarians who believed that, properly managed, these methods held the best promise for the future. There was a growing sentiment that a method was only as good as its management. In *Collection and Disposal of Municipal Refuse,* Hering and Greeley commented, "The natural methods (land dumping, filling, burial) . . . need, in our opinion, more consideration than they have received in the past. Their simplicity and economy heretofore have tended toward neglecting a sufficient study of their efficiency and cost; yet they constitute an important branch of city refuse disposal work, and, as some of them have an extensive application, they need greater study."[27] In 1905, in an *Engineering News* story entitled "The Land Disposal of Garbage: An Opportunity for Engineers and Contractors," the editors noted that experiments in plowing waste into the land in Saint Louis "might well lead engineers and sanitarians to give more attention to the possibilities in the treatment of garbage by burial on agricultural lands. The system is already in quite extensive use, but is generally regarded as more or less a makeshift, as indeed it is when conducted in the ordinary manner."[28] An article in a 1917 issue of *Municipal Journal and Engineer* drew much the same conclusion: "It is possible to deposit garbage and refuse mixed, or even garbage alone if properly treated, on low land without creating a nuisance."[29]

Landfill programs, which never replaced dumping as a primary disposal method in this period, were generally a supplementary means of dealing with inorganic materials. The use of organic wastes alone to fill ravines or to level roads had always been regarded as highly objectionable. When garbage was mixed with large amounts of other materials, the practice was more acceptable, but it rarely provided an adequate means for cities to dispose of all refuse. As utilization and recycling of wastes came into vogue, however, filling low places with refuse or reclaiming marshland and coastal land became desirable for many cities. The "sanitary landfill" was the breakthrough that raised the practice of filling to the status of a primary disposal method, but it was not widely used until after World War II. Using fill for reclamation purposes was becoming more popular, especially as the idea of waste utilization gained support. Davenport, Iowa, situated on the west bank of the Mississippi River, used fill to build up its levies. In Oakland, California, refuse was used as fill along the shoreline of San Francisco Bay.[30]

The best-known reclamation project was begun at Rikers Island in the East River in New York City. The project, begun under Colonel Waring, was controversial. In 1900, the United States War Department, which controlled the harbor lines, forbade the dumping of material behind the enclosing cribwork until a properly constructed seawall was completed. This decree sparked a jurisdictional dispute among city departments: the Department of Street Cleaning, which wanted to use the enclosure for ash dumping; the Department of Correction, which occupied the island and wanted to reclaim more land for its uses; and the Dock Department, which would be required to improve the seawall.[31]

The enthusiasm for waste utilization also produced a renewed interest in feeding garbage to swine. This method became especially popular during World War I, when garbage was fed to hogs to produce more food for the war effort.[32] Some sanitarians remained skeptical about the healthfulness of pork from garbage-fed hogs, but others were not willing to dismiss the method so quickly. Almost everyone agreed that it was ill suited to large cities, especially because it would require a dramatic increase in collections (to ensure the freshness of the garbage) and because the herds of pigs maintained outside the city would have to be enormous.[33] The scale of these problems could be avoided in medium- and small-sized cities, and some did experiment successfully with modern swine-feeding techniques. Samuel Greeley calculated that it would take only seventy-five pigs to dispose of one ton of garbage per day. By properly sorting the garbage and possibly cooking it to avoid spoilage, small cities surrounded by isolated farms could profit from such a method.[34]

Worcester, Massachusetts, operated one of the most highly regarded swine-feeding programs in the country. According to X. H. Goodnough, of the Boston Society of Civil Engineers, Worcester's program was "a remarkable development of that method of garbage disposal." The city maintained a piggery from which, he reported, "a considerable income is derived."[35] Worcester was but one of sixty-one cities and towns in Massachusetts that employed some form of swine feeding. Outside New England, the practice was not as widespread; however, Grand Rapids, Saint Paul, Omaha, Denver, and even Los Angeles (until 1914), used swine feeding as major parts of their disposal programs.[36] Feeding garbage to swine had come a long way from the days when pigs, geese, turkeys, cows, and other animals roamed the streets scrounging for castaway items to eat.

The modification of some primitive methods had not deterred efforts to find modern alternatives to nineteenth-century practices. After a decade or so of practical experience with incineration and reduction—the two "solutions" of the 1890s—engineers and city officials were reluctant to claim that they had found a single best disposal method. Unbridled enthusiasm for the so-called technological wonders from Europe was replaced by healthier skepticism and more deliberate evaluation of the methods. By 1902, experts had begun taking a closer look at incineration and reduction to determine whether they, put to a practical test, really could supplant the older practices.

Of the two methods, incineration managed to retain the greatest number of adherents. The general consensus was that burning waste in furnaces, crematories, or incinerators offered the most sanitary method of disposal and was possibly the most efficient and economical method available. Initially, few people gave more than passing attention to the smoke produced by the incinerators.[37] While many engineers and city officials believed that incineration was in theory a perfect disposal method, they also expressed considerable dissatisfaction with the first generation of American furnaces and crematories installed in the late 1880s. These original crematories (which continued to be built through 1910), had been impulsively adapted from European models, had never lived up to expectations, and were almost without exception branded as failures. Of the 180 furnaces erected between 1885 and 1908, 102 had been abandoned or dismantled by 1909.[38]

In the excitement over the new method, not many bothered to consider whether European-designed equipment would meet American needs. For Europeans—the English in particular—destruction of waste by fire offered many advantages. In England, which lacked cheap undeveloped land, urban dumps had become impractical and costly. Extensive ocean dumping was impractical because of the proximity of neighboring countries. A method had to be devised that reduced waste to the smallest volume possible. Furthermore, because of England's limited energy resources, disposal had to be carried out with minimum fuel consumption and minimum transportation costs. The British "destructors" met these criteria. They burned mixed refuse without the use of additional fuels, and the only by-product of the process was an inert clinker or residue, which could be disposed of with relative ease. In addition, second- and third-generation destructors built before 1910 could be connected to steam boilers for generating heat or electricity. Similar practices on the Continent produced equally impressive results, especially in Germany.[39]

American sanitary engineers were the first group to criticize openly the performance of the American-modified crematories. Although these sanitary engineers had been among the first advocates of the British method, they had become increasingly suspicious of the faulty designs and inefficient operation of the furnaces built in the United States.[40] In 1911, Joseph B. Rider, a consulting engineer in New York City, charged that the early cremation plants had been constructed by rule-of-thumb methods: "What the old horse car is to the modern traction, the ox team is to the automobile, the hour glass is to the chronometer, so are crematories of the past to the modern incinerators or destructor."[41]

Some critics argued that British engineering expertise had not been taken into account in building the crematories in America. Others blamed the excessive use of coal and other fuels to augment the burning, which sharply increased the cost of operation. American furnaces were used primarily to burn organic material instead of mixed refuse, because many cities had adopted source separation to use ash and rubbish for fill or simply disposed of rubbish inexpensively on land. The greater water content in American refuse sent to incinerators thus required higher temperatures for burning. Since coal, fuel oil, and natural gas were abundant in the United States, little attention had been given to retrofitting the British destructors to American conditions. As fuel consumption and costs rose, the temperature at some furnaces was lowered for the sake of conservation. This resulted in waste that was not completely burned and increased gas and smoke emissions. The rising use of external fuel in stoking the crematories added to the cost of incineration and tended to make the method less attractive.[42]

Taken as a whole, criticism of incineration in the United States in the period suggests that the adoption of the British technology was accomplished with little effort to meet special local requirements or to adapt the technology to American cities. Circumstances of space and demography were much different in the United States than in Europe. The American population was dispersed over a larger area and in many regions was substantially less dense. The availability of cheap land in the United States afforded greater opportunity for dumping, making it a practical alternative to other and more expensive or labor-intensive methods. Moreover, energy sources were more plentiful and cheaper in America, which meant that wastes could be hauled greater distances for dumping or fuels could be used for burning.[43]

The frustration of sanitary engineers went beyond faulty designs and improper operation of crematories. They were critical of unscrupulous companies that built and promoted poor-quality equipment, and they particularly resented city officials who adopted the equipment without consulting them. They often felt that their authority had been usurped or their advice circumvented by untrained bureaucrats. Colonel William F. Morse, a sanitary engineer and author of many tracts on waste disposal, led the assault on the "extravagant claims" of the peddlers of the first-generation incinerators. He charged that the "sharp competition of opposing interests [builders of crematories and city authorities] developed mutual misrepresentation and recrimination. Contracts were obtained by personal and political favor, by influential pull, by manipulation and graft, with little regard to the interests of the city or town."[44]

Interestingly, Morse had been the manager of the New York office of Engle Sanitary and Cremation Company, founded in 1886, the first American company to manufacture crematories in quantity. In 1898 and 1899, Morse and Benjamin Boulger built the Morse-Boulger Destructor and in 1902 formed the Morse-Boulger Destructor Company. The company held American rights to Meldrum Brothers' Destructors of Manchester, England, but did not construct equipment under its patents. In 1904, Morse retired from the company and assumed control of Meldrum Destructors and later developed other equipment. Therefore, he had a substantial vested interest in the English-style destructors, which accounts for much of his vehemence against the early promoters of incinerators in the United States.[45]

Although few sanitary engineers were as openly critical of crematory promoters as Colonel Morse was, substantial numbers of them echoed his sentiments that American municipalities had too quickly adopted the new method without adequate investigation and the sound advice of the experts. The 1901 report of the APHA Committee on Disposal of Refuse stated, "No city in America has yet undertaken a systematic comparative test of the several types of destructors that are recommended."[46] Rudolph Hering charged, "In but a few instances did city officials take the initiative in devising proper methods of disposal; hardly one American city can be found where exhaustive preliminary investigations were made and where the solutions suggested and practiced have been as yet entirely satisfactory and final."[47] The ASMI Committee on Refuse Disposal and Street Cleaning noted that some municipalities made mistakes in adopting incineration or reduction plants without necessary investigations: "Again dissatisfaction, in our opinion, in some instances has been brought about by some contractors of plants making guarantees that only technically can be lived up to under special conditions."[48] *Engineering News* asked, "Why do so many of our cities persist in building discredited types of garbage and refuse furnaces?"[49]

In leveling these criticisms, sanitary engineers exonerated themselves of blame, reasserting that disposal, especially by the incineration method, was an engineering problem. E. N. Stacy commented in a typical manner: "The question of refuse disposal is purely an engineering problem and in the majority of cases the committees appointed to investigate and report on the various designs of incinerators available are not competent to make a selection that would prove satisfactory."[50] M. N. Baker said that it would be "hard to name a sanitary or mechanical engineer of national reputation who was ever prominently connected with the design of an American garbage disposal plant of either the cremation or reduction type."[51] Sanitary engineers tenaciously guarded the territory they had come to dominate in the late nineteenth century. As far as they were concerned, refuse collection and disposal were their province.

About the time that the first generation of American crematories were being discredited, the British were well into their second stage of development—moving from low-temperature and slow-combustion furnaces to destructors with artificial drafts operating at higher temperatures and with greater incinerating capacity. These were capable of producing steam for various work purposes. After 1902, serious criticism of the first-generation crematories led to more careful scrutiny of burning waste. Experts began to conduct the first investigations of American crematories, although the resulting debate lasted fifteen or twenty years without resolution.[52] In 1906, North American engineers made the first successful adaption of an English-style destructor in Westmount, Quebec, followed by similar efforts in Vancouver, Seattle, Milwaukee, and West New Brighton, New York.[53] By 1910, many engineers were claiming that a new generation of incineration had arrived.[54] By 1914, approximately three hundred incinerating plants were in operation in the United States and Canada, eighty-eight of them built between 1908 and 1914. About half the plants constructed after 1908 were constructed in the South, which indicates the movement of the technology from the industrial Northeast and the Midwest into areas of increasing urbanization (and the new popularity of incinerators in smaller cities and towns).[55]

English and German projects to convert waste into steam and electrical power eventually prompted similar experiments in the United States. In 1905, New York City began a project to combine a rubbish incinerator with an electric light plant; cost proved a major deterrent to large-scale projects, however. Energy could be more cheaply derived from other sources. Disposal systems that produced energy were difficult to justify in the United States, especially because cheaper practical options, such as land dumping, were available. Ironically, at the point at which American engineering expertise could adapt the British high-temperature destructors to American uses, other factors intervened to frustrate the effort.[56]

Prior to 1920, therefore, incineration failed to maintain a preeminent—

or at least a competitive—place among disposal options. In the first generation of American crematories, inefficient operation and the production of smoke nuisance were prime culprits. In the second generation, costs were more prohibitive. In both cases, viable alternative disposal methods, albeit with problems of their own, also kept incineration from increasing in importance.

The controversy surrounding the incineration method was mild in comparison with the storm over the reduction process. In 1886, a company in Buffalo, New York, had introduced the method into the United States, but it never lived up to its promise. The views of Dr. Quitman Kohnke, city councilman of New Orleans and chairman of that city's Committee on Health, reflected the general disillusionment with reduction. At the meeting of the League of American Municipalities in 1898, he complained, "We have been seduced by the glowing promises of rich rewards which the reduction process has failed to give us."[57] Plants built on the European design in the late 1880s in Buffalo, Milwaukee, Saint Paul, Chicago, Denver, and elsewhere were proving to be failures. Newer plants built on modified patterns in the 1890s were faring little better. By November 1914, only twenty-two of the forty-five reduction plants in the country were in use. Of the twenty-two, nine had changed management and ownership, two had been turned over to municipal operation, and one had burned down and was not replaced. Overall, reduction plants were going out of service much more quickly than were incinerators.[58]

The problems of the reduction process were substantially different from the problems of incineration. Critics of the early crematories continued to favor the burning of waste as a method of disposal, despite the recurrent technical problems. Critics of reduction, however, tended to question the viability of the method itself. Unlike incineration, reduction was a distinctively American process. Although the original plants were applications of the Merz method imported from Vienna, reduction as a practical disposal method never attained success in Austria or other European countries. Except for a plant in Charlottenburg, a suburb of Berlin, the reduction method was not used in Europe during this period.[59] In large measure, reduction was a product of American affluence. As Hering and Greeley stated, "Without doubt, the greater wastefulness of the American people is one reason for this development, as it produced a garbage rich in recoverable elements."[60] Another observer theorized that people of other nations "throw away too little valuable food matter to make the process profitable."[61] As an American method of disposal, reduction had to be evaluated without the benefit of comparative performance in other locations under different conditions.

A comparison of the reduction process with incineration was inevitable. Several critics suggested that, instead of trying to improve the reduction process, cities should abandon it for incineration. A major criticism—not without validity—was that reduction had limited applicability; it could handle only garbage; 70 to 90 percent of the waste had to be disposed of by other means.[62] Cost was also a major consideration. According to one source, the expense of installing a reduction plant varied between four thousand and eight thousand dollars per ton capacity, which was much higher than that of installing an incinerator of equivalent capacity. Some experts also claimed that the cost of disposal was substantially higher (available statistics are not reliable enough to verify that claim).[63] For cities and towns with populations under 100,000, reduction was prohibitively expensive; only in communities in which large volumes of organic waste were produced was reduction economically feasible.[64]

Until about 1905, cities shied away from municipal ownership of reduction plants because of the cost. Ironically, several privately operated reduction plants were showing a profit, largely because they had acquired highly favorable contracts with cities that hired their services; the contracts usually stipulated that the city would pay for collecting and transporting the garbage. At the ASMI convention in 1907, Frederick P. Smith commented, "While it may be true that the contract may be profitable to the operating company, it is doubtful if it is the best method to serve the city's interests."[65] Municipalities were footing the bill for reduction with little control over the operation of the plants and none of the profits.

When city officials realized what was happening, the demand to make reduction a municipal utility increased markedly. Several obstacles stood in their path, however. Some cities were hamstrung by charters that prohibited them from manufacturing goods—reduction produced marketable by-products from waste. City officials also operated at a severe disadvantage because private companies, for obvious reasons, intentionally kept municipal governments ignorant of their expenses and profits, and there were few reliable figures to determine the feasibility of municipally-owned reduction systems.[66] Some cities, nonetheless, tried to gain control of privately-operated facilities or sought funds to build their own. Several Ohio cities led this effort. In fact, municipal ownership and operation of garbage-reduction plants was termed an "Ohio idea." In 1905, Cleveland became the first American city to establish a municipal reduction plant. Soon after the publication of statistics showing the profitability of the Cleveland facility, other Ohio cities—Columbus, Akron, and Dayton (as well as cities in other states, such as Chicago, Schenectady, and Detroit)—built their own plants.

In 1915, Columbus officials claimed that during the four and a half years in which the city owned its reduction plant, the revenue paid the entire operating expenses and fixed costs, including interest and depreciation. In 1914, the plant treated 21,600 tons of garbage at an average cost of $1.86 per ton, and the sale of grease, tankage, and other by-products yielded a revenue of $3.085 per ton. The net earnings for the city in that year were $26,500. Cleveland experienced similar good fortune. In 1905, the city purchased the reduction works from the Newburg Reduction Company. In the first year, it incurred a net loss (or net charge) of $5,243. After 1905, however, there was a net annual gain. In 1910, it amounted to $72,532, or about $1.62 a ton. National figures for 1914 were equally encouraging. Twenty-five reduction plants in the country produced about 60 million pounds of grease and 150,000 tons of tankage having an average market value of $3.5 million. Most of these plants, however, were still in private hands.[67]

The success of municipal ownership in Ohio seemed to give the reduction method renewed credibility. In many locations, however, the problems could not be overcome by a change in control. Over and above the limited application of the reduction method, the high cost of construction and operation, the dependence on a large and constant supply of organic waste, and the unreliability of the by-product market was the stench. Reduction facilities polluted the air with the pungent odor of huge quantities of putrefying wastes, which were "cooked" as part of the process of recovering grease and other by-products. In addition, dark-colored liquids from the compressing process often ran off into nearby streams.[68] Mayor J. J. Williams of Memphis was extremely critical of the reduction process at a meeting of the League of American Municipalities in 1899. "The hopes," he said, "that reduction plants would answer the need have not been realized, and, from a financial standpoint, these plants have proven failures, chiefly because their output commands a very low price in the market. From a sanitary point of view they have proven worse than failures. The air for miles around them is so contaminated that the courts and lawmakers have been appealed to, and have, as a rule, given relief to the sufferers by abating the foul, disease-breeding business. I may mention here that, for no other cause, the laws should prohibit these establishments, because it is degrading and inhuman for human beings to spend their days in such an occupation as assorting the filth of our cities."[69]

Williams's poignant and sensitive criticisms were not taken to heart by those who had faith in the profitability of reduction. Environmental considerations were ignored or rationalized in cities that had decided upon reduction with the hope of recovering revenue. To avoid the perpetual complaints about the smell, plants were built on sites distant from the central city, which tended to increase costs by requiring longer garbage hauls.[70]

Problems of location were acute for cities like New York, in which reduction was economically viable but low-density property was scarce. In 1916 and 1917, a local debate over the placement of a reduction plant grew into a statewide controversy. During negotiations to secure a five-year contract for garbage disposal (to begin January 1, 1917), a decision was made to transport garbage from the boroughs of Manhattan, the Bronx, and Brooklyn to a reduction plant on Staten Island (this plant was to replace plants in Brooklyn which had been closed because of the odor problem). Residents of the island protested and asked Governor Charles S. Whitman to intervene in their behalf. After a preliminary hearing, he referred the matter to the state commissioner of health, who in turn referred it to his deputy, who eventually held a formal hearing. The hearing lasted twelve days, and produced 1,700 pages of testimony. Dr. George C. Whipple, professor of sanitary engineering at Harvard University and a member of Hazen, Whipple, and Fuller, a New York City engineering firm, wrote a summary report based on the testimony. Whipple declared that neither the transportation of waste by scows nor the treatment of garbage by the proposed reduction system would be a health menace. He added, however, that the plant would cause some local nuisance and recommended that officials choose another site. Governor Whitman concurred with Whipple's findings but surprisingly declared that he had no jurisdiction over the matter. His show of concern had turned to indifference or insensitivity to local protests, and contractors began the project without further delay.[71]

The Staten Island controversy highlighted the clash between economic and environmental interests over the reduction method. The attraction of potential revenue led supporters to argue that the drawbacks to reduction were only temporary, readily overcome by technical improvements. Under the new system, they claimed, odors could be eliminated. Many city officials remained suspicious, considering the investment too great and the potential benefits too uncertain. In time the controversy over the reduction method—and other disposal methods as well—made municipal authorities wary of the claims of new-equipment promoters and, indeed, of ready disposal solutions to what had become a major issue for the cities.[72]

While the reduction method failed to attract widespread support in American cities, it did help generate interest in the broader area of utilization of wastes and recycling. By World War I, the perception of waste as a menace had been supplanted in some circles by the notion of "waste as wealth." George E. Dyck, an industrial chemist for the Chicago Bureau of Waste, wrote the following poem in 1916:

> There is wealth in waste,
> Also waste in wealth;
> Save the waste of wealth,
> Turning waste into wealth.[73]

Although hardly Shakespearean, Dyck's quatrain reflected the optimism of many engineers, chemists, city officials, journalists, and sanitarians who saw a way to turn a liability into an asset. In an article in *Cosmopolitan Magazine* entitled "The Chemical House That Jack Built," Theodore Waters extolled the manner in which "every possible substance we use and throw away comes back as new and different material—a wonderful cycle of transformation created by the scientist's skill."[74] Charles Zueblin, the author of *American Municipal Progress* (1916), wrote about the utilization of all kinds of refuse, from street sweepings as fill to the sorting of rubbish for resale.[75] Engineers like Colonel Morse claimed that "almost every article used in a household, after it has been worn out and thrown away, can by the proper agency be turned to some new purpose."[76]

Beyond the parsimony of some householders, interest in waste as wealth was driven primarily by economic rather than by environmental motives. If profits were to be made in recycling and waste utilization, some people were always ready and willing to promote a new method of disposal or a new cleaning device. This view was manifest in a promotional pamphlet circulated by the United States Garbage Reduction Company: "The fortunes of the future will be made from the crumbs that fall from the world's table. The worthless chips of leather of ten years ago make the valuable leatherboard of to-day. The cast-off woollens of thirty years ago now clothe in handsome-looking cloths the poorest in the land. Cotton seed, a waste in the early seventies, is today the source of a substitute for lard, better than the original itself. The flax stalk of the western linseed grower, until recently burned up, has become the fiber from which American linen is made. But greater than all these is the conversion of the greatest menace to public health, the garbage of the townsman, into a useful and valuable commercial product. This is what the United States Garbage Reduction company will do."[77]

Experiments in giving discarded material value multiplied in the early twentieth century. Older techniques were adapted to the new conditions, as in finding innovative ways to use waste materials as fertilizer. In the late 1890s, the United States Department of Agriculture had begun investigating the value of street sweepings as fertilizer. New machines pulverized inorganic materials to be used as fill or fertilizer. Clinker from incineration plants, which had traditionally been used as fill, was tested as a possible improvement on gravel for concrete.[78]

Sorting and reselling items found in rubbish were becoming profitable for some cities, especially because of the growing demand for old rags and used paper by manufacturers of paper products. By the turn of the century, the United States was the world's leading producer of paper and paper goods, with an annual output of approximately 640,000 tons in 1908. Americans were also the leading consumers of paper products. They used about 38.6 pounds per capita a year, as compared with 34.3 pounds used by the English, 29.9 pounds by the Germans, 20.5 pounds by the French, 19 pounds by the Austrians, and 15.4 pounds by the Italians. Because of the increasing demand and the fear of deforestation, manufacturing companies eagerly purchased any material which could be turned into paper pulp quickly and cheaply. In 1913, the United States imported 123,000 tons of rags and 380,000 tons of wastepaper. When foreign sources diminished during World War I, paper manufacturers depended even more heavily on domestic sources to be found in abundance in the cities. In 1916, the United States produced more than 15,000 tons of paper a day, using 5,000 tons of old paper in the process. Demand for paper goods continued to outstrip supply, however, and prices rose steadily. Manufacturers employed innovative technology which allowed them to remove printer's ink from old newspapers through a defibering process. Other techniques were devised to turn old paper into cardboard and pasteboard.[79]

The most promising experimentation took place with conversion of waste into energy, the precursor of modern biomass technology. One approach was to turn refuse into liquid or solid fuels. At a reduction plant in Columbus, Ohio, for example, experiments were conducted in converting garbage into alcohol, to compete with alcohol derived from corn, wheat, and potatoes. In Austin, Texas, E. L. Culver patented a method of turning refuse into fuel bricks called "oakoal" (so named because of the similarity of its burning properties with those of oakwood). According to Dr. William B. Philips of the University of Texas, the fuel bricks contained about the same amount of heat units per pound as the best bituminous lump coal. In the meantime scientists in England developed "coalesine," a fuel briquette made from pulverized refuse. Coalesine gained extensive attention in the United States.[80]

English and German projects to convert waste into steam and electrical power eventually prompted similar experiments in the United States. The writings of W. E. Goodrich and Joseph G. Branch introduced to American engineers the English method of using high-temperature destructors to produce power. Colonel Morse called Goodrich's first book on the subject "a revelation to American engineers and town officials." Goodrich and other English engineers designed destructors to perform a number of power-producing functions. In 1897, a destructor was connected to the steam engines of a sewage pumping station in Hereford. The excess steam generated by the destructor was used to pump the sewage to its final destination. Also in the mid-1890s, a destructor plant was combined with an electricity works. The steam produced by the destructor ran turbines which generated power. By 1912, there were approximately seventy-six combined destructors and electricity works in operation or under construction in the United Kingdom and about seventeen such installations in other countries throughout the world. These works helped generate light, supplied power for traction, and pumped water. In other forms, power generated from waste was connected to gasworks or other municipal operations requiring power. For example, in Nottingham, enough steam was produced by the destructor system to provide one-third of the electricity needed to operate the tramway.[81]

The generation of steam and electrical power through incineration was one of several experiments in ways to obtain wealth from waste.[82] In 1905, New York City began a project to combine a rubbish incinerator and an electric lighting plant. The steam was to run dynamos and provide power to light several city structures, including an East River bridge. Little came of the project, however. Before World War I, there was talk of several incineration plants to be coupled in a similar manner. Only a few were completed. Electric generators were installed in the huge Milwaukee incinerator in December 1913, but, as Samuel Greeley noted, "With this exception no new record of actual steam utilization on a full output basis has been made in this country." A few small-scale projects were developed, such as one in Minneapolis to light and heat a hospital and to light some city streets. In Seattle, power from waste was used for some manufacturing functions; a Rochester, New York, plant furnished power to an adjacent reduction plant; Miami, Florida, generated steam for a municipal hospital to help operate the kitchen and laundry; and Savannah, Georgia, sent power to a steam heater in a nearby waterworks plant.[83]

Cost was a major deterrent to large-scale projects in heat and power generation from waste before World War I, and was competition from other energy sources. A report of the Chamber of Commerce of the United States in 1931 concluded that "only a small number of American incinerators develop steam, the more common practice being to erect plants of cheaper initial cost which are neither designed nor equipped for steam production."[84] Gaining sufficient value from a steam-producing incinerator required utilizing the steam generated day and night, effectively regulating the collection of wastes to ensure proper content for the burning process, calculating the requirements for auxiliary fuel to create the steam, and, of course, determining the market for the energy produced. Although proponents of the method argued the advantages of energy-producing incinerators, it was difficult to overcome the objection that the initial costs of construction were higher than those for traditional crematories. Colonel Morse calculated that the initial cost of a steam-producing plant, which used thirty to seventy-five tons of waste a day, was 15 percent higher than the cost of a simple incinerator. A strong proponent of the new technology, Morse tried to demonstrate to his peers that the investment was worthwhile, but many remained convinced that higher construction costs and potentially higher operating costs (a hotly debated point) made steam-generating incinerators hardly worth the effort.[85]

Most other forms of waste utilization met similar resistance. Although some local experiments proved successful, the idea of obtaining wealth from waste did not produce the kind of profits budding entrepreneurs were looking for and did not find a broad following in American cities. Sanitarian Charles V. Chapin argued that, aside from feeding garbage to swine, he knew of no utilization method "where the value of the products sold yielded a net profit over and above the cost of collection and disposal."[86] Utilization as a means of reducing municipal expenditures for collection and disposal attracted other skeptics. Some engineers, as discussed earlier, claimed that the composition of American refuse, with its allegedly high moisture content, would make the energy-generating processes unsuitable to large-scale incineration and would prove to be very costly.[87]

Utilization failed to make its anticipated impact because it was out of step with certain realities of industrial America. The availability of cheap energy sources—wood, coal, petroleum, and electric power—made the conversion of waste into heat and light seem unnecessary. Private power companies frustrated efforts at extensive experimentation with European innovations. Furthermore, the United States had so many disposal methods available to it that utilization was inevitably compared with them on a cost-benefit basis only. Ultimately, the advent of the sanitary landfill in the post–World War II era undercut several utilization programs for many years by offering a disposal method that seemed to combine both economical and efficient means of disposal.

Interest in the utilization of wastes, though limited in actual practice in the United States, at least pointed to a modification in the way engineers and city officials were beginning to view refuse management. While sanitation practices continued to vary widely from city to city, the nineteenth-century notion that waste collection and disposal were necessary for the health and well-being of the citizens had long since become dogma. Although contemporaries did not fully comprehend the broad ecological implications of efficient and effective refuse management, they were cognizant of the importance of sanitation services in curbing the most obvious nuisances and environmental dangers. As collection and disposal techniques underwent more careful scrutiny and as reformers convinced urbanites that an out of sight, out of mind mentality no longer fit the circumstances, some cities began making the hard choices necessary to provide organized, effective, and consistent service for their citizens. The biggest problem confronting city officials was trying to balance the three pillars of modern refuse management—efficiency, sanitation, and cost. The best-run cities effectively balanced their needs against available resources; the worst-managed cities favored expedience. The perfect system had yet to be devised, a universally conscientious public had not yet been born, and officials with vision were rare. Yet refuse management was never the same after Colonel Waring, the new breed of municipal engineers, and an array of civic reformers.

SEVEN

SOLID WASTE AS POLLUTION IN TWENTIETH-CENTURY AMERICA

After World War I, solid waste collection and disposal underwent some significant alterations, due in part to an evolving environmental, economic, political, and social context, and to noteworthy technical and administrative changes in the services themselves. Through the mid-1960s, refuse management remained primarily local in practice and impact, transformed more appreciably after 1965 by increasing federal programs and regulations. It also emerged as a national environmental issue alongside air and water pollution as land—or third—pollution.[1]

Refuse had become a significant public issue in the nineteenth century as an urban problem. As American cities evolved in form and scale in the twentieth century, solid waste management also changed. The United States was becoming a truly urban nation by 1920. Census figures indicated that for the first time more Americans lived in cities and towns—approximately 51 percent—than in the countryside. By 1960, almost 70 percent of the total resident population occupied urban territory. In addition, the number of urban places themselves increased from 2,722 (2,500 or larger) in 1920 to 6,041 (with some towns under 2,500) in 1960.[2]

Gross population statistics do not tell the whole story. Cities were getting larger and more numerous, but also increasingly complex. Major cities, in particular, were becoming transformed metropolises. Since the 1850s, metropolitan areas had been characterized by an intense concentration of people and activities, an industrial economy, a tendency toward multiple centers (especially by the mid–twentieth century), sharp divisions between work and residence, and more rigid segregation of people based on class and race. After World War II, an increasingly larger share of the urban population could be found in metropolitan areas. In the 1950s alone, the metropolitan population grew nearly five times the non-metropolitan rate. Especially by the mid-twentieth century, the old urban cores had lost or were losing industry, retail trade, and many office functions. Urban regionalization meant stagnant growth at the core as opposed to dynamic growth along the periphery. Suburbanization rapidly increased, characterized by lower population densities (sprawl) and accelerating racial and class segregation. In 1920, the growth rate of suburbs exceeded central cities for the first time. Urban regional networks were replacing the patchwork of more insular cities and suburbs of the previous era.[3]

Few of the thousands of urban places in the United States reached the status of metropolis, but many cities faced growth pains of some kind in the period. Urban sprawl was especially pernicious. *Fortune* editor William H. Whyte Jr. was very concerned about how sprawl influenced the growth of cities in post–World War II America. He wrote:

> The problem is the pattern of growth—or, rather, the lack of one. Because of the leapfrog nature of urban growth, even within the limits of most big cities there is to this day a surprising amount of empty land. But it is scattered; a vacant lot here, a dump there—no one parcel big enough to be of much use. And it is with this same kind of sprawl that we are ruining the whole metropolitan area of the future. In the townships just beyond today's suburbia there is little planning, and development is being left almost entirely in the hands of the speculative builder. Understandably, he follows the line of least resistance and in his wake is left a hit-or-miss pattern of development.[4]

The spate of automobiles, in particular, encouraged sprawl, dispersing work and shopping away from the core, lowering population density, and breaking down uniform patterns of urban development.[5] In 1920, vehicle registration reached eight million, with one car for every five Americans by the onset of the Great Depression. As early as 1922, 135,000 suburban homes in sixty cities were dependent on automobiles for commuting.[6]

Clearly sprawl was the enemy of efficient and economical collection and disposal of solid waste; providing service in this changing urban climate proved to be a serious challenge. More people produced more waste. The length of most collection routes was much greater and further complicated by largely unplanned urban growth in the automobile era. Distances between pick-up points and disposal facilities also were greater. Identifying disposal sites was complicated by competing land uses and by an unwillingness to challenge those with political clout or to take advantage of those who had little or no clout.

Jurisdictional responsibilities were murky not only because of the urban-suburban split, but also because of the rise in the number and type of governmental and quasi-governmental entities in existence, ranging from fringe governments to special districts.[7] In a sarcastic tone, *American City* reported in 1952, "Often a community needs to suffer a near disaster before it will subordinate political boundaries to a metropolitan-area solution of public works problems." It was more typical then to restrict refuse activities "to the boundaries of the parent city and satellite communities."[8]

Between 1920 and 1965, collection and disposal of solid waste clearly favored municipal control, if not necessarily direct municipal operation. Surveys on collection practices conducted during the period were neither comprehensive nor conclusive, but suggest some broad patterns. Table 13 highlights the best-known surveys at the time, and with the exception of the 1929 survey, indicates little significant change for the period with municipal and combined service ranging roughly between 63 and 77 percent for cities of all sizes. Because enforcement of service standards under contract could be difficult, most cities with populations over one hundred thousand had turned to municipal operations by the late 1930s.[9]

While collection and disposal required much lower construction costs than water supply and sewerage services, they utilized a much larger work force.[10] A 1926 Census Bureau survey of the ninety-four American cities with over one hundred thousand people showed that the annual cost per capita was $1.04 for refuse collection and disposal and $1.49 for street cleaning, compared with $.53 for sewers and sewage disposal. By 1934, the figures had changed; refuse collection and disposal rose to $1.14, street cleaning dropped drastically to $.81, and sewers and sewage disposal dropped to $.38. In that same year, refuse collection and disposal represented as much as 7.2 percent of the expenditures of all general departments and as much as 5.1 percent for street cleaning.[11] By the late 1960s, local governments were spending approximately $1.5 billion per year on collection and disposal. Total costs, including municipal service, private service for indus-

TABLE 13

Collection of Residential Waste

Year	*1924*	*1929*	*1939*	*1955*	*1956*	*1964*
Municipal	63%	37%	38%	55%	54%	45%
Contract	25	30	4	15	14	18
Private	10	23	7	11	11	13
Combined					22	
Municipal/Contract			5	8		3
Municipal/Private			31	6		15
Contract/Private			11	2		5
Municipal/Contract/Private			5	3		2
No System	2	16				
Number of Cities Surveyed	96	667	190	908	862	995

Source: "Garbage Collection and Disposal: A Compilation from Questionnaires Returned by 101 City Manager Cities in the U.S. and Canada," *City Manager Magazine* (July 1924): 12-14; E.S. Savas, *The Organization and Efficiency of Solid Waste Collection* (Lexington, MA: D.C. Heath, 1977), 35–39, 43; International City Manager's Association, *Municipal Year Book,* 1957, 327.

trial and other wastes, and self-disposal may have been as high as $5.7 billion.[12]

Because of the high cost of service, several cities with municipal solid waste functions adopted the idea of a service charge for collection and disposal rather than relying on general revenue from property taxes. Proposals abounded for linking the cost of collection to the weight and volume of discards and to hauling distances. Assessments could be based on the number of rooms in a home, the type of residence or business, or the size of water bills. No single approach was acceptable in every community.[13]

In 1956, at least 310 cities with populations exceeding ten thousand had service charges for collection and disposal services. While the first service-charge plan went into practice in 1890, it was not common until the 1940s. The cost per household at that time varied widely from $0.33 to $7.00 per month. This trend followed the user-fee approach commonly employed to underwrite sewer maintenance and treatment plant operations.[14] As part of total solid waste management costs, collection remained the largest percentage. A 1968 survey indicated that the typical community of four hundred thousand spent $2.5 million for collection, but only $900,000 a year for disposal.[15]

Contracted or private service (or some combination) ranged between 35 and 63 percent, and fluctuated a little more sharply than municipal service. The appeal of contracting service was that it was normally cheaper for the city than municipal service, even after adjusting for population density and frequency of collection. The major disadvantage was that it required careful safeguards to ensure that bidding for contracts was competitive and that the service provided was of good quality.[16] Those who favored municipal service, however, were often dubious of contracting because they believed such an arrangement inevitably came with incomplete coverage, unsanitary methods, and high expenses.[17] In reality, collection of municipal refuse was never carried out completely by the cities themselves in this period, especially cities with populations of less than one hundred thousand. The trend away from municipal collection and disposal would be more pronounced beginning in the late 1960s and early 1970s.

In a decidedly local enterprise such as solid waste collection and disposal, municipal responsibility, more regularized management techniques, new technology, and better statistical information did not ensure effective standards of operation or performance. In the late 1960s, there were more than twelve thousand waste haulers, landfill operators, and recyclers in North America—some public, some private. Several enterprises were quite large, others owned only a few trucks. Some workers were associated with unions, some were not. As popular writer Harold Crooks noted, "Each metropolitan area was its own isolated garbage market, and in each of these markets, the garbage industry was based on a modus vivendi worked out over the years by municipal governments, local entrepreneurs, trade unions and other interest groups."[18]

One of the reasons for the lack of standardization had to do with the markets in which public and private refuse people worked. While publicly controlled collection and disposal operations dominated the residential market, they were less influential in the commercial sector and largely absent from the industrial market. By the 1960s, private haulers accounted for a large share of total tonnage and dominated commercial and industrial markets, and in many cases, were exclusively engaged in those ventures. Many of the smaller operations depended on access to municipal incinerators and landfills, while larger companies used private disposal facilities.[19]

In metropolitan areas with fragmented collection and disposal operations, intense competition often resulted in market-sharing arrangements or more coercive methods of control. Underworld elements—involved in the waste industry in some communities—were credited with developing the "property rights system," through which haulers divided the customers and agreed not to compete with each other. As a form of illegal price fixing, participating haulers could not use the law to enforce their system, but relied instead on threats, withholding of service, and price manipulation. For example, the Brooklyn Trade Waste Association was formed—without official approval—and served as a de facto regulatory authority.[20] While New York City had a long history of depending on municipal collection of residential waste, commercial establishments had to contract with private carters.[21] Since the 1950s, at least, the carters had been allegedly controlled by underworld racketeers. Despite the fact that the city's Department of Consumer Affairs was given responsibility to license carters and to set maximum rates, several investigations found that racketeers were still active in the business, and customers were regularly defrauded.[22]

In San Francisco—a large-city exception to municipal collection and disposal—competition was ferocious among independent scavengers until 1920. In that year, the first scavenger cooperative, the Sunset Scavenger Company, was formed. Emilio Rattaro, who started his own scavenger company in 1916, was instrumental in bringing competitors together to form Sunset—a group dominated by Italian Americans of Northern Italian extraction. The bond of ethnic and family ties likely made such an arrangement work. In 1921, a second cooperative was formed—the Scavengers' Protective Association (SPA). While SPA operated primarily in the central city, Sunset served customers more on the urban periphery. In the 1930s, several small companies vied for licenses in various collection districts, but by 1939, Sunset and SPA divided the city into two unified areas. This system stayed in place into the 1960s.[23]

The role of the federal government in local solid waste management was minor before the late 1960s, although the onset of the Great Depression and the New Deal foretold some changes to come. From the standpoint of urban infrastructure development at least, the 1920s began as a promising decade. A number of cities carried out major public works projects, and demand for more continued to rise. This resulted in many new roads, bridges, sewers, schools, and other public buildings, but placed cities deeply in debt. In fact, debt payments began to assume a significant proportion of public spending in these years. At the time increasing levels of debt proved remarkably easy for many communities, since municipalities could borrow at comparatively low interest rates in a favorable market.[24]

The onset of the Great Depression threatened expansion of public works. The accelerating demand for better services and for the repair and replacement of a deteriorating infrastructure in the 1920s ultimately ran well ahead of available resources. The need to manage the debt, rising unemployment, and mounting tax defaults by property owners squelched public and private spending for services. In 1930, of 145 cities with a population of fifty thousand or more, approximately 11 percent of local taxes went unpaid; in 1933, default on taxes rose above 25 percent. In several cases, cities also defaulted on bonds that came due in the 1930s or turned to sinking funds. As a result, the price of municipal bonds plummeted.[25]

With the financial downturn, municipal leaders looked elsewhere for economic relief. State governments offered little support in most cases, and private financial institutions had their own customers and their own solvency to consider.[26] Thus, Washington quickly became the obvious source of hope. President Herbert Hoover's modest anti-Depression measures had little direct impact on restoring public works expenditures to their pre-Depression levels, but they began a pattern of government involvement that was expanded upon during the New Deal. During the Hoover years, public works expenditures dropped dramatically. From 1923 to 1930, public works expenditures had grown from $1.6 billion to $2.85 billion annually; in 1933, they fell back down to $1.6 billion.[27] Of the meager $1.5 billion allotted to help alleviate the economic crisis in July, 1932, only $300 million was appropriated for public works, and of that sum $120 million was earmarked for the federal-state highway program. Less than 1 percent of the funds went to cities.[28]

A variety of New Deal programs during the Franklin Roosevelt administration had a more significant impact on the cities and public works than those developed during the Hoover years.[29] Between 1933 and 1939, the Public Works Administration (PWA) spent $4.8 billion on highways, bridges and dams, airports, sewer and water systems, a variety of public buildings, and other public works projects. More than half of the monies found their way to urban areas, unlike funds allocated by the Reconstruction Finance Corporation in the early 1930s.[30] There were approximately thirty-five thousand different PWA projects throughout the nation, with every county represented save two. The majority of these projects were small, however.

The erosion of revenue during the Depression made cities hard-pressed to maintain adequate collection and disposal services, let alone to invest in necessary capital projects. The Depression also meant that markets for salvageable materials diminished. Cities spent $170 million for sanitation facilities in 1928. By 1932, the outlay had dropped below $50 million.[31] As of March 1, 1939, only 41 governmental refuse projects—compared with 1,527 sewerage projects—received loans and grants. The total federal commitment was approximately $6.6 million for refuse projects and $279 million for sewerage projects, or total project costs of $10.9 million and $466 million, respectively.[32]

Civil Works Administration workers picked up rubbish as part of fire protection programs, but relief funds were not made available for regular collection and disposal services. Federal loans and grants, however, could be used to build incinerators, improve dumpsites, and conduct cost studies.[33] In fact, many of the incinerators in the period were built or enlarged with funds from New Deal programs. In 1933, PWA provided allotments for fifteen incinerators, including New York City's three-hundred-ton Flushing plant and a nine hundred-ton plant in Cleveland. In 1935, PWA financed the construction of about a dozen plants.[34]

The record of PWA and other relief and recovery agencies in the New Deal with respect to public works was decidedly mixed. Many of the projects were tangled in red tape and delays. In some cases, project grants were rescinded because contractors could not be identified quickly enough or the cities failed to provide sufficient matching funds.[35] While the drive for economic recovery increased federal involvement in local affairs in the 1930s, the onset of World War II in the 1940s expanded federal authority and broadened its impact on cities. War mobilization deeply affected private lives—from the call to military service to gas rationing—but it also stimulated the urban economies through the construction of new war industries, defense housing, and the purchase of huge quantities of war materiel and other products.[36] Despite the scale of investment in cities throughout the war and the perpetuation of federal-city partnerships, the further development of public works was now more specifically tied to wartime needs. The Federal Works Agency (FWA) was the heir to the PWA. Established in 1939, it distributed federal grants to states and local governments for highways, public buildings, and other community projects.[37]

While the increasing federal role in urban affairs between the wars influenced the development of public works in general, it had a far less important impact on solid waste collection and disposal in those years. By the late 1960s, that would change, but most immediately, new collection and disposal technologies would have a more profound effect on solid-waste management, especially compared with sanitary services such as water supply and wastewater systems.

Massive quantities of refuse continued to mount in the cities of the interwar years, creating a challenge for solid waste personnel throughout the country. By 1922, a surging recovery from a jarring postwar economic downturn produced an era of real prosperity. World War I was the start of what was to become the modern American consumer and service economy, built upon war-generated investment capital, industrial planning, technical innovation, and rising peacetime demand. Productivity in the 1920s rose twice as fast as population, and the Gross National Product leaped from about $73 billion in 1920 to about $104 billion in 1929 (in 1929 dollars).

Businesses that catered to the consumer market were the most successful of the era. Mass production, high-pressure advertising, and easy consumer credit created democratized materialism. The attractiveness of American consumer goods in the 1920s derived from variety and price. The chemical industry produced an array of new fabrics, kitchen utensils, floor coverings, and cosmetics. With access to confiscated German dye patents and expertise from a team of innovative chemists, DuPont introduced rayon and cellophane. Other synthetics—plastics, Bakelite, Celanese—were produced. Items as wide-ranging as Pyrex cooking utensils, linoleum, lacquers, and antifreeze captured the dollars of the American public. Of all the consumer goods, the automobile was king, making private transportation widespread. By 1930, nine of the twenty leading corporations in the country specialized in consumer goods, compared with one of twenty in 1919. Consumer credit to purchase this array of goods also was on the rise. Between 1919 and 1929, the amount of non-farm consumer credit rose from $32 billion to $60 billion.[38]

The Great Depression obliterated the escalating economic growth of the 1920s, but it only temporarily derailed the consumer trends and habits begun in the 1890s and elaborated in the 1920s. From the late 1940s into the early 1970s, materials consumption in the United States increased at a rate of 4 to 5 percent per year, which was faster than the population growth rate.[39] Generation of solid wastes rose steadily between 1920 and the mid-1960s from 2.7 pounds per capita per day to approximately 4 pounds per capita per day. Between 1955 and 1965 alone, per capita refuse generation increased by 78 percent in New York City; between 1958 and 1968, it increased 51 percent in Los Angeles.[40] In 1965, gross discards from American cities exceeded 103 million tons, of which 6.8 million were recycled, 28.8 million were burned, and 69.6 million were dumped in landfills.[41]

Relatively new materials such as plastics, other synthetic products, and toxic chemicals complicated the collection and disposal process, but so did the rapid increase in paper waste. Paper, plastics, and aluminum in particular grew steadily as a percentage of total municipal solid waste (MSW) in the mid-twentieth century.[42] Paper represented the largest percentage of municipal waste. Yard and food wastes, the major organic discards, steadily declined. Plastics, a new element in the waste stream after World War II, was a relatively small but rapidly increasing portion of the total.[43] The widespread consumption of discardable goods also seriously aggravated the litter problem along roads and highways in urban as well as rural communities.

The unprecedented growth of the packaging industry, a direct response to rampant consumerism after World War II, was largely responsible for the creation of innumerable goods with short useful lives. Packaging took on special importance in the late 1940s because of the rise of self-service merchandising through supermarkets and other consumer outlets. This new direction in marketing required packages that would help sell the products and reduce theft or damage. The use of paper stock rose from 7.3 million tons in 1946 to 10.2 million tons in 1966. For the sake of convenience and significantly less cost of production, nonreturnable bottles and cans replaced standard returnable containers. In 1966, paper and paperboard accounted for 55 percent of packaging materials consumed; glass accounted for 18 percent; metals, 16 percent; wood, 9 percent; and plastics, 2 percent. In 1966, packaging cost the American public $25 billion—3.4 percent of the gross national product—not to mention the cost of collection and disposal. In that year, packaging material amounted to 52 million tons of waste.[44]

In a cultural context, changing products influenced changing habits in an increasingly "throwaway" society. As Susan Strasser stated, "The new consumer culture changed ideas about throwing things away, creating a way of life that incorporated technological advances, organizational changes, and new perspectives, a lifestyle that linked products made for one-time use, municipal trash collection, and the association of traditional reuse and recycling with poverty and backwardness. Packaging taught people the throwaway habit, and new ideals of cleanliness emphasized swift and complete disposal."[45] Given the nature of urban growth after 1920 and the voluminous and varied municipal waste being generated, collection and transportation of refuse to disposal sites were time-consuming operations and extremely costly.[46] The major problems associated with collection revolved around familiar themes, primarily the manner and kind of pickup service provided to the public. Source separation, once viewed as the practice of the future, was largely curtailed—at least in its most comprehensive form. Some cities had different segregation requirements for residential and commercial customers; some required yard wastes to be separated from garbage and other household wastes. A 1968 American Public Works Association study affirmed that 56 percent of the cities surveyed collected mixed wastes, 33 percent collected separated wastes in some form, and the remaining 11 percent used both methods. The major reason for the limited use of source separation was the expense. Multiple collections were expensive, especially in light of the high collection versus disposal costs.[47] To better determine the most efficient and cost-effective collection service, some cities initiated time studies to evaluate the location and standardization of receptacles, the size and types of vehicles essential for service, the length of the haul, and the requirements for secondary transportation.[48]

Refuse collection was characterized more by economies of density than economies of scale. For example, a doubling of the tonnage collected per mile could reduce the average cost by as much as 50 percent. Longer hauls to disposal sites or transfer stations, therefore, could significantly increase costs. This suggests that population diffusion or deconcentration, typical of the modern metropolis, was the enemy of efficient and economical collection.[49]

New transportation technologies were the most significant response to economies of density, especially since little effort was made to promote waste minimization practices among Americans at this time. New equipment more in fashion with the era of the internal combustion engine began to replace hand sweeping and horse-drawn sweeper and sprinkler devices for street sweeping, and horse-drawn tank wagons or can-rack wagons for refuse collection. The city of Boise, Idaho, employed the first practical mechanical sweeper in 1914, and some cities used vacuum devices in the 1920s and 1930s to remove dust from pavements. As greater numbers of cars and trucks clogged the streets, demand for better sweepers rose. By the 1950s, mechanical sweepers had increased both in power and capacity, and replaceable hoppers were introduced that could be set aside and then collected by a special tractor and subsequently dumped.[50]

At first, open trucks replaced many horse-drawn wagons for refuse collection. These open trucks were rather primitive by international standards. Some of the more advanced European collection vehicles in England, Germany, and Switzerland were closed to reduce odors and control dust. Some "dustless" trucks did appear in the United States in the 1930s. By World War II, motorized trucks became the standard. The major limitation of the open and closed trucks was capacity. Open ones held between ten and twenty cubic yards of waste, making frequent dumping necessary. Closed trucks, which held less, also required frequent unloading. Although the trucks were relatively inexpensive, they did not solve the problem of long hauls to disposal sites. Compaction vehicles, considerably more expensive than open or closed trucks, had the advantage of accommodating larger loads, thus reducing the number of hauls on collection routes. They could compress waste to about 30 percent of its curbside volume. Of the collection trucks in use in the late 1960s only about one-third were compactors. However, one study estimated that about 72 percent of private collectors used compacting trucks in 1965.[51]

One answer to the problem of small loads prior to the advent of the compactor trucks was the use of the tractor-trailer system introduced in Chicago in 1922. Horse-drawn trailers filled with refuse were taken to a central point and then moved to disposal locations by motorized tractors. At first regarded as a costly alternative to older methods, the tractor-trailer system proved reasonably economical since the tractors could tow two or three trailers at a time to a disposal site. By 1928, eighty-two cities adopted the method.[52]

Transfer stations—a variant on the tractor-trailer system—were used to centralize wastes for more economic hauling to the final disposal destination. These were points at which collection trucks could unload into larger vehicles or temporary storage facilities. By the 1960s, they could be found throughout the country. Secondary transportation—also a variant on the tractor-trailer system—using large trucks, railroad cars, or barges, was employed also to increase the volume of hauls to disposal locations.[53]

More than any other aspect of solid waste service, labor issues posed serious concerns for collection. The lion's share of collection costs traditionally went toward the wages of crews and their supervisors, at least until the more thorough mechanization of the collection process.[54] The workforce was primarily unskilled or semi-skilled, and the job was highly unattractive. In some communities, employment in refuse collection remained a political sinecure where the head of a municipal collection division did not have the right to hire or fire workers. Instead, the boss of a local committee often had the opportunity to hand out jobs as rewards for political service. To counteract the patronage system to fill posts, about one-quarter of the cities by 1964 adopted civil service or merits systems to obtain employees. Unionization of public and private collection forces increased in the 1960s as well, reaching approximately 30 percent in 1964.[55] Worker grievances, however, were not eliminated by increased unionization, especially in cases where race was an issue either through discriminatory practices or restrictions to union membership. Increasing reliance on mechanized fleets to collect refuse also raised questions about shrinkage of the workforce and the loss of jobs.

Another technology, seemingly unrelated to collection practices, had a noteworthy impact on collection. The in-sink grinder became popular in the 1950s. The idea of the home disposer—or in-sink grinder—can be traced to grinders and shredders utilized at sewage plants in the 1920s and 1930s to convert large solids into fine pulp that could be washed away into the sewer system. The first attempt to grind garbage and dispose of it into sewers took place in Lebanon, Pennsylvania, in 1923. Although the technique was costly and presented some operating difficulties, further experiments followed.[56] In 1935, General Electric adapted the idea of the municipal grinders into the design and manufacture of the "Disposall" for use in the kitchens of private homes.[57] Soon the devices were being publicized in science and women's magazines. While appealing primarily to the suburban homeowner, General Electric and other manufacturers directed their marketing effort at homebuilders, kitchen remodelers, and municipal officials to capture a larger consumer base. Convenience and cleanliness were the bywords of the new device. Its major application, however, awaited the end of World War II.[58]

Although they did not quickly become standard home appliances, the grinders made great inroads in new middle-class subdivisions. A 1968 survey disclosed that there were 63.5 home grinders per 1,000 residences nationwide. The device did not single-handedly solve the collection problem, however, disposing of about one tenth of the total volume of refuse collected. The grinders also faced stiff opposition from those in wastewater treatment because they could clog sewer lines and overload sewage-disposal facilities by transferring the waste problem from one medium to another. The chief engineer of the New Jersey State Department of Health called the device a "creeping menace" to the solution of stream pollution problems. In 1949, opponents succeeded in banning electric disposers in New York, New Haven, Philadelphia, Miami, and much of New Jersey. In Detroit, Denver, and Columbus, legislation sought to increase the use of the device, but usually in new construction only. While the public health benefits of home grinding became more widely accepted by the 1960s, the belief by some proponents that the device could make cities garbage-free was unrealistic.[59]

City officials usually focused on traditional collection and disposal practices, rather than new, promising home technologies to confront the refuse problem. There was no standard disposal method utilized throughout the country in the interwar years, but dumping refuse on land was by far the most typical—especially in smaller communities. Despite persistent criticism of land dumping, approximately 90 percent of cities and towns with populations of less than four thousand relied on open dumps.

Dumping was convenient and simple, but it was notoriously unsanitary, attracting vermin, giving off offensive odors, threatening groundwater supplies, and posing fire hazards. According to the superintendent-engineer for Pittsburgh's Department of Health, "These filthy spots, breeders of disease, and hovels for rats, grew in nuisance owing to increased population and expansion of the municipality, and as a result most cities have done away with dumps by enacting ordinances prohibiting them. Thus they have eliminated a cheap means of disposal, which must be replaced by a moderate-priced method if costs are to be kept near former budgets."[60] This was a serious dilemma most cities faced in these years—eliminate disagreeable disposal sites, but replace them with what?

By the late 1930s, dumps were disappearing from the outskirts of some cities. Ocean or sea dumping was no more tolerable than land dumping, and also was slowly going out of fashion. In 1933, New Jersey coastal cities went to court to force New York City to terminate ocean dumping. Effective July 1, 1934, all discarding of municipal waste at sea came to an end in the New York–New Jersey area as a result of a lower court action against the practice sustained by the U.S. Supreme Court. Oakland, California, had abandoned the practice in 1916 because its barge could not navigate the rough bay waters, but resumed it in 1925. In the 1930s, California prohibited the discharge of garbage into navigable waters or into the Pacific Ocean within twenty miles of shore, and Oakland had to abandon the practice for good.[61]

Although dumping of municipal waste at sea was terminated in 1934, industrial and some commercial wastes were exempt from the ruling. At the end of the 1960s, it was estimated that each year from fifty to sixty-two million tons of waste were dumped into the ocean (60 percent of which was dredging spills produced by harbor-deepening operations, such as those conducted by the U.S. Army Corps of Engineers). In fact, industrial waste dumping in the Northeast more than doubled between 1959 and 1968. New York City continued to dump sewage sludge over an area of several square miles, which was described by critics as a "dead sea of muck and black goo."[62]

Cities achieved limited success with alternatives. A few tried burying their garbage rather than depositing it in open dumps. Excess waste at pig farms was sometimes plowed into the soil.[63] The use of ashes and rubbish—and sometimes garbage—for fill had been practiced for many years, although using organic wastes alone to fill ravines or to level roads was regarded as highly objectionable. The most promising land disposal technology emerging in the 1920s was the precursor to the modern sanitary landfill of the post–World War II era. It essentially combined the practices of filling and open dumping, but with greater attention to the sanitary implications of disposal. The basic principle was to dispose of all forms of waste simultaneously and, at the same time, to eliminate the problem of putrefaction of organic materials by covering the wastes with dirt and/or sweepings. Typical modern sanitary fills, however, were stratified: a layer of garbage was covered with a layer of ashes, street sweepings, or rubbish; then another layer of garbage, and so forth. Alternatively, a layer of mixed refuse might then be covered with a layer of dirt and repeated several times. Chemicals were sometimes sprayed on the fill to retard putrefaction.

The sanitary landfill was a breakthrough that became the primary disposal option in the United States between the end of World War II and the 1980s. Among other things, it was regarded by most experts as the most—or one of the most—cost-effective disposal technologies available.[64] Initial attempts at sanitary fill were tried in Seattle, New Orleans, and Davenport, Iowa, as early as 1910, but they were little more than land-based dumps and did not represent systematic or large-scale disposal. The modern practice began in Great Britain in the 1920s under the name "controlled tipping." However, London and cities in the vicinity simply dumped wastes between houses and covered the piles with street sweepings. The American equivalents to the British practice appeared in the 1930s in New York City, San Francisco, and Fresno, California. In New York, refuse was placed in deep holes, primarily in marshes, and then the holes were covered with dirt. In San Francisco, layers of refuse were deposited in tidelands to produce additional land, but actual trenches were not dug. Fresno employed a trench system and compacted the top layer of dirt.[65]

The Fresno Sanitary Landfill is the oldest true sanitary landfill in the United States, and the oldest compartmentalized municipal landfill in the western states. Located southwest of the city of Fresno, it opened in 1937 and closed in 1987.[66] Jean Vincenz was the man most responsible for developing, implementing, and disseminating the sanitary landfill in the United States. He served as commissioner of public works, city engineer, and manager of utilities in Fresno from 1931 to 1941. Born in 1894 in Enfield, Illinois, he received a degree in civil engineering at Stanford University in 1918 and a degree in public administration from San Diego State College in 1958. After resigning from his positions in Fresno, he became assistant chief of the Repairs and Utilities Division of the Army Corps of Engineers headquarters in Washington, D.C. (1941–1947), and then served as public works director of the San Diego County Public Works Department (1947–1962). In 1960, he was named president of the American Public Works Association.[67]

When Vincenz became commissioner of public works in Fresno, he recommended not renewing the franchise of the Fresno Disposal Company, which operated an incinerator. Prior to developing his sanitary fill, he studied British controlled-tipping techniques, visited several California cities, and consulted with a New York engineer active in developing sanitary fill. Vincenz came to believe that a true sanitary landfill required different elements than those utilized elsewhere, especially systematic construction of refuse cells (trenches), a deeper cover of dirt between layers of refuse, and compaction of both the earth cover and the waste. The trenches and the compaction process were the unique features of the Fresno landfill, although Vincenz argued that compaction was the more important of the two. To his thinking, without compaction and without a real cover, one would attract rats and thus could not claim to have a sanitary landfill.[68]

Sanitary landfills in the major cities of San Francisco and New York got more immediate attention than Vincenz's fill in relatively obscure Fresno. San Francisco began its operations in 1932, initially as an emergency measure, and by 1936 it operated effectively as a primary disposal option for the city. Unlike the Fresno fill, San Francisco's was constructed along tidal flats on the bay. The waste was utilized for reclaiming land eventually used for industrial purposes. Such modifications of the shoreline and leaching problems from the fill eventually raised major environmental concerns, but at the time, the practice was regarded as a success.[69]

The New York landfill began in 1936—a year later than Vincenz's experimental fill—at Riker's Island, which was the site of a city prison. It was much larger than the Fresno enterprise, and although similar in design, it did not have an airtight seal around the whole area. With the expectation of reclaiming additional land, city officials authorized other sites in the 1930s. Not everyone was happy with the decision. Debate broke out on the degree to which the sites were indeed sanitary, and political battles arose over the conduct of the Department of Sanitation in carrying out its disposal policy. The turmoil was not sufficient to undermine the practice in the city.[70]

While there was no mass scramble to build sanitary landfills in the 1930s and early 1940s, momentum slowly shifted in that direction.[71] During World War II, the Army Corps of Engineers experimented with sanitary landfills because it needed to develop a disposal method to handle the great variety and amounts of waste at camps and other installations. At the same time, the corps was unwilling to utilize critical materials on the building of incinerators. While Vincenz was skeptical about extensive adoption of sanitary fills in the army without sufficient supervision and adequate equipment, he built the fills nonetheless. By 1944, 111 posts were using sanitary landfills to dispose of their refuse. By the end of 1945, almost 100 American cities also had adopted the sanitary landfill.[72]

After the war, the Sanitary Engineering Division of the American Society of Civil Engineers prepared a manual on sanitary landfilling, which became a standard guide. It defined sanitary landfilling as "a method of disposing of refuse on land without creating nuisances or hazards to public health or safety, by utilizing the principles of engineering to confine the refuse to the smallest practical area, to reduce it to the smallest practical volume, and to cover it with a layer of earth at the conclusion of each day's operation, or at such more frequent intervals as may be necessary."[73]

During the 1950s and 1960s, the prevailing wisdom among solid waste managers was that sanitary landfilling was the most economical form of disposal, and at the same time, offered a method which produced reclaimed land. A 1961 survey of 250 sanitary landfill sites conducted by the American Society of Civil Engineers showed that completed fills were commonly used for recreational and industrial purposes. Land reclaimed in this manner while suitable for parks, recreational areas, and even parking lots, was not suitable for residential and commercial sites. A dramatic example occurred in the East Bronx. Row houses were erected on filled land in 1959. Within six months, cracks developed in the walls; by 1965, the floors were badly tilted and larger cracks appeared along the brick walls. A year later, the housing commissioner condemned the dwellings and ordered them demolished.[74] Another risk of landfilling was potential groundwater contamination. After World War II, several state departments of health—particularly in California—issued warnings about possible groundwater pollution from sanitary landfills; industrial effluents were of growing concern.[75] Yet throughout the 1960s, the sanitary landfill was spared the harsh criticisms it would receive in the coming decades.

As expectations about disposal changed over time, the role of incineration changed. While sanitary landfilling was in ascendancy in the mid-twentieth century, incineration continued to have a mottled history. Prior to 1920, incineration failed to maintain a preeminent—or at least a prominent—place among disposal options.

In the early 1920s, incinerators were most likely to replace open dumps in suburbs and smaller communities where suitable plant locations were available, where occasional incomplete combustion was not considered objectionable, and where the furnace was not utilized to generate power. At this time, it was widely believed among engineers that incinerators were an acceptable disposal option from a sanitary standpoint.[76] A 1924 report indicated that twenty-seven out of ninety-six of the cities surveyed (29 percent) burned or incinerated their refuse. This compared with sixteen cities (17 percent) that dumped, filled with, or buried refuse; thirty-seven (38 percent) that used refuse as fertilizer or animal feed; two (2 percent) that used reduction; and the remainder that used other methods or no systematic method at all.[77]

By the end of the 1930s, incinerators became a relatively popular disposal alternative. Some statistical summaries of the decade put the total number of incinerators above seven hundred. During the 1930s, however, incineration was responsible for disposing of only about 5 to 10 percent of the refuse produced. The units developed by the standards of forty years later were, as one recent expert noted, "hopelessly in violation of air quality standards," but were not subjected to rigid regulation at the time.[78] In the 1930s, a growing interest in the relationship between air pollution and burning waste was developing among some experts.[79] Nonetheless, the view that incinerators were relatively sanitary was not successfully challenged in this period. More advanced principles of combustion developed in Europe—one way to reduce pollutants—were not yet fully applied to American furnaces. There was a tendency to seek higher temperatures and higher rates of combustion in the American incinerators of the 1930s than in the past.[80]

Part of the problem with the failures of these incinerators was purchasing them from the stock shelf without an engineering study of local factors. Manufacturers were unable or unwilling to supply them because of competitive costs. Some believed that there was an air of mystery surrounding incineration engineering.[81] Incinerators, nonetheless, were gaining a competitive advantage in some urban areas because open dumps were too unpopular to maintain in central cities. They still faced problems of NIMBYism (Not in My Backyard) like the open dumps. One observer rashly noted in 1937: "uninformed and misguided public opinion frequently delays construction of incinerators. Each section of a community wishes the incinerator on another part of town, under the mistaken belief that the production of odors and smoke will be detrimental to health. What is needed is public education to the certainty of odor elimination in the high temperature furnaces of modern design."[82]

More sympathetic to public feelings, a consultant from Detroit stated in 1948, "Public relations are very important. It is very necessary that the residents of the community have an exact knowledge of the various factors contributing to successful plant operation. The importance of the cooperation of the public is a factor most generally neglected by municipalities when undertaking the incineration of refuse for the first time."[83] In a reversal of the trend from the 1920s, incineration in the 1940s began to find favor in larger cities that could afford the technology for disposing of garbage and combustible rubbish, and simultaneously assigning ashes and non-combustibles to landfills. Of the major cities, only New York turned earnestly to landfilling in this period, but even it still relied on incinerators for disposing part of its waste. In cases where incinerators were in competition with landfills, debates focused on cost and convenience more than environmental risk. The extensive use of incineration at army camps and other military installations in the 1940s also broadened this method's appeal.[84]

Incineration continued to develop as a large-city disposal option in the 1950s either because landfill sites were not available and/or hauling costs to distant sites were prohibitive.[85] The technology used for incineration also underwent major changes in the 1950s. Batch-fed furnaces with intermittent discharge of residue was more frequently being replaced by continuously fed, mechanically stoked furnaces with continuous ash removal. Although the demand for economy in city government led to more investigations of heat-recovery systems, relatively few were built or operated successfully in the period.[86] Many engineers still believed that, if properly designed and operated, incinerators need not be offensive. The greatest objection was the traffic of collection vehicles moving to and from disposal sites. But incineration was far from a universal solution to the waste problem. There were reservations about the ability of the method to adapt to rapid urban growth—a key issue in the 1950s—and the lingering concerns about costs.[87] NIMBY issues were too often ignored, and there was little or no public discussion of siting decisions that raised questions about placement in poor or minority communities.

The 1960s witnessed a great number of incinerator abandonments. One estimate suggested that one third of cities with incinerators (about 175) discontinued them in favor of other methods—primarily sanitary landfills. Problems arose, in many cases, because the plants were operated beyond design capacities. Most significantly, the relationship between incineration and air pollution drew increasing attention.[88]

Until the 1960s, knowledge of refuse furnace technology was primarily concentrated in the hands of a few pioneer incinerator builders and some engineers. Increasing recognition of environmental problems linked to burning waste put into question what the experts had been saying about the safety of incinerators. A 1965 study, for example, questioned whether incinerator residue—from a public health standpoint—was suitable for fill. Some argued that the type of control exercised by state health departments over the design and operation of water and sewage treatment plants should be extended to municipal incinerators. In several states and cities, more stringent air pollution control regulations began to appear. Nevertheless, some engineers still insisted that unsatisfactory results from incinerator operation were due to economic compromises rather than a lack of technical expertise.[89]

Waste utilization had long had a place among disposal options, despite the fact that land and sea dumping were the antithesis of conservation.[90] Two major recovery and reuse techniques, swine feeding and reduction, continued to be practiced after World War I, although they showed less promise than sanitary landfilling or incineration. Feeding garbage to swine was somewhat popular during World War I when the country was impelled to increase its food supply, but fell out of favor in the 1920s. During World War II, it saw resurgence. The Depression and the war helped reinvigorate the practice in the late 1930s and 1940s, when droughts in 1934 and 1936 led to severely reduced corn crops and higher pork prices. In 1941, a Department of Agriculture survey of 247 cities (with populations of twenty-five thousand or more) indicated that 27 percent fed their garbage to swine. In many cases, the hog farms and the swine-feeding operations were run privately rather than by the municipalities themselves.[91]

Scientific studies demonstrated that the use of raw garbage as feed was an important factor in the infection of pigs with *Trichinella spiralis*, which could be transmitted to human beings in undercooked pork. Although the morbidity rate for trichinosis was low and mortality was rare, the link between the disease and feeding practices turned several cities toward other methods. However, the practice of feeding garbage to swine continued in some areas of the country until the late 1950s.[92] Between 1953 and 1955, owing to the rapid spread of *vesicular exanthema* (a swine disease which led to the slaughter of more than four hundred thousand pigs), the U.S. Public Health Service and state health departments instituted regulations forbidding the feeding of raw garbage to hogs. Although cooking the waste solved the problem, hog feeding steadily declined by the 1960s because the method became economically prohibitive.[93]

In certain respects, reduction competed with swine feeding for garbage supplies. Because the plants were so expensive to build and operate, they were largely confined to use in cities with populations exceeding one hundred thousand. Aside from their high cost, the market was limited because the plants had to be placed at a distance from the source of waste, due to the horrendous odors emanating from them, and thus hauling charges were high. Reduction plants slightly increased in number after World War I, reaching twenty-four by 1922. But as grease prices dropped in the 1920s, the number of plants declined again. The increased hauling charges, as well as the nuisance value and opportunities for contaminants entering local water supplies, added to the method's limitations.[94]

A few other methods adopted the principles of utilization but with only modest success. Composting processes, carried on primarily in Europe and the Asia subcontinent since the 1920s, were employed to either produce humus or to dispose of refuse.[95] After 1960, approximately 2,600 composting plants were in use outside the United States, 2,500 of which were small plants in India. Yet no American city employed composting of organic waste until the mid-1950s, when a very few plants were built by private companies, which charged municipalities to receive salvageable materials. The inability to sell large quantities of the by-products kept the business from expanding, as did the high cost of transporting compost. Between 1963 and 1964, six out of nine U.S. composting plants closed. By the end of 1967, only three of thirteen U.S. compost operations were active—in Saint Petersburg, Florida; Mobile, Alabama; and Houston, Texas.[96]

Ready markets for cans, bottles, scrap metal, rags, rubber, and paper varied from city to city, and were never sufficiently reliable to attract more attention to rubbish sorting. Private philanthropic organizations had some success with some recovery efforts, but their goals were more social than environmental.[97] Wartime salvage efforts proved much more successful than the peacetime counterparts because public motivation was heightened, and the need to seek markets for recyclables was not a major impediment. For example, wastepaper recycling rates, which stood at about 35 percent during World War II, steadily declined to around 21 percent in 1965.[98]

In World War I, the Waste Reclamation Service of the War Industries Board was modeled after the National Salvage Council in Great Britain. In World War II, the United States again borrowed ideas for other countries in establishing a salvage program. In this case, officials studied Canadian, British, and German systems. Many of the programs of the War Production Board's Salvage Division were based on British practices. The National Salvage Effort relied on cooperation from approximately 1,600 local authorities, which directed the work of numerous volunteer groups, including many women's organizations and the Boy Scouts of America. All kinds of material from scrap metal and rubber tires to paper were collected for the war cause. The paper recycling rate by the end of the war was 35 percent, but soon began to drop when peace was declared.[99] Great enthusiasm for recycling and reuse petered out after the war, as Americans returned to a peacetime society and economy that did everything in its power to forget the monetary restrictions and frugality necessitated by the war effort. A new era of consumption was on its way, and mounds of solid waste were a testament to the anticipated good times.

By the mid-1960s, however, governmental, societal, and environmental forces would play important roles in transforming the solid waste problem from an essentially local issue into a national one. Whether it deserved to be characterized as such or not, concerns of a garbage crisis were widespread. The idea that refuse should be out of sight, out of mind ceased to be a quaint notion from a naïve past, and became fundamentally troubling.

EIGHT

THE GARBAGE CRISIS IN THE LATE TWENTIETH CENTURY

By the mid-1960s, solid waste became a truly national issue. The Lyndon B. Johnson administration passed the first federal regulations. The new laws did not establish federal control of local refuse management, but they did create renewed awareness that the problem of solid waste extended beyond individual city limits and that a new partner would weigh in on how to address that problem. In addition, counties, regional authorities, and special districts complicated the management process. In a more direct way, the rise of private agglomerates in the waste collection and disposal field in this period began to seriously challenge the hard-fought public responsibility for sanitary services achieved years earlier.

Also changing was the almost blind commitment to the sanitary landfill as the dominant disposal option throughout the country. As easily acquired sites for disposal became harder to find, and as refuse was equated with third pollution, dumping solid waste into trenches seemed much less of a panacea than it had appeared to be in the late 1940s and 1950s. Recycling, recovery, and reuse began to receive greater attention as ways to lessen the burden on the landfills and to conserve more virgin materials. However, as solid waste continued to increase in volume, and as new toxic materials entered the municipal waste stream, a manifest unease blanketed refuse management. There was much talk about a garbage crisis by the 1970s, severe enough to spur serious public debate.

The backdrop to the solid waste problem beginning in the late 1960s was regional urban development and looming fiscal problems faced by many cities. The United States had evolved from an urban into a suburban nation. Core cities continued to lose population and have their economic base erode; more self-contained communities—"edge cities" by one description—developed on the periphery; and non-metropolitan growth challenged traditional suburban expansion. Cities in the South and Southwest grew at the expense of older urban centers in the Northeast and Midwest.[1]

The metropolitan complex was home to the majority of Americans after 1970. In 1975, 73 percent of the United States population (213 million) lived in metropolitan counties. From 1980 to 1988, population in metropolitan centers increased twice as fast (9.7 percent) as in non-metropolitan areas (4.5 percent). In 1990, of the total U.S. population of 248.7 million, 78.9 million (31.7 percent) lived in central cities; 79.4 million (31.9 percent) lived in the urban fringe; 28.8 million (11.6 percent) lived outside urbanized areas; and 61.7 million (24.8 percent) lived in rural areas.[2] By 2000, the United States population topped 281.4 million, and 226 million—or more than 80 percent—of its citizens resided in 331 metropolitan (statistical) areas.[3]

The well-accepted division between core city and suburban ring became increasingly ambiguous in metropolitan areas by the twenty-first century. The term suburb was less and less useful in describing what was taking place on the urban fringe.[4] The old central city was only one of several activity centers. Numerous suburbs achieved the status of comprehensive living and working areas. Former bedroom communities that had provided the labor force for central cities were increasingly self-sufficient, attracting a bulk of the new jobs, many new business establishments, and many cultural and recreational facilities once reserved for the downtowns.[5]

Despite anticipated gentrification, population at the core continued to decline.[6] Between 1970 and 1976, central city population dropped by 3.4 percent (to 60.7 million). In 1986, only about 30 percent of Americans lived in central cities; a disproportionate number were poor African Americans and poor Hispanics.[7] The central city "had simply become one more suburb, yet another fragment of metropolitan America serving the special needs of certain classes of urban dwellers."[8] In addition, urban population density declined from 2,766 people per square mile in 1970 to 2,141 in 1990, indicating movement from central cities toward satellite communities and "out-towns."[9]

The scale of urban growth since the 1960s and the shift in population toward the periphery continued to strain solid waste services in fairly obvious ways. Collection had to be carried out over greater distances, but also jurisdictional lines became increasingly blurred, adding an additional layer of perplexity to the process. Moreover, metropolitan growth meant that transfer stations and disposal facilities were difficult to site, due to erratic growth patterns, unrelenting highway construction, new subdivision development, and competing land uses. The cry of "environmental racism" would become louder in the 1980s, when people of color, especially, would charge that landfills and toxic disposal facilities were intentionally placed in the neighborhoods of minorities and the poor to avoid confronting intense opposition from suburbanites and out-migrators who had greater political power than those who believed that they were socially and politically marginalized.

Finding adequate sources of revenue to finance refuse services was an additional concern in a changing urban setting. In the years after 1970, American cities—especially core cities in the East and Midwest—faced what some have called an "enduring plight" characterized by financial stress, the lack of a coherent federal urban policy, and rising social and physical problems.[10] In many cases, the fiscal issues confronting cities were structural. Federal and state governments mandated that cities perform a wide array of functions, and the concentration of poor people at the core also raised the cost of providing many public services, including more public welfare, health care, and hospitals, alongside services such as solid waste collection and disposal. Core cities generally were older than the suburbs, and thus the cost of maintaining or replacing existing infrastructure was high. City residents also had to finance services beneficial to nonresidents, especially commuters and those entering the city to take advantage of cultural and recreational events. Out-migration of middle-income residents and the influx of lower income groups meant that cities had less capacity to raise revenue and to provide satisfactory services.[11] Garbage had to be collected and disposed of no matter who generated it.

Beyond structural problems, fiscal woes were real and widespread. Sources of general revenue were in decline as white flight continued, and the reluctance to increase taxes led city officials to find other means to balance their budgets. To meet service obligations, several departments borrowed from the city's cash flow, intending to pay off the debt with future revenues. Before long, cities were accumulating substantial internal debt.[12]

Cities had long since turned to Washington and their state governments for financial aid. By the 1970s, federal and state funding was an essential component in almost every municipal budget. In 1960, federal grants represented 3.9 percent of the general revenue of cities; in 1977, 16.3 percent. State and federal aid to cities amounted to $42.5 billion ($29 billion state and $13.5 billion federal) in 1977.[13]

Although local dependence upon federal assistance increased in the 1970s, the manner in which those funds were dispersed changed dramatically. Richard Nixon's New Federalism called for greater responsibility of local government in controlling the spending of funds from Washington, and with that change came a shift in financial leverage to suburban communities.[14] The incoming Democratic administration of Jimmy Carter in 1977 continued the local-control approach of the Nixon-Ford years.[15] With the election of Ronald Reagan, however, the New Federalism of the 1970s veered sharply away from any major commitment to cities in the 1980s.[16] The new Republican administration reshuffled priorities, strongly emphasizing federal support for national security and rebuilding of the nation's defense system at the expense of many domestic programs. The Omnibus Budget Reconciliation Act of 1981 reduced federal grants to cities by $14 billion, and cut welfare payments to individuals by $11 billion between 1982 and 1984. In addition, half of the federally subsidized housing units were cut, rents to public housing tenants increased, and job-training programs reduced. The only truly urban program in these years was the promotion of enterprise zones, championed by New York congressman Jack Kemp.[17]

This drastic change in national urban policy in the 1980s—described as a "nonurban urban policy"—combined with the shrinking city tax base to create "havoc among metropolitan governments." This resulted in the closing of schools, parks, and libraries, and forestalled repairs to streets, sewers, and bridges. In retrospect, the dire predictions about the Reagan budget cuts were somewhat exaggerated, since the Democratic Congress restored funds for several programs. In addition, state and local governments established emergency funds and raised taxes to capture lost revenue. Nonetheless, funding for specific urban programs declined 23 percent in 1982 from the projected budget. In addition, state aid did not take up the slack from lost federal support.[18]

Also, in the early 1980s, the public recognized a long-standing trend—deterioration of the existing urban infrastructure and uneven investment in new infrastructure. Government spending on public works, discounting for inflation, rose from $60 billion in 1960 to $97 billion in 1984. This represented a decreasing share of the gross national product from 3.7 percent to 2.7 percent. Between 1970 and 1985, the state portion of spending on infrastructure dropped from 32 percent to 23 percent.[19] In 1981, Pat Choate, a policy analyst at TRW Inc., and Susan Walter, Manager for State Government Issues at General Electric, published *America in Ruins: The Decaying Infrastructure*. The book ignited a wide-scale debate over the state of the nation's public works and future needs. According to Choate and Walter, public facilities were wearing out faster than they were being replaced. The most acute implication of this infrastructure crisis, they argued, was its impact as "a severe bottleneck to national economic renewal." Asserting that the United States was "seriously underinvesting" in public works into the billions of dollars, Choate and Walker painted a dismal picture of existing conditions.[20]

While not everyone embraced the hyperbole of an "America in Ruins" or agreed upon the extent of the infrastructure crisis or the actual costs required to alleviate it, few denied that a massive problem existed. The Public Works Improvement Act of 1984 created the National Council on Public Works Improvement, directed to report to the president and Congress on the state of the nation's infrastructure. In its much-anticipated 1988 report, "Fragile Foundations," the Council took a positive approach, finding "much that is good" about what was being done in the planning, building, operating, and maintaining of the nation's infrastructure. It concluded, however, "on balance, our infrastructure is inadequate to sustain a stable and growing economy. As a nation, we need to renew our commitment to the future and make significant investments of our own to add to those of past generations." The Council's "Report Card on the Nation's Public Works" gave the highest grade (B) to water resources, and its lowest grade (D) to hazardous waste clean-up. Solid waste got a C–; more rigorous testing, monitoring, and development of alternative disposal options was offset by movement toward higher costs.[21] In the 2001 "Renewing America's Infrastructure: A Citizen's Guide" published by the American Society of Civil Engineers, solid waste moved up to a C+, largely on the basis of declining disposal in landfills and increasing waste recovery through recycling.

While solid waste escaped center stage in the infrastructure crisis debate of the early 1980s, talk of a general garbage crisis had been ongoing since the early 1970s.[22] The notion of a crisis—especially in terms of rapidly increasing waste volume and declining landfill space—persisted into the following decades. "Many cities are rapidly approaching a garbage and trash crisis: no place to put it," noted a 1987 study.[23] Other studies, identifying the hub of the crisis in the East, also affirmed its national scope: "The magnitude of New York's dilemma may be exceptional but the general problem itself is not: though reliable nationwide numbers are hard to come by, cities everywhere seem to be running out of places to put their garbage."[24]

Blame for the solid waste predicament was widespread: The nation's affluence inspired a throwaway mentality. Fear of closing landfills led to desperation and more bad decisions in finding alternatives. The landfill crisis was politics pure and simple—NIMBY responses bred an artificial shortage of dumping sites.[25] Whatever the cause, solutions would be difficult to come by. As the *Wall Street Journal* noted: "Like groundwater contamination from septic tanks or smog from commuter patterns, the garbage crisis requires changes in life styles, expectations and local economies. Such changes are hard to mandate."[26]

The idea of a garbage crisis was a convenient, albeit a relatively simplified way, to label a complex set of issues. The notion of "crisis" confers upon the problem relatively tangible, concrete properties, which might be resolved through equally tangible, concrete solutions, such as new technology, effective management, or popular will. In some sense, however, crisis ignores its persistence over time, failing to question whether some waste problems were chronic, recurrent, or temporary. A deeper look at the management of solid waste, collection and disposal practices, the growing governmental regulatory apparatus, and statistics on waste generation may help to clarify "crisis" in the late twentieth century.

Collection of solid waste had long been a burdensome, labor-intensive, and costly municipal enterprise, but not a particularly glaring part of the garbage crisis.[27] Regrettably, this undervalued its importance. Surveys estimated that from 70 to 90 percent of the cost of collection and disposal in the United States went for collection. Conservative estimates placed the annual cost of solid waste management in 1974 at $5 billion, with collection and delivery of wastes to disposal sites accounting for nearly $4 billion.[28]

Refuse collection, as stated earlier, has been characterized more by economies of density than economies of scale. This suggests that population diffusion or deconcentration, as in the case of the modern metropolis, has been the enemy of efficient and economical collection.[29] From the perspective of consumers, frequency of service remained a key concern.

Rising interest in recycling reopened the question of source separation versus mixed refuse collection—a point of contention going back to the late-nineteenth century. Reliance on a mechanized fleet to collect refuse also has had environmental repercussions. As a study by Homer A. Neal and J. R. Schubel noted: "The emissions from the massive fleet of garbage trucks, the traffic congestion aggravated by the trucks, the littering caused by trash blown off the trucks, and the disposal processes themselves all contribute to environmental disturbances which would not have occurred if there were no wastes."[30]

Big cities, by and large, have relied on municipal service because of entrenched bureaucracies and the capacity to budget for collection among other required city services. In the mid-1970s, more than 61 percent of the population living in single-family dwellings had municipal service, and 36 percent had private service.[31] In 1997, seventy of the top one hundred cities had municipal service.[32] As WMX Technologies, Inc., executive (and the former director of solid waste management for Houston) Ulysses Ford stated, "Three functions [of a city's operation] have generally been reserved for the city—police protection, fire protection, and solid waste management, in that order. . . . The common wisdom runs that if a local politician fixes the potholes and ensures that solid waste management and the police and fire departments are well run, he or she can get re-elected."[33]

Labor unions, various city employees, and municipal department managers also have a stake in public-sector approaches to solid waste collection and disposal.[34] Advocates of a public-sector approach to refuse collection and disposal cite what they believe to be the advantages. A 1998 report of the Reason Public Policy Institute in Los Angeles noted that cities have lower capital costs, pay few if any taxes, are exempt from some laws and regulations, and do not need to worry about earning a profit. Some supporters see a trend toward deprivatization, but the evidence is sketchy.[35]

After 1950, increasingly more American communities of various sizes have depended on private service.[36] Dissatisfaction with municipal service led several private collectors—because of their expanded role in residential collection—to claim that they could compete favorably with their public counterparts.[37] The trend in residential collection toward privatization continued steadily in the late 1960s, with the greatest exception in the South where municipal collection remained strong. Advocates of privatization emphasize that the private sector often has a competitive advantage over municipal service, because of its ability to raise capital, but not taxes.[38]

On occasion, questions concerning equitable service delivery under a municipal system have been raised. Bryan D. Jones in his 1980 book, for example, questioned whether municipalities have provided services equally. Relying largely on a Detroit case study, he noted that there may be one of two reasons for the existence of poorer service in African American neighborhoods in Detroit: special collection problems or unsatisfactory performance by sanitation crews in those neighborhoods. His conclusion was, "Less garbage [was] collected because the crews picked up less of it, not because residents discard less."[39] This resulted from lax supervision of the work crews and suggested problems in the service network as opposed to shortcomings in the neighborhoods themselves. Such studies were part of an increasing chorus in the 1980s that began to question more seriously not just the performance of municipal departments, but also private companies. Concern over equal distribution of services and the exposure of poor and minority neighborhoods to a variety of environmental risks were important factors in the emergence of the environmental justice movement.

The initial debates over privatization, however, focused more on quality of service as opposed to equity of service. By 1974, large private firms held contracts in over 300 communities. During the 1970s, when cities of any size were surveyed, less than 42 percent utilized municipal collection exclusively and less than 38 percent had some municipal service.[40] Data from the National Solid Wastes Management Association (NSWMA) published in 1984 indicated that more than 80 percent of the nation's garbage was collected by private companies, and that between 1973 and 1982, communities with private collection increased from 339 to 486. The percentage for the amounts collected may be somewhat misleading, since the pronouncement is unclear as to whether this represents residential collection only or all garbage collection. In addition, the NSWMA is a trade organization that promotes privatization and may have been presenting the proverbial half-filled glass.[41]

Despite the trend toward more private collection, debate continued to rage over its value and efficiency. Public works directors and their employees came down decidedly on the side of municipal service. Public/private partnerships had been in operation since the nineteenth century, however, and municipal officials did not balk at them as long as cities maintained control of the contracts or franchises.[42] Nevertheless, in the 1970s and 1980s, some social scientists made a strong case that private collection was cheaper and more efficient, recommending increased privatization of the service.[43]

In some respects, collection was a special case, since not all municipal services could easily be privatized. In addition, some studies noted that in very small cities—less than one thousand people—privatization produced diseconomies that could be corrected by increasing the scale of the operations through a contract or a municipally operated system.[44] In a 1998 article in *Waste Age*, Laith B. Ezzet suggested that exclusive municipal collection was likely to remain strong in the largest cities, with limited potential growth for private solid waste companies in that market. This was the mirror image of the success of private companies in the field of commercial collection.[45] In cases where cities were trying to maintain or regain control of collection service, "managed competition" was a new catch phrase. However, it is not clear if the attempt to level the playing field could buck the privatization trend.[46]

With increased privatization also came consolidation of the solid waste industry. "Historically," financial consultant Anne Hartman noted in *Waste Age*, "the solid waste control industry as a whole has been fragmented, undercapitalized, and unsophisticated." The industry, she added, was now "assuming a new stature, and new management methods are being brought to bear on the longstanding and relatively simple problem of disposing of wastes."[47] Three companies that dominated the U.S. industry in 1980 had been formed as recently as the late 1960s and early 1970s—Waste Management, Inc., Browning-Ferris Industries (BFI), and SCA Services of Boston. While they only handled 15 percent of the nation's solid waste, they had extensive influence in the industry because of their capital, their size, and their management teams.[48]

Into the 1980s, consolidation in the industry was producing record growth. In 1987, the six largest public U.S. waste management companies reported annual revenues of $5 billion from their solid and hazardous waste services.[49] Growth, however, also brought charges of antitrust and mismanagement of disposal sites, including alleged violations of federal environmental statutes. CEO of Browning-Ferris and former EPA administrator William D. Ruckelshaus conceded that "Free markets are not perfect," but also asserted that "they are efficient allocators of services." He also observed, "Private enterprise is trying to mesh with a set of public processes, and there is sand in the gears."[50] In response to the variety of charges,

Browning-Ferris, Waste Management, and the other agglomerates sought to bolster their image, and in some cases attempted corporate makeover campaigns. Indeed, Ruckelshaus's selection as CEO was an attempt to give BFI a high profile and public legitimacy. Criticism of the waste giants did not disappear, and, in some cases, periodic decreases in stock value reflected unease with the consolidation and conduct of the solid waste industry.[51]

Consolidation and vertical integration accounted for much of the industry's growth in the 1990s, as some companies became international businesses. In 1996, the top six companies posted revenues of more than $19 billion.[52] In 1995–1996, large numbers of consolidations took place; twenty-eight of *Waste Age*'s top one hundred companies were acquired or merged with other top one hundred companies. The acquisitions, in addition, were largely part of a plan for vertical integration in which companies control the waste stream and then dump the waste into their own landfills. As John T. Aquino noted, "big has bought small, big has bought big, and small may be merging with small."[53] Analysts cited a number of reasons for why smaller haulers were willing to sell to bigger companies. Most prominent was the notion that government regulations, especially compliance in matters of personnel, safety, and the environment, had driven up the cost of doing business. In addition, traditional family-owned companies often felt unprepared to deal with an industry becoming increasingly more complex. In some cases, larger companies were willing to maintain the original name of the smaller hauling company and involve the original owners in management. And, of course, profit opportunities encouraged several small companies to sell out.[54]

In March, 1998, USA Waste Services, Inc.—at the time, the third-largest waste disposal company in the country—acquired the ailing industry leader, Waste Management, Inc., to become the largest waste disposal corporation. The "new" WMI now controlled more than 20 percent of the national waste stream. Of the top one hundred companies ranked by revenue in 1997, fifteen were acquired in 1998, and consolidations continued along the whole spectrum of the industry.[55] *Waste News* reported on January 12, 1998, "Industry observers see no fast finish for the current consolidation trend in the trash-handling and scrap metal markets."[56] Indeed, the very next year, Allied Waste Industries, Inc., of Scottsdale, Arizona (the second largest non-hazardous waste management company in the United States in 2004), announced that it would acquire Houston's Browning-Ferris Industries, then ranked number two behind WMI.[57] Several lingering questions remained by the turn of the century: Would consolidations ever end in the solid waste industry, and how predictable was the industry's future? Could the public sector compete in a world of increasingly powerful waste companies despite the chronic and continual fragmentation of the industry?

The transformation of the solid waste industry since the 1960s and the consolidations in the 1980s and 1990s did not stabilize the industry or make it immune from continued criticism. As journalist Barnaby J. Feder noted in 1990, "Waste handlers have provided skeptics with every conceivable reason to distrust them, including technological fiascos, bid rigging and other antitrust violations, mismanagement and links to organized crime." Exposés on the waste agglomerates and recurrent stories of price-fixing and mob involvement dogged the industry.[58] Critic Harold Crooks asserted, "The fragmented and disorganized nature of the garbage business left most firms extremely vulnerable to the hazards of the open market, and so like other big-city industries in similar situations, it often resorted to coercive means of self-protection."[59] Even after the rise of the new agglomerates, the industry was not free of its rough and tumble history, its intense competition, its internal battles, and its brushes with government agencies and the courts. Yet as Crooks also noted, somewhat ironically, "the mobster has become a dinosaur in the waste disposal industry. Financial muscle had replaced the physical kind."[60]

As agglomerates emerged to transform the private solid waste business, the role of the federal government in solid waste management began to modify the regulatory environment in which collection and disposal took place. Acknowledging the poor state of solid waste research, the American Public Works Association, the major national organization of public works professionals, and the U.S. Public Health Service sponsored the first National Conference on Solid Waste Research in Chicago in December 1963. Soon thereafter, the APWA established the Institute for Solid Wastes (1966) in recognition of the special significance of the issue.[61]

In a message on conservation and restoration of national beauty in 1965, President Lyndon Johnson called for "better solutions to the disposal of solid waste" and recommended federal legislation and research and development funds to assist state governments in creating comprehensive disposal programs. Soon after Johnson's speech, Congress passed the Solid Waste Disposal Act as Title 2 of the 1965 amendments to the Clean Air Act.[62] By recognizing the mounting quantities and changing character of refuse as a national problem, the act was the first piece of legislation to involve the federal government in solid waste management. Its primary thrust was to "initiate and accelerate" a national research and development program and to provide technical and financial assistance to state and local governments and interstate agencies in the "planning, development, and conduct" of disposal programs. The act also was meant to stimulate the promulgation of guidelines for collection, transportation, separation, recovery, and disposal of solid wastes.[63]

The Solid Waste Disposal Act focused on demonstration projects to develop new methods of solid waste collection, storage, processing, disposal, and the reduction of unsalvageable wastes. By 1967, the government awarded approximately $9 million on various projects.[64] The new law noted the inability of current methods to deal with solid waste and the failure of resource recovery programs to convert waste economically into usable by-products. However, the act was incomplete in its assessment of the waste problem and did not mandate a regulatory authority to deal with broader issues related to refuse. In addition, the primary focus was on disposal of waste, not collection or street cleaning.[65]

Although the Solid Waste Disposal Act was a significant step in the direction of federal involvement, there was little reliable data on the extent of the solid waste problem. President Johnson, with the advice of his Scientific Advisory Committee, initiated the National Survey of Community Solid Waste Practices (1968)—the first truly national study of its kind in the twentieth century. Of the approximately six thousand communities taking part in the survey, less than half provided estimates of the amounts of waste produced or collected. Although the samples chosen were incomplete and the execution imperfect, the survey helped to fill the data gap and led to other statistical compilations.[66]

Initially the enforcement of the 1965 act fell to the United States Public Health Service (USPHS) and to the Bureau of Mines (BOM) in the Department of the Interior. The USPHS was given responsibility for municipal wastes, while the BOM supervised mining and fossil-fuel waste from power plants and industrial steam plants. With the creation of the Environmental Protection Agency (EPA) in 1970, responsibility for most refuse activities was transferred to it. In the 1970s, the Office of Solid Waste (OSW) acquired the authority to conduct special studies of problems related to solid waste, award grants, and publish guidelines.[67]

To refine the 1965 act, Congress passed the Resource Recovery Act in 1970, which shifted the emphasis of federal involvement from disposal to recycling, resource recovery, and the conversion of waste to energy. It created the National Commission on Materials Policy to develop a national policy on materials requirements, supply, use, recovery, and disposal. Another feature of the 1970 legislation was the stipulation that a national system be implemented for storing and disposing of hazardous wastes.[68]

Although the federal legislation was incomplete and federal interest in solid waste issues inconsistent, both the 1965 and 1970 acts caused the states to become more deeply involved in what had been local issues related to solid waste collection and disposal. At the time of the passage of the 1965 act, there were no state-level solid waste agencies in the country, and only five states had employees assigned to any phase of solid waste management. Responding to federal legislation, pressure, or incentives, states began to enact solid waste management statutes and to designate agencies as solid waste management offices. Four and a half years after the first grant of technical assistance from the federal government was awarded, forty-four states had active programs. The most significant immediate result was the development of solid waste management plans—a prerequisite for receiving federal funds. Often built on the plans of counties and/or municipalities, they stressed disposal issues more than collection. The state programs, however, demonstrated little uniformity.[69]

Competition also existed among various governmental entities in the form of jurisdictional rivalries. As one report noted, "Solid waste disposal is rapidly becoming an intrastate, as well as interstate, matter."[70] A variety of new arrangements emerged out of the realization that the garbage problem was infrequently contained within the city limits or within a particular jurisdiction. Some areas tried interlocal agreements, countywide systems, multicounty corporations, or intrastate planning bodies. Often born out of necessity, these arrangements spoke to the complex problems of collection and disposal.

The question of "Who is in charge?" is particularly significant because of the commitment of both public and private entities to integrated waste management systems. Integrated waste management is a relatively common-sense notion adopted by the EPA that stresses reliance on, as James R. Pfafflin and Edward N. Ziegler noted, "a hierarchy of options from most desirable to least desirable," with source reduction on the high end and with the sanitary landfill on the low end.[71] The EPA's *The Solid Waste Dilemma: An Agenda for Action* (1989) recommended "using 'integrated waste management' systems to solve municipal solid waste generation and management problems at the local, regional, and national levels. In this holistic approach, systems are designed so that some or all of the four waste management options (source reduction, recycling, combustion and landfills) are used as a complement to one another to safely and efficiently manage municipal solid waste."[72]

In setting such an ambitious goal, the EPA and others were treating integrated waste management as a solution to the garbage problem. Proponents may not have given sufficient attention to the fact that attempts to implement integrated waste management came with the risk of sharpening the traditional rivalry between public and private forces seeking to control collection and disposal systems. The EPA's promotion of the concept suggested a significant new role for the federal government not only in setting a national agenda for solid waste, but in setting an expanded regulatory role.

The Resource Conservation and Recovery Act (RCRA) passed in 1976, and reauthorized as the Hazardous and Solid Waste Amendments of 1984, added significantly to the federal role in solid waste management. It was the first comprehensive framework for hazardous waste management in the United States, and completely changed the language of the Resource Recovery Act, redefining solid waste to include hazardous waste.[73] RCRA continued provisions on solid waste and resource recovery; ordered the EPA to require "cradle to grave" tracking of hazardous waste and controls on hazardous waste facilities; closed most open dumps; and set minimum standards—including size and location—for waste disposal facilities through rules and regulations promulgated through the states.

The EPA did not give major attention to carrying out the provisions dealing with hazardous wastes until publicity over Love Canal in 1978.[74] According to the National Council on Public Works Improvement 1987 infrastructure report, the original RCRA bill "was based on the assumption that the free enterprise system would reduce the solid waste problem by being responsible for producing and procuring the material made from recovered materials. Furthermore, as the cost of disposing of solid waste at landfills increased over time, there would be increased financial benefit to industry to either reduce the quantity of waste being generated, recycle significant portions of the waste or use the waste to generate energy."[75]

The enlarged federal role in solid waste management in the 1960s and 1970s faced a serious setback in the 1980s. After the election of Ronald Reagan, substantial cutbacks of the EPA's budget, administrative disruptions in the agency, and greater attention toward hazardous waste cleanup (begun in the 1970s) diminished the emphasis on municipal solid waste management issues until the end of the decade.[76]

The inclusion of the OSW within the EPA initially had provided a base for federal solid waste programs, but not without controversy. Formerly housed in the Health, Education, and Welfare Department (HEW), its role was to provide technical assistance and grant money to state and local governments. As such, it did not necessarily attract the best and the brightest within the EPA, who wanted to go where there were more opportunities. Between 1973 and 1976, the staff decreased from 225 to 174 employees. Despite a small hazardous waste group within the OSW besieged with a series of complex problems to confront, several officials still preferred to concentrate on hazardous wastes and deemphasize solid waste management issues. In the mid-1970s, the EPA had proposed a drastic cutback in the federal solid waste program, but Congress and state and local groups balked. The EPA backed away from its more extreme position and announced its willingness to continue to develop and promote resource recovery systems and technology. But when the immediate threat of the energy crisis had passed, and belt-tightening provisions cut into the EPA's budget in the Reagan years, the agency all but turned away from the municipal solid waste issue.[77]

With public attention alerted to a pending garbage crisis, and with the strengthening of the provisions of the Resource Conservation and Recovery Act (RCRA) affecting the MSW in the mid-1980s, the EPA was back in the middle of the municipal solid waste debate—along with other federal agencies such as the Office of Technology Assessment. The issuance of *The Solid Waste Dilemma* was a clear statement of the EPA's desire to set the tone for attacking the crisis. Speaking about the EPA's new role, especially in light of the establishment of an Office of Pollution Prevention, director William Reilly stated, "That is going to be the hallmark of this place—pollution prevention and reduction at the source."[78]

But the return of the EPA engendered several responses. There were some who applauded the resolve of the federal government to join the battle over the garbage crisis. Others argued that the boost in regulatory authority was not matched with action, especially since the primary focus of the RCRA was to be on issues of hazardous waste as opposed to municipal solid waste. William Kovacs, who was chief counsel of the House of Representatives subcommittee on transportation and commerce with primary jurisdiction over solid and hazardous waste during the enactment of the RCRA, observed that the "EPA's implementation of RCRA can only be described as tardy, fragmented, at times nonexistent, and consistently inconsistent."[79]

In the EPA's defense, however, Congress had enacted the RCRA as a broad, unwieldy monster of a law. As one study of the EPA noted: "Those experts (on the Congressional subcommittee who helped write the legislation) were not very expert. In particular they lacked a clear vision of the 'problem' that they wanted EPA to solve. As a result the agency was instructed to establish an exceedingly ambitious program with little guidance about how that was to be done. By also specifying an utterly unrealistic time schedule, Congress put EPA in a classic 'no win' situation."[80] For those living under the new standards—especially local and state governments, various waste disposal authorities and private businesses—such challenges at the national level often produced confusion over the type and extent of compliance with new rules, and sometimes suspicion and cynicism over federal intrusion into local affairs. For optimists, the expanded role of the federal government in the solid waste issue could mobilize Americans to action. For pessimists, it simply added one additional layer of bewilderment.[81]

The vast quantities of wastes generated in the United States contributed significantly to the problems of collection, and ultimately, disposal. Although estimates vary widely, government and government-sponsored studies (generally the ones with lower estimates) stated that between 1970 and 2000 municipal solid waste increased from 121.1 million tons per year to 231.9 million tons; amounts of wastes per capita increased from 3.3 to 4.5 pounds per day.[82]

Population increases, as much as changes in consumption patterns, were responsible for the steady rise in aggregate discards.[83] The 1992 study of the Franklin Associates, Ltd., which analyzed trends in MSW generation between 1972 and 1987 for the federal government, found that during the fifteen-year period under study the United States population grew by 16 percent (from 209.9 million in 1972 to 243.8 million in 1987) and per capita waste discarded also grew by 16 percent. The report concluded that "perhaps as much as one-half of the growth in trash over the 15 years was caused simply by our growing population."[84]

Although per capita waste-generation rates may have flattened, or even declined in some studies, annual per capita waste generation remains high in the United States especially in comparison with other countries. For example, figures for the late 1980s show rates as high as 6.4 pounds

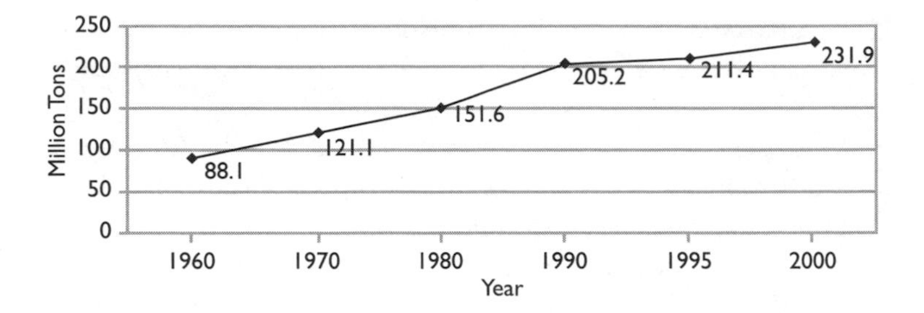

FIGURE 5. Generation of MSW, 1960–2000, National Total Per Year.
Source: U.S. Environmental Protection Agency, Office of Solid Waste and Emergency Response, *Municipal Solid Waste in the United States: 2000 Facts and Figures, Executive Summary* (Washington, D.C.: Environmental Protection Agency, June 2002), p. 2.

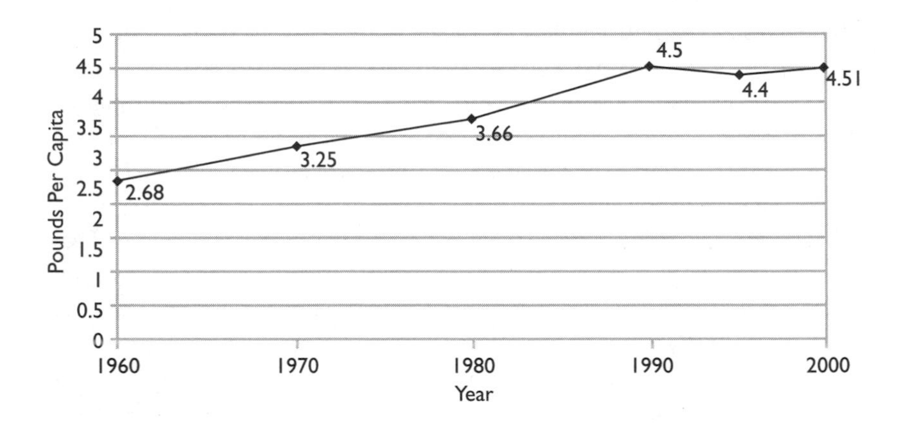

FIGURE 6. Generation of MSW, 1960–2000, Per Capita Per Year.
Source: U.S. Environmental Protection Agency, Office of Solid Waste and Emergency Response, *Municipal Solid Waste in the United States: 2000 Facts and Figures, Executive Summary* (Washington, D.C.: Environmental Protection Agency, June 2002), p. 2.

per capita per day for Los Angeles, 5.8 for Philadelphia, and 4.0 for New York City compared to Tokyo at 3.0, Paris and Toronto at 2.4, and Rome at 1.5.[85]

While the quality of waste has been a critical factor in the garbage problem, it has not been the only problem. Composition of municipal waste called into question the rationale for consumption of scarce resources, and at the same time led to demands for quick and efficient disposal of society's discards. The waste stream after 1970 included a complex mix of hard-to-replace as well as recyclable materials, and a vast array

TABLE 14

Type of Discards in MSW, 1970–1999 (by percent)

Year	*1970*	*1980*	*1985*	*1990*	*1995*	*1999*
Paper/paperboard	36.3	36.6	37.4	37.5	38.6	38.1
Glass	10.4	10.0	8.0	6.7	6.1	5.5
Metals	11.6	9.7	8.6	8.3	7.5	7.8
Plastics	2.5	5.2	7.1	8.3	8.9	10.5
Rubber/leather	2.6	2.9	2.3	2.4	2.9	2.7
Textiles	1.6	1.7	1.7	2.9	3.5	3.9
Wood	3.3	3.3	5.0	6.3	4.9	5.3
Food waste	10.5	8.8	8.0	6.7	10.3	10.9
Yard waste	19.0	18.4	18.2	17.9	14.0	12.1
Other	2.2	3.4	3.6	3.1	3.3	3.2

Source: U.S. EPA, *Characterization of Municipal Solid Waste in the United States: 1992 Update*, prepared by Franklin Associates, Ltd. (Washington, D.C.: EPA, 1992); U.S. Department of Commerce, U.S. Census Bureau, *Statistical Abstract of the United States: 2001* (Washington, D.C.: Department of Commerce, 2001), p. 217.

Note: There are a few anomalies in this data and other data on waste composition because surveys vary in technique and interpretation of the evidence.

of toxic substances. How best to collect these materials, let alone how to dispose of them, challenged traditional solid waste practices.

As table 14 shows, paper has made up the largest percentage of municipal waste since 1970. Plastics, a new element in the waste stream since World War II, represented a relatively small but rapidly increasing portion of the total. In recent years, approximately 30 percent of plastics were used in packaging.[86] Yard and food wastes, the major organic discards, have steadily declined. Inorganic materials were scant in terms of percentage, but include many household chemicals with potentially serious environmental implications.[87]

The nature of the waste stream suggests important front-end problems. Of all the discarded items, paper, plastics, and aluminum have grown steadily as a percentage of the total municipal solid waste (MSW).[88] The use of these materials reflects a substantial increase in packaging waste and a wide variety of additional uses for paper. Few would disagree that rampant consumerism, especially since World War II, has been a major contributing factor to the solid waste problem. The conclusion that excessive packaging is the whole problem misstates the case, however. Between 1958 and 1976, packaging consumption rose by 63 percent. But rapid growth in aggregate packaging waste did not continue into the 1980s and

1990s. For example, between 1970 and 1986, packaging as a component of MSW increased only 9 percent, while durable items increased 35 percent; clothing/footwear, 88 percent; non-durables, 300 percent; newspapers/magazines/books, 40 percent; and office and commercial paper, 69 percent.[89]

A more critical issue with packaging was changes in composition. Paperboard cartons, plastic jugs, and plastic jars replaced glass milk bottles and other glass containers; aluminum beverage cans replaced steel cans; and plastic grocery bags began to replace paper bags. In 1985, 47 billion pounds of plastics were produced in the United States. Of the 39 billion pounds consumed domestically, 33 percent was used for packaging, and over half of the discarded plastics were from packaging. Estimates suggest an annual growth rate in plastic waste at about 10.3 percent per year. The dependence on plastics required extensive use of petrochemical and other feedstocks, resulting in discards of substantial bulk—if not weight—in landfills, posed a health hazard if burned in incinerators, and demanded more complex recycling methods than glass or paper. Additional problems arose if packages were composites of several materials.[90]

In the late 1980s, McDonald's polystyrene (PS) "clamshell" package became a symbol of wastefulness, resource depletion, and disposal problems for proponents of recycling and other environmental advocates. In 1987, Vermonters Organized for Cleanup called for a ban on the use of PS packaging, which ultimately led to a national grassroots campaign aimed at McDonald's. While the hamburger giant's initial response was to fall back on a public relations response, it bent to the demand to phase out using PS containers and to substantially reduce its solid waste production. In 1991, McDonald's USA announced its Waste Reduction Action Plan to manage solid waste. Several local governments also proposed and passed ordinances to ban PS from cups and related items, and such campaigns had some success. However, recycling and reuse solutions sometimes became an alternative to actual resource minimization efforts. Ordinances alone could not reverse the use of some plastics for packaging, and certainly inspired an increasing defense of the value of its products by the packaging industry.[91]

In the case of paper, not only did the amount of discards increase, but the uses also changed. While some forms of packaging were no longer dependent on paper products, other uses were growing. Office paper and commercial printing papers began mounting up, while newsprint as a percentage of total MSW remained constant. The rise in the use of computer printouts and facsimile paper, paper for direct mail advertising, and the output from high-speed copiers was particularly significant. The anticipation of a "paperless office" because of computers and sophisticated electronic communications was offset by these new uses. The potential for recycling paper was great, although paper and paperboard of various types were often mixed in landfills and thus had to be recycled at the source. Even with source separation, effective techniques for converting all but newsprint and clean, white paper into new products were not yet viable in the market. The problem with paper was further complicated because it added significantly to the volume of waste in a greater proportion than weight. Since most projections about aggregate discards are calculated in weight, paper's contribution to the disposal problem was often underestimated, even after the paper was compacted.[92]

The sheer volume and increasing complexity of the municipal waste stream made disposal the focal point of solid waste management in the late twentieth century. Ironically, the technology once hailed as the clean, efficient, and inexpensive panacea to disposal problems—the sanitary landfill—became the central symbol of the garbage crisis. Shrinking landfill sites in the eastern United States, more stringent federal regulations, revelations about environmental risks, competition from recycling programs, and rising public criticism turned a disposal cure-all at mid-century into an emerging pariah by the beginning of the twenty-first century.

The great symbol of the end of the sanitary landfill era was Fresh Kills Landfill, owned by New York City and operated by the Department of Sanitation.[93] Located on a salt marsh on the western shore of Staten Island in 1948, the largest landfill in the world spreads over more than 2,100 acres with four mounds ranging in height from ninety to five hundred feet. It is so big, as the oft-repeated observation goes, that as the highest point on the Eastern Seaboard south of Maine, it can be seen from space. Before it closed, barges from nine marine transfer stations operated around the clock, six days a week, to deliver approximately 11,000 tons of refuse daily, or approximately 2.7 million tons of solid waste and incinerator residue annually.[94]

Beyond its inclusion in the *Guinness Book of World Records*, Fresh Kills was fraught with controversy almost from its inception. Residents of Staten Island never forgave the decision makers in Manhattan who singled them out to carry the burden of New York City's waste disposal. In addition, like all landfills of its time, Fresh Kills has few environmental controls including no bottom liner to protect the surrounding area from leachate (toxic liquids)—estimated at four million liters per day entering New York Harbor. Ironically, despite its size it did not even accommodate half of the city's daily solid waste production.[95] It was, without a doubt, at the core of every public battle over solid waste disposal in New York City from the closure of the Brooklyn Landfill in the early 1990s to the intense debates over incineration as an alternative to dumping.

Celebrations all over Staten Island broke out in March 2001, when a garbage barge from Queens departed for Fresh Kills on the last of four hundred thousand such trips to deposit what was to be the last load at the landfill. New York governor George E. Pataki, New York City mayor Rudolph W. Giuliani, and Staten Island borough president Guy V. Molinari were on hand to mark the occasion. "This is a glorious day for Staten Island," Molinari stated. The landfill was "the most notorious environmental burden in Staten Island history."[96] The euphoria was short lived, however. In the wake of the shocking attacks of September 11, 2001, and the destruction of the twin towers of the World Trade Center, Fresh Kills was opened to receive most of the estimated 1.2 million tons of debris from Lower Manhattan. And as several people pointed out, Fresh Kills also became the final resting place for remains of many of the victims of that tragic event.[97] What had been a reviled disposal site became hallowed ground as well.

The controversy over Fresh Kills during its more than fifty-year history was not enough to undermine the support for landfilling as the key disposal option in the United States. By the 1970s, however, solid waste professionals and others began to doubt the adequacy of the sanitary landfill to serve the future needs of cities. The initial discussions centered on the problem of acquiring adequate space. The reason that sanitary landfills were attractive in the United States in the first place—availability of cheap and abundant land—was the very issue turned on its head to discredit them. Siting new landfills became problematic, especially in the East. Many communities there simply did not set aside land specifically designated for waste disposal facilities, or physically or environmentally marginal land was no longer available. The strong consensus favoring sanitary fills also meant that other disposal options received much less attention, and in many cases had languished, creating a vacuum that could not be easily and quickly filled with viable alternatives.[98]

Landfill siting was treacherous business because of citizen resistance and increasingly rigid environmental standards. A great deal has been made of the NIMBY syndrome, which exposed growing popular skepticism toward the environmental soundness of landfills. Equally important, NIMBY received wide press coverage when attempts were made to site landfills beyond the inner city along the urban fringe, where the population was not characteristically poor and also predominantly white. Sometimes groups were pitted against each other. In Contra Costa County, California, all of the proposed sites for new landfill space were to be located on the blue-collar east side of the county. The organizer of WHEW (We Have Enough Waste), an east county coalition group, stated, "Whenever there are undesirable things to be located, if there is a dump or a jail, where do you look? East county! Is it fair that east county citizens be the recipients of all the garbage for the rest of the county?"[99]

Even the once passive or politically neutral neighborhoods were fighting back, unwilling to provide dumping space for the whole city. In the 1980s, grassroots organizations that had been dealing with toxins also extended their range of interests beyond hazardous waste facilities to landfills and incinerators. For example, the Citizen's Clearinghouse for Hazardous Wastes (CCHW) based in Falls Church, Virginia, and headed by Love Canal activist Lois Gibbs, established the Solid Waste Organizing Project in 1986 to stop unsafe and unsound disposal practices, to promote alternatives such as recycling and composting, and to change industry practices to reduce and eliminate products that add to the solid waste problems. The CCHW claimed victories for local grassroots efforts from the Eagle Mountain dump proposal in California to Sparta, Georgia.[100]

However, siting landfills moved beyond battles in the suburbs by the 1980s. The environmental justice movement that emerged by mid-decade—promoted by former civil rights activists and a variety of grassroots organizations—sought to confront problems perceived to be rooted in "environmental racism." The building of landfills in inner-city minority neighborhoods—along with environmental risks such as lead poisoning or exposure to pesticides—became lightning rods for protest. Some in the movement connected class and race, but many others viewed racism as the prime culprit. Rev. Benjamin F. Chavis Jr., former head of the NAACP, is credited with coining the term "environmental racism" during his tenure as executive director of the United Church of Christ's Commission for Racial Justice (CRJ). He stated, "Millions of African Americans, Latinos, Asians, Pacific Islanders, and Native Americans are trapped in polluted environments because of their race and color. Inhabitants of these communities are exposed to greater health and environmental risks than is the general population."[101] Rev. Chavis became interested in the connection between race and pollution in 1982 when residents of Warren County, North Carolina—predominantly African American—asked the CRJ for help in resisting the siting of a PCB (polychlorinated biphenyl) dump in their community. The protest proved unsuccessful, but the event proved to be a cause celebre, which built momentum for future protests and further articulation of the promotion of social and environmental justice.[102]

In reality, landfill siting issues drew condemnation as being environmentally racist before Warren County. In *Margaret Bean et al. v. Southwestern Waste Management Corp. et al.* (1979), the first lawsuit of its kind, a class-action suit was filed against the city of Houston, the state of Texas, and Browning-Ferris Industries for the decision to site a municipal landfill in Northwood Manor (a middle-income neighborhood whose population was 82 percent African American). The plaintiffs claimed that the decision was in part motivated by racial discrimination. The court ruled that the decision was "insensitive and illogical" but that the plaintiffs had not demonstrated "discriminatory intent" on the basis of race, making cases of this type difficult for plaintiffs to win. Similar suits met with the same result, but the issue of siting landfills based on race and class gave ammunition to the emerging environmental justice movement.[103]

Changing tactics by waste companies or municipal authorities to tout the alleged safety of new waste disposal facilities or to promote them as economic opportunities—jobs, support for schools and parks, and so forth—present communities with real dilemmas. Rejecting a facility holding the opportunity for economic productivity or embracing such a facility without certainty of its potential health and environmental risks imposes difficult choices. Environmental justice advocates brand economic incentive packages from waste companies as nothing more than "job blackmail," although waste companies would deny this harsh description.[104] "Some critics argue that environmental justice is a creature of liberal public interest groups," stated Michael Gross in a 1999 issue of *Waste Age*, "and that it deprives minority communities of high-paying, stable employment."[105] Sociologist David Pellow observed to the contrary that apart from proximity to waste disposal sites, people of color, the poor, and immigrants can experience workplace hazards at waste disposal sites. In recycling facilities placed in low-income neighborhoods in Chicago, for example, "the people who actually do the work got more than they bargained for" in the form of overwork, exposure to various environmental risks, and low pay.[106]

Growing concern that sanitary landfills are not as sanitary as their name implies gave citizen groups and others weapons to resist new sitings. Birds, insects, rodents, and other animals that frequent landfills carry pathogens back to people. Even in the most pristine facility, this problem is not wholly eliminated. Only one of six existing MSW landfills in the 1980s was lined. If not properly lined, landfills pose a threat to groundwater and surface water, especially through leachate. Only one of twenty had a leachate collection system, and only about 25 percent had the capability to monitor groundwater. Contamination from various hazardous materials and incinerator ash found its way into the landfills from homes and factories.[107] Garbologist William Rathje noted, "The potentially toxic legacy of landfills that may long ago have been covered over by hospitals and golf courses illustrates one of the terrible ironies of enlightened garbage management: an idea that seems sensible and right is often overtaken by changes in society and in the contents of its garbage."[108]

Until the promulgation of strict federal laws in the early 1990s, more than 75 percent of MSW ended up in landfills. However, the amounts directed to landfills still remained high at the start of the twenty-first century.[109] Conversely, the number of sanitary landfills has declined rapidly since the 1980s. Figures vary widely on what constitutes a landfill for purposes of statistical evaluation, but all estimates clearly indicate a substantial shrinking of the total number of landfills in the United States. One study set the number of municipal landfills at 15,577 in 1980, another reported a drop to less than 8,000 by 1989. Between 1990 and 2000 the number of landfills declined from approximately 7,300 to 2,200 (some assessments are even as low as 1,967). About eighteen years of landfill capacity overall remains in the United States, but the distribution of landfills is uneven. Predictably, the Northeast continues to lose sites, but in the West and South, declines have been sharp as well.[110]

The steep reduction in landfill sites is attributable to a number of factors including mounting quantities of waste products; limited available land; NIMBY and NIABY (Not in Anybody's Backyard) pressure; competition from other disposal options (increasingly from recycling and to a lesser extent from incineration in the 1990s); and sweeping changes in federal law. Of particular significance for municipal solid waste was Subtitle D of the Resource Conservation and Recovery Act (RCRA), which passed Congress in 1976 and was reauthorized as the Hazardous and Solid Waste Amendments of 1984.

Subtitle D provided for the development of environmentally sound disposal methods and the protection of groundwater with new landfill standards. To achieve sound methods of waste disposal and to encourage the conservation of resources, RCRA outlined plans for federal financial and technical assistance to state and local jurisdictions for the development of solid waste management plans that led to cooperation among federal, state, and local governments and private industry. The 1984 amendments substantially tightened standards with respect to landfills. According to analyst William Kovacs: "Simply put, the premise of RCRA is that the solid waste problem is a result of our industrial society. As such, the true costs of disposal should be borne by those benefiting from the products that generate the waste. Therefore, once the subsidy of the cheap landfill is removed, other more advanced technologies will develop."[111]

Under RCRA, the EPA had acquired extensive regulatory authority over municipal solid waste, especially in the design and operation of landfills and incinerators. Congress directed the EPA through RCRA to develop landfill criteria that included a prohibition on open dumping. The EPA issued the criteria in 1979. (Landfills also were subject to stricter state regulations and financial liability under the Comprehensive Environmental Response, Compensation, and Liability Act [CERCLA] for cleanup of contaminated sites.) The 1984 amendments attempted to tighten standards with respect to landfills. On September 11, 1991, the EPA announced the long-awaited revised Subtitle D regulations for MSW landfills with an October 9, 1993, effective date (extended for some sites). The new rules set minimum national standards, and with few exceptions, any landfill that accepted municipal solid waste fell under the revised criteria. These included location restrictions, revised operating and design requirements, groundwater monitoring, and closure and post-closure care. The cost of meeting the Subtitle D criteria for existing and new landfills was extremely high and resulted in further constriction in the landfill supply nationwide, thus favoring the development of larger regional landfills to lower per unit costs or moving communities toward alternative disposal methods. In addition, landfills were more frequently becoming privately held because the costs were too high for communities to bear them alone. In 1984, only 17 percent of the landfills in the United States were privately owned; by 1991, that figure increased to more than 50 percent.[112]

One solution to the dearth of landfills in one location—due to the inability or unwillingness to meet stricter regulations, lack of sites, and NIMBY concerns—was exporting wastes. This provided at least a near-term option for cities that believed few choices were available to them. Yet interstate transfers did not solve the landfill shortage, and in fact, they sometimes stimulated new local protests in rural America. In the not-so-distant past, interstate transfer was an action that smacked of desperation by those who had given up on purely local solutions. In the late 1980s, however, three northeastern states—New Jersey, Pennsylvania, and New York—exported eight million tons of garbage per year, much of it to the Midwest where landfill sites were more plentiful and where tipping fees were lower.[113]

In the 1990s, interstate movement of solid waste became commonplace, particularly because it proved to be relatively inexpensive and a simple alternative to other disposal options. In 1990, the National Solid Wastes Management Association conducted the first survey of interstate movement of nonhazardous solid waste. In 1995, all states either imported or exported solid waste, with 252 "interactions" (movement of solid waste between two states or countries) occurring. This represented a 24 percent increase over the 1992 figure (203 interactions) and 91 percent more than in 1990 (132 interactions).[114]

In 2000, 32 million tons of MSW crossed state lines, an increase of 13 percent since 1998. New York, by far, was the biggest net exporter with 6.3 million tons, followed by New Jersey, Maryland, Missouri, and Illinois. Imported waste shipments were concentrated in the Midwest and along the East Coast, with Pennsylvania the biggest net importer by a wide margin (9.2 million tons compared with Virginia, in second place, with 3.7 million tons).[115]

Several factors account for the movement of waste across state lines, and they suggest some basic characteristics of recent solid waste disposal conditions. Numerous cities still relied on the closest and most readily available disposal capacity. In some cases, this meant crossing state lines. "Waste does not recognize state lines," stated Bill Sells, federal relations director of the Environmental Industry Associations, which represents private waste companies. "It goes to the nearest logical facility that will satisfy the requirements of the generator."[116] Saint Louis and Kansas City, for example, delivered waste across state lines because disposal sites were closest there, and transport was not restricted by geographic barriers. Institutional changes, such as the regionalization of service delivery and the consolidation of the solid waste management industry, reduced the significance of viewing state lines as impenetrable boundaries. In the case of the latter, large, integrated private companies sought to develop cost-effective service by taking a regional approach to collection and disposal. As landfills became larger and fewer, available options for dumping waste narrowed. Of course, large waste companies often ship solid waste to their own landfills across state lines rather than to a rival's facility.[117]

The trend toward interstate transport of solid waste does not suggest that state governments lacked concern about the movement of such materials across state lines—or that regulations to control the transport were ignored. State and local governments throughout the country passed flow-control ordinances that designated where MSW was taken for processing, treatment, or disposal. Because of flow controls, certain designated facilities—such as incinerators—could maintain monopolies on local sources of solid waste or recoverable materials. Relationships between private solid waste companies and municipal authorities, therefore, were further strained by the flow control issue. In 1992, Congress directed the EPA to review the issue as it related to municipal solid waste management. The EPA found that thirty-five states, the District of Columbia, and the Virgin Islands directly authorized flow control. Four other states authorized it through various administrative mechanisms, such as a solid waste management plan.

The agency concluded that flow controls were efficient tools for solid waste management, but not essential for developing a new management capacity or achieving recycling goals. The courts thought differently in the 1990s (although the reverse was true for courts in the 1980s). For example, the U.S. Supreme Court ruled in May 1994 that flow-control ordinances—like one passed in Spokane County, Washington, to ensure a supply of waste for incineration—violated the Interstate Commerce Clause of the Constitution. The court's view was that the movement of interstate refuse was understood to be "commerce." The *Carbone* decision, as it is known, was one of several Supreme Court rulings on solid waste matters since 1990. For its part, Congress continued to examine the merits of flow control, while supporters hoped for a reversal of the 1994 decision.

Battle lines were even being drawn along state borders. In February 1999, inspectors from several states began an East Coast garbage truck inspection blitz. On one day alone, 417 trucks were stopped in Maryland, the District of Columbia, and New Jersey. Of those, 37 were ordered off the road because they might be carrying potential hazards. The legislative debate over flow control and the battles in the courts and in individual states made it unclear as to whether local governments would be able to rely on flow control to direct refuse to designated facilities in the twenty-first century.[118]

The changing role of the landfill in the disposal mix of the late twentieth century was at the heart of the debate over what to do with the relentless production of solid waste, from perplexing problems over land use to flow controls. Of the available alternative disposal methods, incineration fell in and out of favor in this period. It emerged as a disposal option that met relatively specific needs rather than completely replacing existing methods.[119]

In 1966, the total number of operating incinerators in the United States was 265, and by the end of 1974, one expert estimated that only 160 plants were in use. The growing skepticism about landfills and the energy crisis of the early 1970s, however, turned attention back to incineration. The renewed promise of raising steam to produce energy plus the development of refuse-derived fuel (RDF) shifted interest away from the question of disposal to the question of energy generation.[120]

Two types of waste combustion facilities—both with roots in the late nineteenth and early twentieth centuries—were in use in the 1970s. The first only burned waste to reduce the volume to be landfilled. For the most part, this type of plant was built before 1975 and was relatively small in size (with a capacity of less than one thousand tons per day). It was typically a mass-burn unit, combusting waste with very little or no preprocessing. The second type was the waste-to-energy facility (WTE) that burned the waste but utilized the resulting hot flue gases to produce steam for the production of electricity or for sale directly to users of thermal energy. This second type became popular in the mid-1970s as one of several alternatives to fossil fuels. The WTE plant could employ mass-burn techniques or could be a RDF facility. The RDF plant often turned combustible substances into pellets which burned hot and evenly and were generally combusted in specially designed boilers.[121]

Resource recovery plants got a modest boost in 1979 because of the Public Utility Regulatory Policies Act (PURPA). The law provided guaranteed markets for electricity sales, despite the high cost of generating it from garbage. The potential for energy generation made resource recovery systems intriguing to several cities, not because they filled a gap left by reduced landfill sites, but because they held out the possibility of some return on investment. Federal legislation, however, did not produce market forces that could give resource recovery plants a strong competitive role in the disposal arena. Sales of steam and RDF did not prove easy, though.[122] As fears over an energy crisis faded, incineration—including

WTE—faded as well. According to some experts, "recent WTE project cancellations suggest that the future of WTE is most difficult to predict of all MSW management options in the United States."[123]

The economic feasibility of incineration has been a chronic concern among engineers and city officials. Improvements in ash disposal practices and air pollution control equipment added substantial expense to construction and operation costs. Yet incineration disposal fees, considered quite high in the recent past, became increasingly competitive in parts of the United States, where tipping fees at landfills were also high. Economic feasibility, however, was not the only reason why incinerators had a difficult time competing with sanitary landfills. There were several time-specific problems related to, but not exclusively dependent on, cost, including the assumptions that: incinerators could not overcome their environmental liabilities; their value met only specific disposal needs; and the production of usable by-products did not outweigh other liabilities.

The cost of adding pollution controls was viewed as prohibitive over the years. And into the late 1980s, incinerators did not appear to be compatible with the emerging philosophy of resource reuse, despite the promise of recovering metals from incinerator residue. Mass burning, in particular, caused alarm over the production of highly toxic dioxin and furans, acid gas, and heavy metal emissions.[124] WTE plants were not exempt from this criticism. Opponents worldwide claimed that they were essentially unreliable—not able to produce energy in sufficient quantity to offset costs—and that they also produced dioxin emissions. Indeed, in its May 2000 report, the EPA for the first time stated that dioxin emitted from incinerators and other types of combustors caused cancer in humans.[125]

Much as in the case of landfills, grassroots groups and environmental justice advocates fought the siting of incinerators. Environmentalist Barry Commoner referred to the new generation of incinerators as "dioxin-producing factories," and through his leadership, citizens' groups in New York City managed to delay the development of an elaborate resource recovery program there.[126] Concerned Citizens of South-Central Los Angeles (CCSCLA) were successful in the 1980s in scuttling a plan to build a huge incinerator in a predominantly black, inner-city neighborhood. Battles over proposed incinerator facilities increased substantially during the mid- to late-1980s.[127] A 1990 Greenpeace report charged that communities with existing incinerators had populations of people of color 89 percent higher than the national average, and 60 percent higher in communities with proposed incinerators.[128]

Cities found themselves between the proverbial rock and a hard place. Some bureaucrats and industry leaders, however, did adopt methods to confront NIMBYism. The board of California Waste Management, for example, hired the political consulting firm, Cerrell Associates, who developed a plan for dealing with public opposition to waste-to-energy facilities. The Cerrell Report essentially outlined how to identify communities least able, or least likely, to protest. This was an effort to manage "community acceptance" of the waste facilities.[129]

By the mid-1980s, the environmental debate over disposal had pushed aside enthusiasm for waste-to-energy facilities. Promoters of WTE had to make the dual case that the system they embraced could generate sufficient energy to offset construction and operating costs, and to do it without serious threats to the air or groundwater. As the debate intensified, it was apparent that public standards were changing as well. What had been acceptable in the 1970s was no longer acceptable by the mid-1980s.[130] By one count, only 134 waste-reduction-only (38) and waste-to-energy (96) facilities were operating in the United States in 1988, although the total number had been reported as substantially higher. Of that number only 18 were new WTE plants starting up in that year.[131]

The decade of support that the EPA and the Department of Energy had provided for planning, research, demonstration, and commercialization of incinerators also was undermined by federal budget cuts and a shift in interest away from MSW to hazardous and toxic wastes as popularized by Love Canal (1978).[132] Furthermore, the nature of financing incineration systems was changing, which affected local management decisions. Some investment firms limited availability of bonds. The 1986 Tax Reform Act reduced the investment tax credit. It also was unclear if PURPA made much of an impact on WTE development, especially since the law was never popular among electric utilities and some municipalities suffered because of unfavorable long-term contracts.[133]

In the early 1990s, although somewhat apparent as early as 1988, conditions began to show a reverse in the trend toward less frequent use of incinerators by municipalities. Further constriction of sanitary landfills was a major reason. Somewhat of a counterweight were national standards for municipal waste combusters, promulgated by the EPA in 1991, which placed serious restrictions on existing incinerators. The 1990 Clean Air Act also limited incinerator emissions. The hope was that cleaner technologies ultimately would prevail to meet federal standards and would encourage more efficient energy production through new WTE units.

This point was made clear in the Department of Energy's 1991 National Energy Strategy forecasts.[134] Figures for the late 1990s and 2000 suggest yet another renewed interest in incinerators built upon shrinking landfills, the promise of cleaner operation, and the hope of energy recovery. *BioCycle* reported 119 incinerators in operation in 1999 and 122 in 2000.[135]

In recent years, promoters of WTE developed new strategies to increase its use. Efforts were made to introduce WTE facilities into communities through a merchant approach, whereby companies in the private sector identified communities that appeared to have certain needs that could be effectively met in a project-development approach. Among the existing merchant facilities was Wheelabrator Environmental Systems' plant operated in Millbury, Massachusetts, since 1988.[136] Although this strategy met with enthusiasm in a few communities, citizen groups—especially those associated with the environmental justice movement—feared that such an approach was just another form of "job blackmail," and thwarted efforts to site such facilities on financial, environmental, and social grounds.[137]

An alternative approach for communities interested in changing their disposal options was to issue a request for proposal (RFP) for a new solid waste management facility. Rather than have the private firm take the initiative, communities could solicit offers from a variety of companies stating their specific requirements. Local officials then could evaluate the proposals and determine which, if any, would serve their needs.[138]

The continuing debate over incineration was a classic confrontation between those with faith in technology to overcome social and economic problems versus those who were suspicious of the technical fix. In theory, waste-to-energy offered two major advantages: an efficient means to reduce the volume of wastes and the production of a valuable by-product. Proponents did not deny the potential environmental hazards associated with burning waste, but were inclined to believe that while older facilities usually had failed to meet new emissions standards, newer WTE facilities could stand the performance test.[139]

The long-standing debate over sanitary landfills versus incineration took a new turn in the 1980s. Recycling emerged as a widely accepted alternative disposal strategy. Once regarded as a grassroots approach to source reduction and a remonstration against over consumption, the solid waste management community took recycling seriously as a way to stretch existing resources and to generate less waste that would otherwise be dumped or burned. As an Oregon engineer told the American Institute of Chemical Engineers, "[O]nce ridiculed as an ineffective hobby of environmentalists," recycling is now regarded as "an essential component of solid-waste management and a cost-effective way to reduce dependence on landfills."[140]

The new commitment to recycling, however, raised numerous questions. What kind of incentives should be used to get the compliance of households, businesses, haulers, and manufacturers? What about mandatory recycling laws? Should there be a clear policy of government procurement of recycled products? How much attention should be paid to recycling literacy? And the biggest question of all: Can markets be found and developed to take the increasing volume of recyclables?[141]

The groundwork for the resurgence of recycling in the 1980s began in the late 1960s, although forms of recycling and reuse were present earlier. In the 1960s and early 1970s, drop-off centers for recyclables could be found in a few cities, such as Berkeley, California. In 1968, Madison, Wisconsin, may have been the first city to begin curbside recycling of newspapers. In that same year, the aluminum industry in the United States began to recycle discarded aluminum products.[142] In 1971, Oregon was the first state to enact a bottle bill, placing a five-cent refund on every beer and soft-drink container unless it could be reused by more than one bottler (in which case it would have a two-cent refund value) and outlawed pull-tab cans.[143]

Recycling centers in several communities experienced modest success, but the inconvenience of dropping off materials at a centralized location severely limited participation. Curbside collection programs, while not inexpensive to run, have been the most effective method of gathering recyclables, achieving the highest diversion rates from the waste stream of any option available. From the modest beginnings in Madison and a few other locations, citywide and regularly routed curbside programs increased to 218 by 1978. (Of these,178 collected newspapers only.) They were located primarily in California and in the Northeast.[144] In 1989 it was estimated that there were 1,600 full scale and pilot curbside recycling programs in the United States, with participation rates estimated at 49 to 92 percent.

As the 1990s began, recycling was a growth industry, and by mid-decade markets for recovered materials were on the rise. By 1991, the number soared to 2,711 programs; forty-seven of the fifty largest cities had such programs in 1992. With the increase in curbside programs came the development of materials recovery facilities (MRFs) operating in twenty-four states in the early 1990s. EPA estimates for 1996 fix curbside collec-

TABLE 15

Generation and Recovery of Materials, 1980–2000 (in millions of tons)

	1980	*1990*	*1995*	*2000*
Waste generated, total	151.5	205.2	211.4	231.9
Paper/paperboard	54.7	72.7	81.7	86.7
Ferrous metals	11.6	12.6	11.6	13.5
Aluminum	1.8	2.8	3.0	3.2
Other nonferrous metals	1.1	1.1	1.3	1.4
Glass	15.0	13.1	12.8	12.8
Plastics	7.9	17.1	18.9	24.7
Yard wastes	27.5	35	29.7	27.7
Other wastes	31.9	50.7	52.4	57.1
Materials recovered, total	14.5	33.6	54.9	69.9
Paper/paperboard	11.9	20.2	32.7	39.4
Ferrous metals	0.4	2.6	4.1	4.6
Aluminum	0.3	1.0	0.9	0.9
Other nonferrous metals	0.5	0.7	0.8	0.9
Glass	0.8	2.6	3.1	2.9
Plastics		0.4	1.0	1.3
Yard wastes		4.2	9.0	15.8
Other wastes	0.6	1.8	3.2	16.5

Source: U.S. Department of Commerce, U.S. Census Bureau, *Statistical Abstract of the United States: 2001*, p. 218; U.S. Environmental Protection Agency, Office of Solid Waste, *Municipal Solid Waste in the United States: 2000 Facts and Figures, Executive Summary* (June 2002), p. 7, Internet, www.epa.gov.

tion programs at more than 7,000, serving about half of the U.S. population. Estimates for 1999 were 9,349 with 139, 826 (or 52 percent) served.[145] A 1993 Congressional Research Service Report stated that "In the urban states of the Northeast and Pacific Coast . . . curbside programs are now so common that areas without them are becoming the exception rather than the rule." By later in the decade, this was also true for other parts of the country with large urban populations.[146]

A major goal for most communities and the nation in general was to increase the recycling rate, which stood at 10 percent in the late 1980s. The EPA's 1988 draft report of *The Solid Waste Dilemma: An Agenda for Action* called for a national recycling goal of 25 percent by 1992. In the late 1990s, the EPA raised the goal to at least 35 percent of MSW by 2005, and it called for reducing the generation of solid waste to 4.3 pounds per capita per day. In 1998, thirty states estimated that their recycling and/or waste reduction rates were 25 percent or greater. Some communities, however, continue to claim achievements beyond that national standard, and conflicting claims have led to ongoing debate over the extent of success achieved in raising the standard.[147]

At the very least, implementing recycling programs put municipal policymakers on the side of conservation of resources and gave concerned citizens a way to participate in confronting the solid waste dilemma in specific, and environmental problems in general. The perception of recycling as an answer to the nation's disposal problems gave it strong momentum in the 1980s. Its linkage to waste reduction and waste minimization as a means to conserve resources and reduce pollution attracted new followers. A powerful argument for recycling was that the environment could be protected from many pollutants by reducing the amount of waste that is produced, thus reducing the need to find ways to treat the unwanted residues.[148]

As figure 7 shows, however, relative increases in the amounts of waste recovered have yet to keep up with rates of generation. But figure 8 also shows that recycling and composting rates have been steadily increasing since the 1960s.[149]

Skeptics of the new faith in recycling—regarded as "anti-recyclers" by proponents—raised concern about the cost of recycling, the lack of stable markets, and what they believe to be inaccurate portrayal of the success of many recycling programs.[150] Cautionary notes even appended many enthusiastic reports, and it was clear that rates of recycling in the United States, especially for paper and glass, were relatively low in comparison with other industrialized nations.[151] Economic hard times at the end of the twentieth century, after a record-setting period of prosperity in the 1990s, adversely affected recycling like many other areas of the economy and provided additional fuel for the debate.[152]

Critics of recycling, especially from conservative think tanks and industry trade groups, believe the notion of a garbage crisis is vastly overstated, are not harsh critics of landfilling nor incineration, and generally resist the notion of a managed economy in which recycling programs are subsidized and where markets should achieve the most efficient level of recycling and reuse. Proponents of recycling bristle at the "economics only" argument, asserting that recycling makes environmental sense by also reinforcing sound resource use practices.[153]

Leading the backlash was writer John Tierney in an article appearing in the June 30, 1996, issue of the *New York Times Magazine*, and reprinted

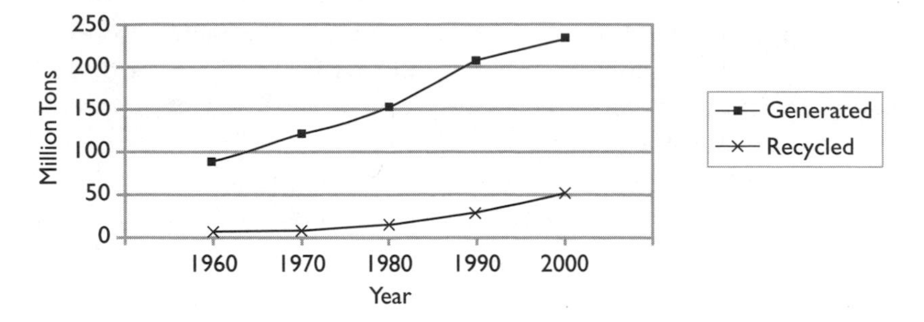

FIGURE 7. Materials Generated and Recovered by Recycling, 1960–2000.
Source: U.S. Environmental Protection Agency, Office of Solid Waste, *Municipal Solid Waste in the United States: 2000 Facts and Figures, Executive Summary* (June 2002), p. 2.

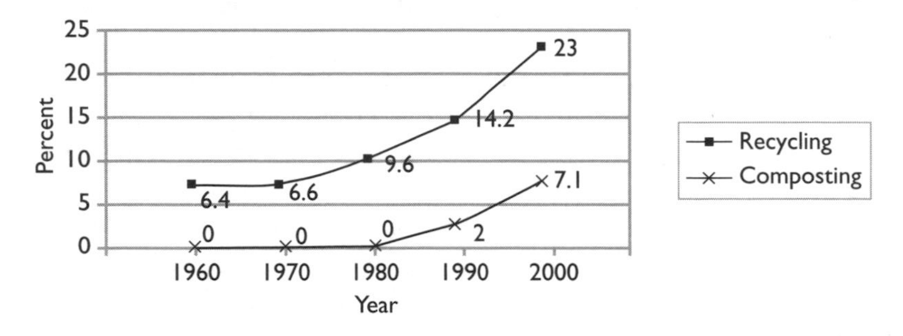

FIGURE 8. Waste Recovery by Recycling and Composting, 1960–2000.
Source: U.S. Environmental Protection Agency, Office of Solid Waste, *Municipal Solid Waste in the United States: 2000 Facts and Figures, Executive Summary* (June 2002), p. 2.

in more than twenty newspapers. In "Recycling Is Garbage," Tierney wrote that as a result of the *Mobro* incident in 1987, "The citizens of the richest society in the history of the planet suddenly became obsessed with personally handling their own waste." He argued that as a result of concern about the filling up of landfills, Americans saw recycling as their only option, but failed to recognize that the crisis of 1987 was a false alarm and that mandatory recycling programs "aren't good for posterity." In offering short-term benefits to a few groups—politicians, public relations consultants, environmental organizations, and waste-handling corporations—recycling diverted money from "genuine social and environmental problems." He concluded, "Americans have embraced recycling as a Transcendental experience, an act of moral redemption."[154]

Supporters of recycling were outraged, and the *New York Times Magazine* was inundated with letters. They argued that Tierney was selective in his use of data. He characterized the recycling craze as impulsive, rather than as part of a longstanding concern about waste disposal. He criticized recycling supporters for not being grounded in the realities of cost-benefit, but did not provide careful documentation of recycling's failures and successes. He depicted advocates of recycling as representing a clearly identifiable social or political fringe with a single ambition of doing something to relieve guilt. And he painted the history of recycling as if it already had played itself out, and thus its results could be definitively evaluated. Failing to view recycling as a burgeoning, rather than a mature, disposal option weakened Tierney's critique. However, just as the sanitary landfill came to be viewed as a disposal panacea, the euphoria over recycling needed to be tempered with a strong record of tangible results in its future. Tierney may have gone too far, but recycling was not above criticism, any more than the sanitary landfill, only to be later discredited.[155] Recycling, therefore, was more than the core of an argument over waste disposal methods, as the centerpiece in a debate over lifestyle and political, economic, and social values.

In recent years, as public attention to recycling increased, interest in littering has subsided. In the 1970s, however, concern over a litter epidemic had inspired—temporarily at least—a significant civic response rivaling other aspects of the refuse problem. Cities posted fines for littering along streets and highways; radio and television public-service announcements promoted antilittering campaigns; civic groups underwrote the cost of trash cans and placed them at strategic points throughout the cities and along roadsides; and old-fashioned cleanup campaigns reappeared in many areas. New York City initiated "Project Scorecard," a program designed to measure the cleanliness of the city's streets and sidewalks and to evaluate the litter problem. One of the most innovative ideas was the bottle bill. Beverage and container companies and other opponents fought this approach to the litter problem, but the idea first started in Oregon in 1972 soon spread to other states. In 1975, although few states had actual bottle laws, more than one thousand pieces of restrictive container legislation had been introduced at state and local levels throughout the country.[156]

The best-known national campaign against littering was directed by Keep America Beautiful, Inc. (KAB). Founded in 1953 as an antilitter association, KAB grew into a national organization that promoted several programs related to the environment. It has had strong links with the beverage container industry, and proved to be more interested in emphasizing individual and community responsibility for littering than in directing a critical eye at packaging products. KAB made "litterbug" a household word and, in so doing, brought the litter problem to public attention. In the 1970s, the litterbug campaign was dropped because of its limited and almost juvenile appeal and was replaced by a more sophisticated and comprehensive program. The program's main thrust was to inspire action on the local level through its Clean Community System (CCS). By the spring of 1980, there were 181 CCS communities in the United States, representing a population of over 21 million in thirty-five states. KAB employs a behavioral-based systems approach to the litter problem, designed to educate citizens about the sources of litter and to encourage them to promote effective antilitter ordinances and efficient solid waste management practices. KAB also conducted national and international campaigns. It produced television commercials featuring a "Crying Indian" (Iron Eyes Cody), who surveyed the damage done to America by littering, shedding a tear for the sad state of affairs: "People Start Pollution, People Can Stop It." The message was a play on the "Indian as a conservationist" theme, and although it was almost as elementary as the litterbug campaign, it made its case to millions of television viewers.[157]

By the standards of the 1990s and into the twenty-first century, the antilitter campaign seemed almost frivolous in dealing with the growing concern over a garbage crisis. Yet it was one more way to bring the issue of solid waste down to a personal level. In a world where environmental problems seemed beyond individual control, garbage has always been real and immediate. If solving the so-called garbage crisis too often seemed bewildering for everyday citizens, acknowledging solid waste as a nuisance, risk, or burden in some visceral way was never beyond human imagination.

CONCLUSION

The refuse problem in urban America between 1880 and 2000 had many dimensions. It was, foremost, an environmental problem of no less importance than air, water, or noise pollution. It also stimulated vigorous debate over the extent and limits of individual versus community responsibility. It became a significant focal point for municipal-reform efforts. Refuse, the seemingly mundane and oft-neglected residue of human activity, came into the public consciousness significantly during the late nineteenth century and raised several uncomfortable questions about health, aesthetics, and the quality of urban life.

The rampant increase in solid wastes was a central feature of environmental problems in American cities in the late nineteenth and early twentieth centuries. For the first time in the nation's history, cities confronted garbage and rubbish in quantities far beyond the capacities of traditional collection and disposal practices. More important, the perception of the problem also began to change. Refuse had been considered a nuisance that brought temporary discomfort or inconvenience rather than a serious environmental danger. When the waste was removed from the range of human senses, it ceased to be an eyesore, a bad odor, an obstacle to traffic, or a bothersome annoyance. Few individuals cared what became of the waste after it had been discarded. An out of sight, out of mind viewpoint, however, was inappropriate and impractical in an era transformed by urban growth and economic change. City life especially made individual disregard obsolete. As the mounds of solid waste piled up on every street corner and in every alley of the nation's cities, urbanites were obliged to address the problem. Beginning in the 1880s, city leaders and many of their constituents began to take the garbage nuisance more seriously. Individual acts of neglect could become communitywide dangers, especially after it became common knowledge that there was a direct relationship between disease and waste. With the realization that inadequate or inappropriate collection and disposal threatened the physical city as well as the public health came the awareness that refuse was a serious environmental problem. A change in attitude was a necessary first step in seeking a solution.

The threat to health undercut the notion of refuse as simply a nuisance. This was also true for other kinds of pollution—tainted water supplies, billowing coal smoke, unnecessary noise—that accompanied rapid urbanization and industrialization. In the effort to deal with refuse, environmental sanitation came to be regarded as the most effective safeguard against the ravages of communicable diseases. Despite its flawed theoretical base, environmental sanitation mobilized urbanites to deal forthrightly with garbage and rubbish in the late nineteenth century. The perception of refuse as a danger to health led the way to an examination of other unwanted properties of that blight which might further undermine urban living conditions or the surrounding physical environment. Existing methods of disposal, such as sea dumping, filling with untreated wastes, swine feeding, and open burning came under scrutiny. Experts like George Waring experimented with collection methods such as source separation to try to improve disposal practices. During this period, some effort was made to link the refuse problem with various other physical and social phenomena of the emerging urban-industrial age. Overcrowded residences—especially in tenement districts—and human and transportation congestion in business districts were viewed as conditions obviously contributing to the refuse problem. Sometimes, these observations were carried too far, especially when foreign-born or working-class people were blamed for the problem or when "cleanliness" was simplistically equated with "civilization." There was some peripheral concern about a connection between refuse and resource conservation. The experiments with the reduction process, recycling programs, and sanitary landfills came out of that concern. Austin Bierbower, in a 1907 issue of *Overland Monthly,* made one of the clearest statements on this subject, albeit revealing some of his own stereotypes: "Nowhere in the world is there such a waste of material as in this country. In our eagerness to get the most results from our resources, and to get them quickly, we destroy perhaps as much as we use. Americans have not learned to save; and their wastefulness imperils their future. Our resources are fast giving out, and the next problem will be to make them last. In passing the alleys of an American city, a foreigner marvels at the quantity of produce in the garbage boxes. The thrifty Germans would have saved this; and there is no excuse for letting it spoil in these days of cold storage and quick transportation."[1]

This glimmer of conservationist thinking was not central to the environmentalism of refuse collection and disposal in the late nineteenth and early twentieth centuries, however. Primary attention was given to eradication of waste rather than to the origins of that waste. Reflections on the broader implications of the refuse problem, resource conservation in particular, were ignored mainly because an immediate, practical solution to waste disposal seemed imperative to prevent the ravages of disease and the encumbrance of mounting piles of rejectamenta. Despite the limitations on their environmental perspective, contemporaries recognized the risk of land pollution and sought to control it.

In the years since the 1880s, the refuse problem and the perception of refuse as an environmental hazard underwent substantial evolution. The changing composition of waste—synthetics, new chemical compounds, less organic material, more paper products—presented different challenges and required more appropriate methods of collection and disposal. The scale of metropolitan growth and the regional and national significance of cities placed even greater significance on controlling the problems associated with solid wastes. Widespread consumerism and the rise of a throwaway culture further heightened the challenges and repercussions of third pollution in modern America.

Some valuable lessons were drawn from the enduring experience with the refuse problem. The coining of the term "third pollution" attests to the significance accorded to the refuse problem amongst general environmental conditions of the city. The study of urban ecology further suggests that pollution must be understood as part of the urban process. As demographic, geographic, climatic, economic, political, and social conditions change over time, pollution problems such as solid waste also change. Density and distribution of population, geographic variables, fluctuations in weather, and ratios of residential to commercial establishments are more often considered before new collection and disposal systems are implemented. This is not to say that municipal officials design sanitation services based exclusively on environmental factors, only that they no longer neglect such factors in making their assessments. Unfortunately, the process of siting disposal facilities is often more political than scientific as environmental justice advocates and others rightly claim.

As we acquire a more holistic view of the urban environment in specific and the larger physical world in general, the front end of the refuse problem, namely waste generation, becomes more critical. Finding ways to manage the flow of garbage, trash, and discards of all types once they enter the waste stream is only half the battle. Nineteenth-century Americans did little more than deal with their environmental problems as they became evident. It also was typical to view multiple environmental problems distinct from one another. Refuse, noise, smoke, and dirty water appeared to have separate causes, requiring separate and specific solutions. Little attention was given—or, in some respects, is being given today—to limiting the use of virgin materials or conserving resources in any sustainable manner. The sheer volume of wastes, of course, has made it difficult to do more than cope with the daily deluge, and recent efforts at recycling only begin to address the complex issue of waste reduction.

Successfully managing with the refuse problem required more than a change in attitude about the threat of pollution. Implementing programs to eradicate waste had to be devised. Yet it was extremely important that refuse came to be perceived as an environmental problem that threatened the entire community. This awareness—plus the desire to enhance the power of city governments—enabled urbanites to consider collection and disposal as municipal responsibilities. The scale of urbanization in the late nineteenth century made private action increasingly impractical, or at the very least city authorities were unwilling to strengthen their supervision of private companies. A community perspective may not have been necessary for some services, but for refuse, sewerage, the acquisition of pure water, and so forth, any other approach at the time was considered unreliable or unworkable. Consequently the development of municipal waste services was not a radical notion but a pragmatic one. Problems that were most threatening (such as fire) or solutions that offered the most immediate return (such as transportation) received the most direct attention.[2]

Recognized first as a health hazard and later as a technical problem, refuse—it was believed—could not be left to individuals or franchisees to dispense with, but must be the responsibility of those trained in sanitation, public health, and engineering. Individuals with such training had already been attracted to municipal-government service to deal with health problems, sewerage, and water supplies. These activities had effectively linked technical experts with municipal authorities in efforts to protect some of the basic features of the urban physical environment (more elusive forms of pollution, such as smoke and noise, came under municipal authority later and in a much less successful way).[3]

Transforming refuse collection and disposal into municipal functions also grew more generally out of cities' efforts to establish "home rule." Municipal authorities actively discouraged state legislatures from interfering in sanitation services and especially tried to gain or maintain control over city health departments. Criticism of contracted services also represented an attempt by city officials to head off any challenge to their growing authority. The controversy over contracted versus municipally operated programs was secondarily a debate over the operation of a communitywide system of collection and disposal. Sometimes the question of control (or responsibility) was confused with the question of operation, but there was little doubt that cities would ignore the refuse problem any longer.

Colonel Waring introduced modern refuse management to the United States, drawing on European precedents, past American practices and experiments, and his own experiences as a sewerage and drainage engineer. Waring's emphasis was on developing a total sanitation system that took into account social, political, and economic considerations, as well as technical and organizational improvements. None of the individual parts of his program were particularly new or innovative—with the possible exception of the White Wings and the Juvenile Street Cleaning League, which were promotional masterpieces. His real accomplishment was uniting a commitment to municipal responsibility with the application of technical and organizational expertise, the rallying of civic action, and the cultivation of good public relations. Waring's effective coalition of municipal leaders, technical experts, and activists demonstrated that a community-wide sanitation program could work. That was his legacy to the twentieth-century city.

Waring's example led to the institutionalization of more professional street-cleaning and refuse services throughout the nation's cities in the twentieth century. Street-cleaning practices benefited from improved pavements and more sophisticated sweeping and flushing devices. In street cleaning and refuse collection and disposal, however, improved administrative and organizational techniques were as important as, if not more important than, technical advances. These techniques included better scheduling of tasks to be completed, better routing of workers and vehicles, and more competent cost keeping and budgeting methods. The gathering of statistics about past and present practices, especially by sanitary engineers, added to a more comprehensive knowledge of the waste problem, which also contributed to more successful programs. The emphasis on "the one best way" of doing things was meant not to pay homage to the fashionable efficiency ethic of the day but to solve a pressing community problem. The efforts of Waring and the promotion of home rule and municipal socialism guaranteed that a mechanism for dealing with the collection and disposal of refuse was firmly implanted in the urban American bureaucracy in the late nineteenth and early twentieth centuries. Yet the success of a program of solid waste management depended on more than a new department of street cleaning or engineering. Street cleaning, collection, and disposal are expensive services. The level of funding determined the choice of disposal methods, the quality and extent of collection practices, and the frequency of cleaning. The degree to which a city maintained a financial commitment to its sanitation services determined its effectiveness.

The practices established in the early twentieth century have been remarkably influential. Collection and disposal practices have been modified by motorized vehicles and other technical innovations, but the administrative and organizational functions of modern public works departments are largely refinements of past practice. Of course, questions of responsibility and control have been and continue to be adjusted to meet the demands of the metropolis and regional urban development. Jurisdictional boundaries for providing service may be set along county or district lines rather than restricted to city limits. The magnitude of the task often requires a greater degree of cooperation among cities, counties, and a variety of regional authorities. These changes in jurisdiction have led to a rethinking about which services municipal governments should provide and which might be best accomplished through privatization.

The role of the private sector in collection and disposal has grown markedly since the 1950s. While municipal control over the disposition of municipal solid waste has not been completely eroded—particularly in the major cities—the sense of crucialness of these public services has changed. Along with the difficulties created by competing governmental entities and an array of complex regulations, budgetary considerations, growing demands for social services, and the rising influence of solid waste conglomerates have altered the climate of municipal control. The role of the federal government in solid waste management also has become more proactive in recent years as the implementation of Subtitle D, several environmental regulations, the role of the courts in dealing with flow control, and the modest commitment to environmental justice principles suggest.

Sanitation reformers of the late nineteenth and early twentieth centuries succeeded in convincing city officials and the public that the refuse problem could no longer be ignored (a major accomplishment of civic reform) and that it could be controlled through the mechanism of municipal government (a major accomplishment of Waring and other sanitary engineers). Operating primarily in the public realm outside municipal government, civic reformers encouraged compliance with sanitation ordinances, promoted citizen involvement in cleanup projects, and lobbied for better collection and disposal methods. Working primarily away from the public gaze within municipal government, sanitary engineers set about to shape a workable refuse-management program. Civic reformers articulated the problem in lay terms; sanitary engineers focused on finding solutions. Having an external and an internal sphere, both a popular and a professional dimension, refuse reform had an excellent chance for success. Refuse reform was particularly effective when the two dimensions were united, as in the ways of Colonel Waring. It was least effective when the two failed to achieve some link.

Sanitation reform provided two major legacies. It contributed to the institutionalization of refuse management, and it offered a departure point for a more sophisticated environmentalism among American city dwellers. Much has been made of the contributions of the conservation movement of the early twentieth century to the ecology movement of the 1960s and beyond. Too little has been made of the contributions of sanitation reform in particular and urban environmental reform in general to modern environmentalist thought and action. Certainly there were limits to urban environmental reform at the turn of the century. There were no omnibus environmental organizations, only specialized groups interested in particular forms of pollution or specific environmental issues—antismoke groups, noise-abatement leagues, sanitation groups, and so on. Few people recognized the important interrelationships among all forms of pollution or any other physical threats to the city. Even city planners, who seemed best qualified to merge the myriad reform interests into a general program for change, contributed little of substance.[4] Nevertheless, contemporary environmental reformers rejected the notion that pollution of any kind was an unavoidable by-product of industrialization, that a tradeoff had to be made between a clean environment and material progress. There was never a rejection of the economic benefits of industrialization and urbanization, but an assertion that material progress had meaning only if concern for the quality of the physical surroundings was not abandoned in the process. Refuse reformers in particular did not question an economic system and a society that generated the most abundant waste materials in the world. They rarely made the connection between affluence and quantity of refuse. Instead, they contended that the physical environment should not be sacrificed to high productivity and consumerism. This was, of course, a compromise, but a compromise acceptable to a great number of people. A more incisive ecological perspective, one that would be less conciliatory to the rampant exploitation of natural resources, would have to find its place in a later period of American history.

The modest environmentalist goals of sanitation and urban environmental reform also encompassed a faith in professional expertise, technique, and scientific method in solving pollution problems. A society that grew powerful because of its mastery of machines, acquisition of vast resources, and massive production of goods was not likely to abandon a faith in and a dependence on technology and scientific method to help curb the excesses of those activities. Yet the civic phases of refuse reform, for instance, were not totally dependent on the technical fix or scientific expertise; they were accompanied by aesthetic considerations and a humanism that demonstrated great faith in the ability of citizens to rally to the civic call. American urbanites had not totally abandoned the natural world for a manufactured one; they simply tried to adapt to onrushing economic change without accepting the inevitability of defilement of the physical world. The echoes of this brand of environmentalism can be heard in more modern times. The refuse problem in American cities posed a challenge to urbanites that could not be resolved in a generation or a century. Credit should be given to civic reformers, engineers, and city officials who confronted the mounds of waste and sought practical solutions. The experience of the late nineteenth and early twentieth centuries provided guidelines for effective methods of collecting and disposing of waste. It also offered broader guidelines for comprehension of the intricacies of the urban environment.

Nonetheless, refuse reform was not an ongoing feature of twentieth-century municipal life. One would be hard-pressed to identify many lively civic organizations in the 1920s, 1930s, 1940s, or 1950s comparable to Progressive Era groups. This is not to say that the waste problem was ignored, but technical solutions through the efforts of sanitary engineers—soon to become environmental engineers—seemed to attract increasing credibility through mid-century. By the late twentieth century, however, environmental voices rose more loudly to confront solid and hazardous wastes, manifest in a revitalized demand for recycling and reuse and the criticisms over siting waste facilities by the environmental justice movement. Embedded in the call for new recycling programs was an intimacy between the average citizen and the environment—the idea that personal action could make a difference. In a complex world where global warming and ozone depletion seemed beyond the control of the individual, recycling offered a way to participate in environmental reform, if only in a modest way. On the other end of the spectrum, recycling and reuse of materials was linked to the notion of sustainable development—a change in habits and mindset that could stretch resources without harming the world around us. For its part, environmental justice advocates questioned an environmental movement that appeared to give too little attention to everyday urban problems, of which mounting piles of solid and hazardous wastes were important features. Supporters of environmental justice also raised the uncomfortable notion that even in a society that over the years had established processes and procedures for collecting and disposing of its waste, some people paid more dearly than others for third pollution. Refuse reform from the nineteenth century to the present underwent vast changes, but continued as a reminder of a lingering problem that at best had been managed but not resolved.

Over the years and around the world, solid waste has been plentiful, unwieldy, and polluting. The United States, unfortunately, has been and remains a leader in many categories associated with municipal solid waste generation, although the waste problem is hardly restricted to one country. While composition of solid waste varies considerably throughout Europe, organic material and paper currently dominate the waste stream. These categories account for between 50 and 80 percent of residential waste materials. In Eastern Europe, organic materials are more plentiful than glass, plastics, and metals. Overall figures suggest substantially less use of packaging material in Europe as a whole than in the U.S.

Affluence is a strong dictator of waste volume and variety throughout the world. For example, in high-income economies in Israel, Saudi Arabia, and the United Arab Emirates, abandoned automobiles, furniture, and packaging are openly discarded. In Asia, paper and plastics are generally greatest in Tokyo and Singapore, while very low in Beijing and Shanghai (due in part to recovery and recycling). On the Indian subcontinent, organic and inert matter dominates waste disposal. The same is true for African and Latin American cities, where waste is high in organic material.

While the United States has sought to implement an integrated solid waste management system, western Europe leads the world in such an endeavor. Governments in all western European countries are required to design systems around integrated models with waste prevention at the core. While the public sector has been at the center of these programs for years, private companies carry out waste management services. Even in eastern Europe, private companies are making headway. In East Asia and the Pacific, women often manage garbage in the household, pay for collection service, separate recyclables, and sell items to private waste collectors. In many developing nations, collection of waste still remains an imposing task for individuals, governments, and private companies. In some cities, such as Katmandu, Nepal, there is no formal waste collection service of any kind. In Mexico City, the national government controls collection and disposal arrangements, and does not allow private contractors to operate. Throughout Latin America, collection coverage is reasonably good in large cities, although squatter settlements rarely receive adequate collection. Like Europe, cities in industrialized countries of Asia, Australia, New Zealand, Hong Kong, Japan, and Singapore have waste collection that is mechanized and capital-intensive. In the poorest countries, collection rates may not exceed 50 percent and may not extend to the poor.

Disposal poses many of the same problems throughout the world as in the United States. There is a clear split between northern and southern Europe. In some northern European countries, landfill practices parallel current U.S. experiences with approximately half of the waste finding its way into trenches. In Greece, Spain, Hungary, and Poland, virtually all collected waste goes into the ground. Many of the landfills in Europe are of the small, uncontrolled municipal type, but efforts are being made to shift toward larger regional varieties. Unlike in the U.S., NIMBYism rarely influences the siting of landfills. Landfilling has been the cheapest and most typical form of disposal in East, South, and West Asia, much of Africa, and in the Pacific, but some countries have seen sharply rising costs in recent years. Landfilling is on the rise in Latin America and the Caribbean, especially near large cities. Ocean dumping, although banned or restricted in many places, is still common. Incineration, furthermore, has a very checkered history internationally.

The U.S. experience with recycling, while vastly changed from years past, falls short of recycling and recovery efforts in other parts of the industrialized world. Germany and Denmark, in particular, have aggressive recycling policies. Denmark, for example, recycles about 65 percent of its wastes. One of the unique aspects of recovery of materials in western Europe is the pervasive idea of "producer responsibility" for proper disposal of packaging and other products. In some western European countries, packaging reduction goals had been set at 75 percent by 2002. Recycling and recovery in other parts of Europe and the world is much more uneven. Throughout Latin America and the Caribbean, materials recovery is extensive with recycling programs in all large cities and most moderate-size communities. In Asia, a large portion of household organic waste is fed to animals. However, the large composting plants once so prevalent in developing countries in Asia—including those pioneered in India—are out of use or not working at full capacity. In many countries in the developing world, informal waste picking—like in nineteenth-century New York—is widespread with many informal networks of pickers, buyers, traders, and recyclers in the place of formal public systems or private companies. Many of these networks arise as necessities in areas of poorly paid or unemployed people and in areas where resources are scarce.[5]

The solid waste problem, therefore, is part of life—ancient and modern. Success in managing solid waste, let alone reducing it, can be measured in many ways. The experience in the United States grows out of its own history and culture, as much as out of the inherent reality that garbage will always be with us. Underlying this mundane subject are many perceptions of our urban environment and many questions about things that we value (what we choose to keep) and what we do not (what we throw away).

NOTES

Preface

1. Daniel C. Walsh, "Urban Residential Refuse Composition and Generation Rates for the 20th Century," *Environmental Science and Technology* 36 (November 2002): 4936–42. See also Kirk Johnson, "Finding Surprises in the Garbage," *New York Times*, November 22, 2002.

2. Michael Thompson, *Rubbish Theory: The Creation and Destruction of Value* (New York: Oxford University Press, 1979), 11.

3. See Martin V. Melosi, "The Fresno Sanitary Landfill in an American Cultural Context," *Public Historian* 24 (Summer 2002): 17–35.

Introduction

1. Lewis Mumford, *The City in History: Its Origins, Its Transformation, and Its Prospects* (New York: Harcourt, Brace and World, 1961), 75. See also E. S. Savas, *The Organization and Efficiency of Solid Waste Collection* (Lexington, MA: D.C. Heath and Co., 1977), 11–13; Charles G. Gunnerson, "Debris Accumulation in Ancient and Modern Cities," *Journal of the Environmental Engineering Division, ASCE* 99 (June 1973): 229–43.

2. George Rosen, *A History of Public Health* (New York: MD Publication, 1938), 25–27; Fred B. Welch, "History of Sanitation" (paper presented at the first general meeting of the Wisconsin Section of the National Association of Sanitarians, Inc., Milwaukee, WI, December 1944), 39; Savas, *Organization and Efficiency*, 11–12; Edward S. Hopkins, ed., *Elements of Sanitation* (New York: D. Van Nostrand Co., Inc., 1939), 104; Frederick Charles Krepp, *The Sewage Question* (London: Longmans, Green, and Co., 1867), 7. For information about early street paving, see "The Early History of Street Paving and Street Cleaning," *Engineering News* 36 (July 1896): 47–48.

3. Benjamin Freedman, *Sanitarian's Handbook: Theory and Administrative Practice* (New Orleans: Peerless Publishing Co., 1957), 2; Savas, *Organization and Efficiency*, 12. See also "Early History," 47; Welch, "History of Sanitation," 39.

4. Savas, *Organization and Efficiency*, 13; Mumford, *City in History*, 130.

5. Welch, "History of Sanitation," 41; Savas, *Organization and Efficiency*, 13–14; Hopkins, *Elements of Sanitation*, 104; Rosen, *History of Public Health*, 48. See also Katie Kelly, *Garbage: The History and Future of Garbage in America* (New York: Saturday Review Press, 1973), 16–18.

6. Rosen, *History of Public Health,* 56–57.

7. "Early History," 47–48; Rosen, *History of Public Health,* 57–58.

8. Savas, *Organization and Efficiency,* 14–15; Mumford, *City in History,* 290–92; "Early History," 48.

9. Rosen, *History of Public Health,* 122–23; Savas, *Organization and Efficiency,* 16–17; Charles Singer, et al., eds., *The Industrial Revolution, 1750 to 1850,* vol. 4 of *A History of Technology* (New York: Oxford University Press, 1958), 505–6; Kelly, *Garbage,* 20.

10. For bibliography on the Industrial Revolution, see Peter N. Stearns and John H. Hinshaw, *The ABC-CLIO World History Companion to the Industrial Revolution* (Santa Barbara, CA: ABC-CLIO, 1996), 299–310.

11. Eric C. Lampard, "The Social Impact of the Industrial Revolution," in *The Emergence of Modern Industrial Society, Earliest Times to 1900,* ed. Melvin Kranzberg and Carroll W. Pursell Jr., vol. 1 of *Technology in Western Civilization* (New York: Oxford University Press, 1967), 305.

12. Mumford, *City in History,* 447.

13. Asa Briggs, *Victorian Cities* (New York: Harper and Row, 1963), 18.

14. Eric E. Lampard, "The Urbanizing World," in *The Victorian City: Images and Realities,* ed. H. J. Dyos and Michael Wolff, (London: Routledge and Keegan Paul, 1973), 1:4, 10–13, 21–22; H. J. Habakkuk and M. Postan, *The Industrial Revolutions and After: Incomes, Population, and Technological Change,* vol. 6 of *The Cambridge Economic History of Europe* (Cambridge: Cambridge University Press, 1966), 274; Mumford, *City in History,* 461–65.

15. Habakkuk and Postan, *Industrial Revolutions and After,* 276.

16. Briggs, *Victorian Cities,* 20, 144; Lampard, "Urbanizing World," 19–22; Mumford, *City in History,* 462–63, 476–79.

17. Lampard, " Urbanizing World," 43–45; Lampard, "Social Impact," 317.

18. R. M. Hartwell, "The Service Revolution: The Growth of Services in Modern Economy," in *The Industrial Revolution, 1700–1914,* ed. Carlo M. Cipolla, vol. 33 of *The Fontana Economic History of Europe* (London: Harvester Press, 1976), 364–67; Savas, *Organization and Efficiency,* 19–20. See also Bill Luckin, *Pollution and Control: A Social History of the Thames in the Nineteenth Century* (Bristol: Adam Hilger, 1986); Anthony S. Wohl, *Endangered Lives: Public Health in Victorian Britain* (Cambridge, MA: J.M. Dent, 1983).

19. On Chadwick and the "sanitary idea," see Martin V. Melosi, *The Sanitary City: Urban Infrastructure in America from Colonial Times to the Present* (Baltimore: Johns Hopkins University Press, 2000), 43–57; Christopher Hamlin, *Public Health and Social Justice in the Age of Chadwick: Britain, 1800–1854* (Cambridge: Cambridge University Press, 1998). See also J. C. Wylie, *The Wastes of Civilization* (London: Faber and Faber, 1959), 50ff.; Briggs, *Victorian Cities,* 19; Rosen, *History of Public Health,* 214–17, 252–59; Anthony N. B. Garvan, "Technology and Domestic Life, 1830–1880," in Kranzberg and Purcell, *Emergence of Modern Industrial Society,* 555–56; Welch, "History of Sanitation," 43–45.

20. US Bureau of the Census, *Characteristics of the Population,* vol. 1 of *Census of Population:* 1960, pt. A, 1–15, table 8; David R. Goldfield and Blaine A. Brownell, *Urban America: From Downtown to No Town* (Boston: Houghton Mifflin Co., 1979), 13–21.

21. See Melosi, *Sanitary City,* 17–42.

22. Carl Bridenbaugh, *Cities in the Wilderness: The First Century of Urban Life in America, 1625–1742* (New York: Ronald Press, 1938), 18, 85–86. See also Carl Bridenbaugh, *Cities in Revolt: Urban Life in America, 1743–1776* (New York: Alfred A. Knopf, 1955), 32–33, 239–40.

23. Sam Bass Warner Jr., *The Private City: Philadelphia in Three Periods of Its Growth* (Philadelphia: University of Pennsylvania Press, 1968; reprint 1975), 16; Charles E. Rosenberg, *The Cholera Years: The United States in 1832, 1849, and 1866* (Chicago: University of Chicago Press, 1962), 112–13, 184; Bessie Louise Pierce, *The Beginning of a City, 1673–1848*, vol. 1 of *A History of Chicago* (Chicago: University of Chicago Press, 1937; reprint 1975), 204–5, 338–39; Constance McLaughlin Green, *Washington: Village and Capital, 1800–1878* (Princeton, NJ: Princeton University Press, 1962), 211. See also John Duffy, *The Sanitarians: A History of American Public Health* (Urbana: University of Illinois Press, 1990), 13, 30–33; Stanley K. Schultz, *Constructing Urban Culture: American Cities and City Planning, 1800–1920* (Philadelphia: Temple University Press, 1989), 43, 119–21, 140.

24. John Duffy, *A History of Public Health in New York City, 1625–1866* (New York: Russell Sage Foundation, 1974), 180–93; 356–75; Sidney I. Pomerantz, *New York, an American City, 1783–1803: A Study of Urban Life* (Port Washington, NY: Ira J. Friedman, 1938; reprint, 1965), 270–76, 295–96, 344; John B. Blake, *Public Health in the Town of Boston, 1630–1822* (Cambridge, MA: Harvard University Press, 1959), 18, 100–4, 156–63; Bayrd Still, *Milwaukee: The History of a City* (Madison: State Historical Society of Wisconsin, 1948), 103, 239–40; Bridenbaugh, *Cities in the Wilderness*, 85; Bridenbaugh, *Cities in Revolt*, 30–32, 240.

25. Blake, *Public Health*, 157–58.

26. Bridenbaugh, *Cities in the Wilderness*, 18, 165–67, 239, 321ff; Blake, *Public Health*, 15–16, 103–4, 209–10; Duffy, *History of Public Health in New York City*, 180–93, 356–90; Pomerantz, *New York, an American City*, 251, 269–76, 295–96, 344; Bridenbaugh, *Cities in Revolt*, 31–33, 240; Green, *Washington*, 93–94, 211, 254–55; Still, *Milwaukee*, 239–40; William E. Korbitz, ed., *Urban Public Works Administration* (Washington DC: International City Management Association, 1976), 9–96; Ernest S. Griffith, *A History of American City Government: The Colonial Period* (New York: Oxford University Press, 1938), 261–91; American Public Works Association, *Street Cleaning Practice*, Rodney R. Fleming, 3d ed. (Chicago: APWA, 1978), 2–3 (hereafter cited as APWA); APWA, *Street and Urban Road Maintenance* (Chicago: Public Administration Service, 1963), 5–6.

27. US Bureau of the Census, *Characteristics of the Population* (Washington, DC: 1961), pt. A,. 1–14, table 8. See also, Blake McKelvey, *American Urbanization: A Comparative History* (Glenview, IL: Scott Foresman, 1973), 73, 104; Zane L. Miller and Patricia M. Melvin, *The Urbanization of Modern America: A Brief History*, 2d. ed. (San Diego: Harcourt Brace Jovanovich, 1987), 72, 79; Howard P. Chudacoff and Judith E. Smith, *The Evolution of American Urban Society*, 4th ed. (Englewood Cliffs, NJ: Prentice-Hall, 1994), 90.

28. Sam Bass Warner Jr., *The Urban Wilderness: A History of the American City* (New York: Harper and Row, 1972), 55–112, Maury Klein and Harvey A. Kantor, *Prisoners of Progress: American Industrial Cities, 1850–1920* (New York: Macmillan Co., 1976), 68–108.

29. Sam H. Schurr and Bruce C. Netschert, *Energy in the American Economy, 1850–1975* (Baltimore, MD: Johns Hopkins University Press, 1960), 57–83; Joseph M. Petulla, *American Environmental History: The Exploitation and Conservation of Natural Resources* (San Francisco: Boyd and Fraser Publishing Co., 1977), 149–51, 189–91. See also David Stradling, *Smokestacks and Progressives: Environmentalists, Engineers, and Air Quality, 1881–1951* (Baltimore: Johns Hopkins University Press, 1999).

30. Ellis L. Armstrong, Michael C. Robinson, and Suellen M. Hoy, eds. *History of Public Works in the United States, 1776–1976* (Chicago: APWA, 1976), 410; Robert R. Russell, *A History of the American Economic System* (New York: Appleton-Century-Crofts, 1964), 183;

Raymond W. Smilor, "Cacophony at 34th and 6th: The Noise Problem in America, 1900-1930," *American Studies* 28 (Spring 1977): 28–29; Mumford, *City in History*, 459. For additional references, see Martin V. Melosi, ed., *Effluent America: Cities, Industry, Energy, and the Environment* (Pittsburgh: University of Pittsburgh Press, 2001), 18–21.

31. Martin V. Melosi, "Environmental Crisis in the City: The Relationship between Industrialization and Urban Pollution," in *Pollution and Reform in American Cities, 1870–1930*, ed. Martin V. Melosi, (Austin: University of Texas Press, 1980), 6–9.

32. Ibid., 9–10.

33. Ibid., 11.

34. Jane Addams, *Twenty Years at Hull-House* (1905; reprint, New York: New American Library, 1961), 209.

35. Jacob Riis, *How the Other Half Lives* (1890; reprint, New York: Hill and Wang, 1957), 9–10.

36. David Brody, "Slavic Immigrants in the Steel Mills," in *The Private Side of American History*, ed. Thomas R. Frazier (New York: Harcourt Brace Jovanovich, 1975), 133; Addams, *Twenty Years at Hull-House,* 207–9; Riis, *How the Other Half Lives,* 8.

37. Thomas C. Cochran and William Miller, *The Age of Enterprise: A Social History of Industrial America* (New York: Macmillan Co., 1942; reprint, New York: Harper, 1961), 262; Roy Lubove, *Twentieth-Century Pittsburgh: Government, Business, and Environmental Change* (New York: John Wiley & Sons, 1969), 18; Klein and Kantor, *Prisoners of Progress,* 314–28. See also Melosi, *Sanitary City*, 62–72, for responses to epidemic disease in this period.

38. For a more thorough discussion of these pollution problems and city-service development, see Melosi, *Sanitary City*; Melosi, *Pollution and Reform.*

ONE. Out of Sight, Out of Mind: The Refuse Problem in the Late Nineteenth Century

1. G. T. Ferris, "Cleansing of Great Cities," *Harper's Weekly,* January 10, 1891, 33.

2. The Reverend Hugh Miller Thompson, "Disposal of City Garbage at New Orleans," *Sanitarian* 7 (November 1879): 545.

3. Boston Street Department, *Annual Report* (1891), 119–20.

4. "Disposal of Refuse in American Cities," *Scientific American,* August 29, 1891, 136; John McGaw Woodbury, "The Wastes of a Great City," *Scribner's Magazine,* October 1903, 392; Henry Smith Williams, "How New York Is Kept Partially Clean," *Harper's Weekly*, October 13, 1894, 973.

5. Rudolph Hering and Samuel A. Greeley, *Collection and Disposal of Municipal Refuse* (New York: McGraw-Hill, 1921), 13; H. de B. Parsons, *The Disposal of Municipal Refuse* (New York: John Wiley and Co., 1906), 27.

6. These figures are meant only to demonstrate the relatively large quantities of urban refuse produced in the United States, not to describe European conditions. See Hering and Greeley, *Collection and Disposal,* 13, 28, 70; Rudolph Hering, "Disposal of Municipal Refuse," *Transactions of the American Society of Civil Engineers* 54 (1904): 265–308.

7. Franz Schneider Jr., "The Disposal of a City's Waste," *Scientific American*, July 13, 1912, 24.

8. Joel A. Tarr, "Urban Pollution: Many Long Years Ago," *American Heritage*, October 1971, 65–69, 106. For a newer version of this essay, see "The Horse—Polluter of the City," in *The Search for the Ultimate Sink: Urban Pollution in Historical Perspective,* ed. Joel A. Tarr,

(Akron: University of Akron Press, 1996), 323–33; "Disposal of Refuse," 52; "Clean Streets and Motor Traffic," *Literary Digest* September 5, 1914, 413; Clay McShane and Joel A. Tarr, "The Centrality of the Horse in the Nineteenth-Century American City," in *The Making of Urban America*, ed. Raymond A. Mohl (Wilmington, DE: Scholarly Resources, 1997), 105–7, 121.

9. APWA, *History of Public Works in the United States, 1776–1976*, 164 (hereafter cited as APWA); Duffy, *History of Public Health in New York City*, 127; Tarr, "Urban Pollution," 67–69; "Clean Streets and Motor Traffic," 413; McShane and Tarr, "Centrality of the Horse," 111–12, 122.

10. "Clean Streets and Motor Traffic," 413; McShane and Tarr, "Centrality of the Horse," 124.

11. Howard D. Kramer, "The Germ Theory and the Public Health Program in the United States," *Bulletin of the History of Medicine* 22 (May-June 1948): 233–47. In response to the disastrous yellow fever epidemic of 1878, health reformers were able to secure from Congress the first national quarantine act and the creation of the National Board of Health. Neither proved effective. In 1883, Congress eliminated the appropriation of the National Board of Health and it collapsed. See John Duffy, "Social Impact of Disease in the Late 19th Century," in *Sickness and Health in America: Readings in the History of Medicine and Public Health*, ed. Judith Walzer Leavitt and Ronald L. Numbers (Madison: University of Wisconsin Press, 1978), 399–400. See also Melosi, *Sanitary City*, 43–72; Christopher Hamlin, *Public Health and Social Justice in the Age of Chadwick: Britain, 1800-1854* (Cambridge: Cambridge University Press, 1998).

12. Kramer, "Germ Theory," 233–47; James H. Cassedy, *Charles V. Chapin and the Public Health Movement* (Cambridge, MA: Harvard University Press, 1962), 39–61; Duffy, *History of Public Health in New York City*, 91–111; Henry I. Bowditch, *Public Hygiene in America* (Boston: Little, Brown Co., 1877), 29–41.

13. For sources dealing with public health problems in nineteenth-century cities, see Melosi, ed., *Effluent America*, 122–23; Duffy, *Sanitarians*; Nancy Tomes, *The Gospel of Germs: Men, Women, and the Microbe in American Life* (Cambridge, MA: Harvard University Press, 1998).

14. Gen. Emmons Clark, "Street-Cleaning in Large Cities," *Popular Science Monthly*, April 1891, 748.

15. Henry B. Wood, "Street Work in Boston," *Journal of the Association of Engineering Societies* 11 (August 1892): 433–34.

16. "City Refuse Disposal," *Engineering News* 28 (October 6, 1892): 325. See also John S. Billings, "Municipal Sanitation in Washington and Baltimore," *Forum* 15 (August 1893): 727–37.

17. Henry Smith Williams, "The Disposal of Garbage," *Harper's Weekly*, September 1, 1894, 835.

18. City of Boston, Joint Special Committee on the Disposing of City Offal, *Report of the Joint Special Committee on the Disposing of City Offal* (1893), 17.

19. In the late nineteenth and early twentieth centuries, the efforts of many cities to move away from state interference in their affairs resulted in frequent demands for municipal "home rule." This movement took many forms, including efforts to increase the appointive power of mayors and to gain control of various service departments. See Ernest S. Griffith, *A History of American City Government: The Conspicuous Failure, 1870–1900*

(Washington, DC: University Press of America, 1974), 215; Griffith, *A History of American City Government: The Progressive Years and Their Aftermath, 1900–1920* (Washington, DC: University Press of America, 1983), 124–25, 128; Jon C. Teaford, ed., *The Unheralded Triumph: City Government in America, 1870–1900* (Baltimore: Johns Hopkins University Press, 1984), 105, 122; Charles N. Glaab and A. Theodore Brown, *A History of Urban America*, 174–76.

20. *Engineering News* 23 (January 4, 1890): 13.

21. *Engineering News* 23 (February 15, 1890): 156. See also Washington DC Health Department, *Report of the Health Officer* (1882), 10; Saint Louis Health Department, *Annual Report* (1894), 9; (1895), 9.

22. Chicago Department of Health, *Annual Report* (1892), 15–16; *Engineering News* 33 (January 3, 1895): 1. See also George E. Hooker, "Cleaning Streets by Contract: A Sidelight from Chicago," *Review of Reviews* 15 (March 1897): 437–41.

23. City of Newton, Mass., *Report of the Board of Health upon the Sanitary Disposition of Garbage and Other Municipal Waste, and the Reorganization of the Department* (1895), 3–4.

24. US Department of the Interior, Census Office, *Report on the Social Statistics of Cities, Tenth Census, 1880*, comp. George E. Waring Jr.

25. Good examples of this phenomenon can be found in Sam Bass Warner Jr., *Streetcar Suburbs: The Process of Growth in Boston, 1870–1900* (Cambridge: Harvard University Press, 1962).

26. US Department of the Interior, Census Office, *Report on the Social Statistics of Cities, Tenth Census, 1880.*

27. Clay McShane, "Transforming the Use of Urban Space: A Look at the Revolution in Street Pavements, 1880–1924," *Journal of Urban History* 5 (May 1979): 283.

28. Cassedy, *Charles V. Chapin*, 44–45; Wilson G. Smillie, *Public Health: Its Promise for the Future* (New York: Macmillan Co., 1955), 351–52; William F. Morse, "Methods of Collection and Disposal of Waste and Garbage by Cremation" (Paper presented at the Sanitary Convention of State and Local Boards of Health of Pennsylvania, Erie, PA., 1892), 3.

29. See US Department of the Interior, Census Office, *Report on the Social Statistics of Cities, Tenth Census, 1880.*

30. Ibid. See also Smillie, *Public Health*, 352.

31. American Public Health Association (hereafter cited as APHA), *A Half Century of Public Health*, ed. Mazyck P. Ravenel (New York: APHA; reprint New York: Arno Press and New York Times, 1970), 190–91; Hering and Greeley, *Collection and Disposal*, 2.

32. G. W. Hosmer, "The Garbage Problem," *Harper's Weekly*, August 11, 1894, 750.

33. "Disposal of Refuse in American Cities," 136. For an extensive listing of contemporary periodical articles dealing with the refuse problem, see Robert C. Brooks, *A Bibliography of Municipal Problems and City Conditions* (1901; reprint New York: Arno Press and the New York Times, 1970).

34. Ferris, "Cleansing of Great Cities," 33.

35. John S. Billings, "Municipal Sanitation: Defects in American Cities," *Forum* 15 (May 1893): 305.

36. New York City, Citizens' Committee of Twenty-one, *Statement and Report of the Citizens' Committee of Twenty-one Respecting the Efforts to Procure Reform in the System of Cleaning the Streets of the City of New York* (1881).

37. Citizens' Association of Chicago, *Annual Report* (1880), 16–18; Chicago Department of Health, *Annual Report* (1892), 29–31. See also *Engineering News* 33 (January 3, 1895): 1.

38. Mary E. Trautmann, "Women's Health Protective Association," *Municipal Affairs* 2 (September 1898): 439–43; Duffy, *History of Public Health in New York City*, 124, 130, 132.

39. New York Ladies' Health Protective Association, *Memorial of the New York Ladies' Health Protective Association to the Hon. Abram S. Hewitt on the Subject of Street Cleaning,* (1887), 4–5. See also Suellen M. Hoy, "'Municipal Housekeeping': The Role of Women in Improving Urban Sanitation Practices, 1880–1917," in Melosi, *Pollution and Reform*, 173–98; Hoy, *Chasing Dirt: The American Pursuit of Cleanliness* (New York: Oxford University Press, 1995).

40. See Philadelphia Department of Public Works, Bureau of Street Cleaning, *Annual Report* (1893), 58; Mrs. C. G. Wagner, "What Women Are Doing for Civic Cleanliness" *Municipal Journal and Engineer* 11 (July 1901): 35; Edith Parker Thomson, "What Women Have Done for the Public Health," *Forum* 24 (September 1897): 46–55.

41. City of Newton, Mass., *Report of the Board of Health,* (1895), p. 5; Philadelphia, Bureau of Health, *Annual Report* (1892), 18–19; Chicago Department of Public Works, *Annual Report* (1889), 23–24; Detroit Board of Health, *Annual Report* (1882), 115.

42. See Joel A. Tarr, "From City to Farm: Urban Wastes and the American Farmer," *Agricultural History* 49 (October 1975): 598–612.

43. Washington DC Health Department, *Report of Health Officer* (1889), 31. See also Boston Board of Health, *Annual Report* (1883–84), 56; Baltimore Department of Street Cleaning, *Annual Report* (1882), 11; (1887), 19–20.

44. Hugh Miller Thompson, "Disposal of City Garbage," 546.

45. New York City, Department of Street Cleaning, *Annual Report* (1886), 23; "City Refuse Disposal," *Engineering News* 28 (October 6, 1892): 325; Chicago Department of Public Works, *Annual Report* (1882), 25.

46. Henry Smith Williams, "How New York Is Kept Partially Clean," 974. See also Hosmer, "Garbage Problem," 711; New York State Assembly, Committee on the Affairs of Cities, *Report to the Assembly of the State of New York, April, 1880, as to the Present System of Street Cleaning in the City of New York, and the Means Whereby a More Efficient and Economical Method of Doing the Work May Be Secured*, 50; "The Disposal of Garbage and Other City Refuse," *Engineering News* 32 (July 26, 1894): 72–73. See also Steven H. Corey, "Garbage in the Sea," *Seaport* 25 (Winter-Spring 1991): 18–23.

47. Lawrence H. Larsen, "Nineteenth-Century Street Sanitation: A Study of Filth and Frustration," *Wisconsin Magazine of History* 52 (Spring 1969): 239–40.

48. Statistics Gleaned from US Department of the Interior, Census Office, *Report on the Social Statistics of Cities, Tenth Census, 1880.*

49. See "Street Department Notes," *Municipality and County* 1 (April 1895): 155; Boston Street Department, *Annual Report* (1891), 128. See also US Department of the Interior, Census Office, *Report on the Social Statistics of Cities, Tenth Census, 1880.*

50. Clark, "Street-Cleaning in Large Cities," 748.

51. Boston Board of Health, *Annual Report* (1887), 36.

52. McShane, "Transforming the Use of Urban Space," 279–82.

53. US Department of the Interior, Census Office, *Report on the Social Statistics of Cities in the United States, Eleventh Census, 1890,* comp. John S. Billings, 18.

54. Boston Board of Health, *Annual Report* (1891), 120–121.

55. Chicago Department of Public Works, *Annual Report* (1894), 11. See also "Cost of Street Cleaning in Various Cities," *Engineering News* 33 (April 11, 1895): 247.

56. F. W. Hewes, "Street Cleaning," *Harper's Weekly,* March 9, 1895, 233–34. See also *Engineering News* 23 (January 4, 1890): 12–13; "Street Cleaning Statistics," *Municipality and County* 1 (April 1895): 153; Duffy, *History of Public Health in New York City,* 51–69.

57. See Stanley K. Schultz and Clay McShane, "To Engineer the Metropolis: Sewers, Sanitation, and City Planning in Late-Nineteenth-Century America," *Journal of American History* 65 (September 1978): 389–411.

58. J. Berrien Lindsley, "On the Cremation of Garbage," *Journal of the American Medical Association* 11 (October 13, 1888): 514; George Baird, "Destruction of Night-Soil and Garbage by Fire," APHA, *Public Health: Papers and Reports* 12 (1886): 120.

59. S. S. Kilvington, "Garbage Furnaces and the Destruction of Organic Matter by Fire" in APHA, *Public Health: Papers and Reports* 14 (1889): 170.

60. Walter Francis Goodrich, *Refuse Disposal and Power Production* (Westminster: Archibald Constable and Co., 1904), 3, 9–10; Chamber of Commerce of the United States, Construction and Civic Development Department, *Refuse Disposal in American Cities* (Washington DC, 1931), 15; William F. Morse, "The Disposal of the City's Waste," *American City,* May 1910, 23; W. Howard White, "European Garbage Removal and Sewage Disposal," *Transactions of the American Society of Civil Engineers* 15 (December 1886): 869.

61. William Mayo Venable, *Garbage Crematories in America* (New York: John Wiley and Co., 1906), 88; "Garbage-cremation," *Science* 12 (December 7, 1888): 265–66; William F. Morse, "The Utilization and Disposal of Municipal Waste," *Journal of the Franklin Institute* 158 (July 1904): 25–42.

62. See "City Refuse Disposal," *Engineering News* 28 (October 6, 1892): 325; George H. Rohe "Recent Advances in Preventive Medicine," *Journal of the American Medical Association* 9 (July 2, 1887): 5–6; Thomas H. Manly and Douglas H. Stewart, "The Economical and Efficient Disposal of the Household Garbage of New York," *Sanitarian* 34 (February 1895): 106; Chicago Department of Health, *Annual Report* (1888), 11; New York City Department of Street Cleaning, *Annual Report* (1887–88), 17–18.

63. Boston Health Department, *Annual Report* (1890), 76–77.

64. See Martin V. Melosi, "Technology Diffusion and Refuse Disposal: The Case of the British Destructor," in *Technology and the Rise of the Networked City in Europe and America,* ed. Joel A. Tarr and Gabriel Dupuy (Philadelphia: Temple University Press, 1988), 207–22.

65. See Douglas H. Stewart, "The Gold in Garbage," *Journal of the American Medical Association* 25 (September 21, 1895): 484–86; "The Merz System of Garbage Utilization in Four American Cities," *Engineering News* 32 (November 1, 1894): 354–59; "The Report of the New York Garbage Commission," *Engineering News* 32 (November 29, 1894): 452; Charles V. Chapin, *Municipal Sanitation in the United States* (Providence, RI: Providence Press, 1901), 703–706. See also "The New York Garbage Deodorizing Plant at Riker's Island," *Engineering News* 32 (August 2, 1894): 90; "Garbage Utilization at Cincinnati and New Orleans," *Engineering News* 36 (November 8, 1896): 236–38; "The Utilization of New York City Garbage," *Scientific American* 77 (August 14, 1897): 102; Saint Louis Department of Health, *Annual Report* (1895–96), 31–32; Chicago Department of Health, *Biennial Report* (1897–98), 28-29; "Disposal of Garbage," *City Government* 5 (August 1898): 67; "Recent Refuse Disposal Practices," *Municipal Journal and Engineer* 37 (December 10, 1914): 848–49.

For reduction, incineration, and other methods of disposal in this period, see Daniel J. Zarin, "Searching for Pennies in Piles of Trash: Municipal Refuse Utilization in the United States, 1870–1930," *Environmental Review* 11 (Fall 1987): 207–22.

TWO. The "Apostle of Cleanliness" and the Origins of Refuse Management

1. George A. Soper, "George Edwin Waring," *Dictionary of American Biography* (New York: Charles Scribner's Sons, 1936), 19: 456. See also *New York Times*, October 28, 1898, 1; October 29, 1898, 7; October 30, 1898, 1, 18; October 31, 1898, 1.

2. *New York Times*, November 23, 1898, 6.

3. Soper, "George Edwin Waring," 456; Albert Shaw, *Life of Col. Geo. E. Waring, Jr.: The Greatest Apostle of Cleanliness* (New York: Patriotic League, 1899); Richard W. G. Welling, Miscellaneous Papers, New York Public Library, Manuscript Division; *Newport News*, November 23, 28, 1898; *New York Times*, November 23, 1898, 6; November 24, 1898, 7.

4. Shaw, *Life of Col. Geo. E. Waring, Jr.;* Helen Gray Cone, "Waring," *Century Illustrated Monthly Magazine*, February 1900, 547; "George E. Waring," *Outlook*, November 5, 1898, 564–65; *Newport News*, November 22, 1898; November 25, 1898; "Colonel Waring," *Nation*, November 3, 1898, 326–27. See also William Potts, "George Edwin Waring, Jr.," *Charities Review*, 1898, 461–68; W. P. Gerhard, "A Half Century of Sanitation," pt. 2, *American Architect and Building News* 63 (March 4, 1899): 67.

5. *National Cyclopedia of American Biography* (New York: James T. White & Co., 1929), 6:157; *New York Times*, October 30, 1898, 2; Neil FitzSimons, "Pollution Fighter: George Waring" (manuscript), 2.

6. George E. Waring Jr., *The Elements of Agriculture* (New York: D. Appleton and Co., 1854), 279–85. See also Martin V. Melosi, *Pragmatic Environmentalist: Sanitary Engineer George E. Waring, Jr.* (Washington DC: Public Works Historical Society, 1977), 6.

7. Laura Wood Roper. *FLO: A Biography of Frederick Law Olmsted* (Baltimore, MD: Johns Hopkins University Press, 1973), xiv.

8. Ibid., xiii–xiv, 139,148, 462. See also Soper, "George Edwin Waring." 456; Albert Fein, *Frederick Law Olmsted and the American Environmental Tradition* (New York: George Braziller, 1972), 28–29.

9. "The Military Element in Colonel Waring's Career," *Century Illustrated Monthly Magazine*, February 1900, 544–47; Soper, "George Edwin Waring," 456; Roper, *FLO*, 162.

10. George E. Waring. Jr., *Whip and Spur* (Boston: J. R. Osgood and Co., 1875), 67–68.

11. George E. Waring Jr., "Education at West Point," *Outlook*, August 6, 1898): 825.

12. *Newport News*, December 6, 1898. See also *National Cyclopedia of American Biography*, 6:157.

13. *National Cyclopedia*, 6:157; George E. Waring Jr., *Sewerage and Land-Drainage*, (New York: D. Van Nostrand Co., 1885), 108. Additional works on drainage and sewerage written by Waring in the 1870s and 1880s include *The Sanitary Condition of City and Country Dwelling Houses* (New York: D. Van Nostrand Co., 1877); "House Drainage and Sewerage" (Paper presented to the Philadelphia Social Science Association, 1878); "The Draining of a Village," *Harper's Magazine*, June 1879, 132–35; "Recent Modifications in Sanitary Drainage," *Atlantic Monthly*, July 1879, 56–62; "The Sewering and Draining of Cities" (Paper presented to the APHA, Nashville, TN, 1879); "Suggestions for the Sanitary Drainage of Washington City," *Smithsonian Miscellaneous Collections* 26 (1880): 1–23; "Storm-water in Town Sewerage" (Paper presented to the APHA, New Orleans, LA, 1881);

"Sanitary Drainage," *North American Review*, July 1883, 57–67; *How to Drain a House* (New York: Henry Holt and Co., 1885); *The Disposal of Sewage, and the Protection of Streams Used as Sources of Water Supply* (Philadelphia: W. J. Dorman, 1886).

14. For the fullest treatment of the separate system and Waring's association with it, see Joel A. Tarr, "The Separate vs. Combined Sewer Problem: A Case Study in Urban Technology Design Choice," *Journal of Urban History* 5 (May 1979): 308–39.

15. George E. Waring Jr., "The Memphis System of Memphis and Elsewhere" (Paper presented to the APHA, Mexico City, 1893); Waring, "Sewering and Draining"; Soper, "George Edwin Waring," 456; William Henry Corfield, *The Treatment and Utilization of Sewage*, 3d ed. (London: Macmillan Co, 1887), 199; Harold W. Babbitt, *Sewerage and Sewage Treatment* (New York: John Wiley and Sons, 1922; reprint 1953), 3; Detroit Board of Health, *Annual Report* (1882),115; Glaab and Brown, *History of Urban America*, 165–66; Tarr, "Separate vs. Combined Sewer Problem," 315–18.

16. Leonard Metcalf and Harrison P. Eddy, *Sewerage and Sewage Disposal* (New York: McGraw-Hill, 1922; reprint 1930), 11–12. See also James H. Cassedy, "The Flamboyant Colonel Waring: An Anti-Contagionist Holds the American Stage in the Age of Pasteur and Koch," *Bulletin of the History of Medicine* 36 (March-April 1962): 168–70. For more details on the Memphis system, see Melosi, *Sanitary City*, 153–60.

17. See George E. Waring Jr., *Earth-Closets: How to Make Them and How to Use Them* (New York: Tribune Association, 1868); Cassedy, "Flamboyant Colonel Waring," 165; Roper, *FLO*, 320, 324

18. See US Department of the Interior, Census Office, *Report on the Social Statistics of Cities, Tenth Census, 1880;* George E. Waring Jr., "The National Board of Health," *Atlantic Monthly*, December 1879, 732–38. An interesting sidelight on Waring's work on the tenth census is that he selected George Washington Cable, the Creole novelist, as an assistant on the project. See Charles Philip Butcher, *George Washington Cable* (New York: Twayne Publishers, 1962), 60–63. Arlin Turner, *George W. Cable: A Biography* (Durham, NC: Duke University Press, 1956), 108–12.

19. Waring's popular works include *Whip and Spur: A Farmer's Vacation* (Boston: J. R. Osgood & Co., 1876); *The Bride of the Rhine* (Boston: J. R. Osgood and Co., 1878); *Tyrol and the Skirt of the Alps* (New York: Harper and Brothers, 1879); *Ruby* (Boston: J. R. Osgood and Co., 1883); *Vix* (Boston: J. R. Osgood and Co., 1883); *The Saddle-Horse* (New York: Orange Judd Co., 1881).

20. Justin Kaplan. *Lincoln Steffens: A Biography* (New York: Simon and Schuster, 1974), 73.

21. Duffy, *History of Public Health in New York City*, 108–9; Roper, *FLO*, 462; Theodore Roosevelt *Theodore Roosevelt: An Autobiography* (New York: Charles Scribner, 1924), 168; Richard Skolnik, "George Edwin Waring, Jr.: A Model for Reformers," *New York Historical Society Quarterly* 52 (October 1968): 357.

22. George E. Waring Jr., "The Cleaning of the Streets of New York," *Harper's Weekly*, October 29, 1895, 1022. For a recent, detailed account of solid waste management in New York City, including Waring's tenure, see Steven Hunt Corey, "King Garbage: A History of Solid Waste Management in New York City, 1881–1970" (PhD diss., New York University, 1994).

23. *Engineering News* 33 (January 3, 1895): 8. See also Martin V. Melosi, "'Out of Sight, Out of Mind': The Environment and the Disposal of Municipal Refuse, 1860–1920,"

Historian 35 (August 1973): 626–27; "Hitch Your Wagon to a Star," *Garden and Forest* 9 (June 24, 1896): 251–52.

24. George E. Waring Jr., *The Causation of Typhoid Fever* (Cambridge, MA: Riverside Press, 1878), 7; Waring, *Sewerage and Land-Drainage,* 21.

25. "Sewer gas" was a term applied to the odoriferous gases emanating from decomposing matter.

26. Waring, "Suggestions for Sanitary Drainage," 12. See also Waring, "Storm-water in Town Sewerage"; Waring, *How to Drain a House,* iii–vi, 1–11, 69–71; Waring, "Sanitary Drainage," *North American Review,* July 1883, 57–67; Waring, "Village Sanitary Work," *Scribner's Monthly,* June 1887, 176–77; Cassedy, "Flamboyant Colonel Waring," 166; Glaab and Brown, *History of Urban America,* 165.

27. George E. Waring Jr., "Partial Purification of Sewage," *Engineering News* 31 (January 4, 1894): 15–16; Waring, *Modern Methods of Sewage Disposal* (New York: D. Van Nostrand Co., 1894), 1–19, 30–33, 41–47, 54–57, 214–15, 239–43; Waring, *Disposal of Sewage.*

28. Waring, *Sewerage and Land-Drainage,* 22.

29. George E. Waring Jr., "Out of Sight, Out of Mind," *Century Illustrated Monthly Magazine,* 47, n.s. 25, April 1894, 939.

30. George E. Waring Jr., "The Sanitary Condition of City and Country Dwelling Houses" (Paper presented to the APHA, Boston, MA, 1876), 5–21.

31. George E. Waring Jr., *Village Improvements and Farm Villages* (Boston: J. R. Osgood and Co., 1877), 17. See also Waring, "Village Improvement Associations," *Scribner's Monthly,* May 1877, 97–98.

32. George E. Waring Jr., "The Sanitary Condition of New York," pts. 1 and 2, *Scribner's Monthly,* May 1881, 64–75; June 1881, 179–89.

33. George E. Waring Jr., *Street Cleaning and the Disposal of a City's Wastes: Methods and Results and the Effect upon Public Health, Public Morals, and Municipal Prosperity* (New York: Doubleday and McClure Co., 1898), 15–18. See also Waring, "The Cleaning of a Great City," *McClure's Magazine,* September 1897, 911–15; E. Burgoyne Baker, "The Refuse of a Great City." *Munsey's Magazine,* April 1900, 84; Skolnik, "George Edwin Waring, Jr.," 359–60. For more detailed descriptions of Waring's sometimes unusual political views, see Waring, "Government by Party," *North American Review,* November 1896, 587–94; Waring, "The Drink Problem in New York City Politics." *Outlook,* October 15, 1898, 436–40.

34. *New York Times,* April 28, 1895, 19; October 31, 1897, 3; November 29, 1897, 6; "Col. Waring on Street Cleaning," *City Government* 4 (June 1898): 223; "Hon. William S. Andrews and Col. George E. Waring, Jr., on the Cost of Street Cleaning in New York City," *Municipal Record and Advertiser,* June 26, 1897, 15.

35. Waring had a penchant for embroiling himself in political controversies. For instance, he created a furor when he declared that the Grand Army of the Republic was an aggregation of "pension bummers." See *New York Times,* April 23, 1895, 2.

36. Waring, "Cleaning of a Great City," 916; Waring, *Street Cleaning,* 19–20; Waring, "Street Cleaning," *City Government* 3 (October 1897): 118.

37. Waring, "Street Cleaning," 118; Waring, *Street Cleaning,* 19–43; Baker, "Refuse of a Great City," 84.

38. "Hitch Your Wagon to a Star," 251–52; Waring, "Street Cleaning," 118: Waring, "Cleaning of a Great City," 917; Potts, "George Edwin Waring, Jr.," 465; Elizabeth Fee and Steven H. Corey, *Garbage! The History and Politics of Trash in New York City* (New York: The

New York Public Library, 1994), 37. Fee and Corey note that Waring also established a New York School for Street Cleaners "with classrooms full of hand trucks, brooms, and garbage cans," 39.

39. George E. Waring Jr., "The Labor Question in the Department of Street Cleaning of New York." *Municipal Affairs* 1 (September 1897): 515–24; Waring, *Street Cleaning* 24–43; Charles Zueblin, *American Municipal Progress*, rev. ed. (New York: Macmillan Co., 1948), 76; *New York Times*, July 28, 1895, 5; Skolnik, "George Edwin Waring, Jr.," 360–61.

40. Sometimes the faith in hand sweeping got a little out of control. In an effort to increase the number of sweepers in his force within his budgetary limits, Waring recommended this exploitative plan to the members at the Good Government Club: "Another suggestion I wish to make to you is that the city employ, instead of 2,500 men, 5,000 persons to clean the streets. The present force constitutes a sort of labor aristocracy, each member of which makes about $780 a year, probably the best pay for common labor in the world. I see no objection to this, excepting that the city has to pay it. Now I suggest that we employ 5,000 men, women, and boys to do this work. Good drivers and experienced sweepers would receive about their present pay while the 1,500 boys and the 11,000 women would make about 50 cents a day. There are plenty of foreign women who would be glad to work three or four hours in the early morning at sweeping, and plenty of boys who would do the cart work in the day time. I would employ only one person in every family so that 5,000 families would be helped each receiving about $160 a year from the work." *New York Times*, June 8, 1895, 7

41. Waring, *Street Cleaning*, 37–42; Zueblin, *American Municipal Progress*, 75–76; Waring, "Street Cleaning," 118–19; Waring, "The Cleaning of the Streets of New York," *Harper's Weekly*, October 26, 1895, 1024; Waring, "Cleansing of Cities and Public Health," *Engineering Magazine*, February 1895, 810; Waring, "The Relations of Good Paving to Street Cleaning," *Engineering Magazine*, February 1897, 781–85; Skolnik, "George Edwin Waring, Jr.," 365.

42. John Brisben Walker, "Great Problems in Organization: The Street-Cleaning Work of Colonel Waring in New York," *Cosmopolitan Magazine*, December 1898, 235. See also Potts, "George Edwin Waring, Jr.," 466.

43. Waring, "Cleaning of a Great City," 921. See also Waring, "Cleaning of the Streets," 1022–23.

44. Waring, *Street Cleaning*, 91.

45. Waring, "The Disposal of a City's Waste," *North American Review*, July 1895, 52.

46. Waring, "Disposal of a City's Waste," 51–56; Waring, "Cleaning of a Great City," 919. Waring, *Street Cleaning*, 43; Woodbury, "Wastes of a Great City," 388–89. For good background on the history of resource recovery and source separation in the United States before World War II, see Suellen M. Hoy and Michael C. Robinson, *Recovering the Past: A Handbook of Community Recycling Programs, 1890–1945* (Chicago: Public Works Historical Society, 1979).

47. "The Fouling of the Beaches," *Harper's Weekly*, July 2, 1898, 663; Waring, "Disposal of a City's Waste," 53–54; Waring, "Cleaning of a Great City," 918–19.

48. Waring, "The Utilization of City Garbage," *Cosmopolitan Magazine* 24 (February 1898): 406–8.

49. Waring, "Disposal of a City's Waste," 51–56.

50. Ibid., 51. See also "Garbage," *Municipality and County* 1 (April 1895): 164; Waring, "Cleaning of a Great City," 917–18; Zarin, "Searching for Pennies," 216–217.

51. Hering and Greeley, *Collection and Disposal of Municipal Refuse,* 299

52. Waring, "Utilization of City Garbage," 408–11; Waring, "The Cleaning of a Great City," 919.

53. Waring, *Street Cleaning,* 68–73; Waring, "Disposal of a City's Waste," 56; Waring, "Cleaning of a Great City," 919–20; "The Delehanty Dumping-Scow," *Harper's Weekly,* October 24, 1896, 1051; Waring, *Draining for Profit and Draining for Health,* 150.

54. *New York Times,* October 11, 1896, Sunday Supplement, 2.

55. Skolnik, "George Edwin Waring, Jr.," 365–66; Cassedy, "Flamboyant Colonel Waring," 171.

56. Waring, "Cleaning of a Great City," 922.

57. Daniel Eli Burnstein, "Clean Streets and the Pursuit of Progress: Urban Reform in New York City in the Progressive Era" (PhD diss., Rutgers University, 1992), 192, 214, 247.

58. Waring, "Cleaning of a Great City," 922.

59. Waring, *Street Cleaning,* 177–86; Baker, "Refuse of a Great City," 90.

60. Waring, *Street Cleaning,* 186.

61. Reuben S. Simons, "The Juvenile Street Cleaning Leagues of New York," *American City,* October 1910, 163–66; Burnstein, "Clean Streets," 227, 240.

62. Waring, *Street Cleaning,* 181. The rest of the stanzas are as follows:

> No longer will you see a child fall helpless in the street
> Because some slippery peeling betrayed his trusting feet;
> We do what we are able to make our sidewalks neat;
> And we will keep right on.
>
> And all the people far and near, in sunshine or in rain,
> Rejoice to see our cleaner streets, and find the reason plain:
> We children take a hand to keep our thoroughfares so clean;
> And we will keep right on.

63. Ibid., 184; See also 177–83; Potts, "George Edwin Waring, Jr.," 467.

64. Daniel Burnstein, "George E. Waring, Jr. and the Perceived Nexus of Personal and Environmental Conditions in Progressive Era Urban America" (Paper presented at the American Society for Environmental History Conference, March 2000), 3.

65. Simons, "Juvenile Street Cleaning," 163–66; Skolnik, "George Edwin Waring, Jr.," 366–67; Zueblin, *American Municipal Progress,* 76.

66. See Woodbury, "Wastes of a Great City," 387–90.

67. *American Municipal Progress,* 75–76, 82; Delos F. Wilcox, *The American City: A Problem in Democracy* (New York: Macmillan Co., 1904; reprint, 1906), 118, 224; George A. Soper, *Modern Methods of Street Cleaning* (New York: Engineering News Publishing Co., 1909), 165; John A. Fairlie, *Municipal Administration* (New York: Macmillan Co., 1901; reprint, 1906), 258–59; "Tammany and the Streets," *Outlook,* October 20, 1900, 427–28; "Street Cleaning," *Outlook,* October 20, 1900, 426–27; "The Disposal of New York's Refuse," *Scientific American,* October 24, 1903, 292–94; Woodbury, "Wastes of a Great City," 387; Baker,

"Refuse of a Great City," 81, 89–90; Duffy, *History of Public Health in New York City*, 125–26; "Military Element," 544–47.

THREE. Refuse as an Engineering Problem: Sanitary Engineers and Municipal Reform

1. "The Sanitary Engineer—A New Social Profession." *Charities and the Commons (Survey)* 16 (June 2, 1906): 286.

2. Schultz and McShane, "To Engineer the Metropolis," 389–411. Schultz and McShane primarily emphasize the role of the engineer as technical expert and administrator, rather than as an environmental generalist. See also Martin V. Melosi, "Sanitary Engineers in American Cities: Changing Roles from the Age of Miasmas to the Age of Ecology," in *Civil Engineering History: Engineers Make History*, ed. Jerry R. Rogers, et al. (New York: American Society for Civil Engineers, 1997), 108–22.

3. Welch, "History of Sanitation," 45.

4. Duffy, *History of Public Health in New York City*, 91.

5. Kramer, "Germ Theory," 233–47. See also Cassedy, *Charles V. Chapin*, 39–45, 96, 141; Barbara G. Rosenkrantz, *Public Health and the State: Changing Views in Massachusetts, 1842–1936* (Cambridge, MA: Harvard University Press, 1972), 75, 103, 177–82; Rosen, *History of Public Health*, 233–50.

6. Kramer, like some other historians of public health and medicine, assumed that the shift away from environmental sanitation by health workers was a good sign: "Many municipal matters that had concerned earlier sanitarians, such as street cleaning, had gravitated into other hands, where they really belonged." See Kramer, "Germ Theory," 246. For an excellent discussion of the transition from the filth theory to the germ theory, see Tomes, *Gospel of Germs*. Tomes makes an effective argument that the transition was not as abrupt as we might think. In fact, contagionist and anticontagionist approaches were sometimes melded together for several years until the germ theory clearly won out in the twentieth century.

7. Charles V. Chapin, "The End of the Filth Theory of Disease," *Popular Science Monthly*, January 1902, 239.

8. Charles V. Chapin, "Sanitation in Providence," in *Proceedings of the Providence, Rhode Island, Conference for Good Government and the Thirteenth Annual Meeting of the National Municipal League* (Philadelphia: National Municipal League, 1909), 326.

9. P. M. Hall, "The Collection and Disposal of City Waste and the Public Health," *American Journal of Public Health* 3 (April 1913): 314–15.

10. Ibid., 316–17. See also "Control of Garbage Disposal," *Municipal Journal and Engineer* 30 (May 3, 1911): 633.

11. Edward D. Rich, "State Health Departments and Municipal Refuse Disposal," *American Journal of Public Health* 8 (February 1918): 135–36. See also "Disposal of Garbage," *City Government* 5 (August 1898): 66; Benjamin Lee, "Preventive Medicine in Pennsylvania," *Sanitarian* 40 (February 1898): 99–100; William Hay McLain, "A Sanitary Method of Garbage Collection," *American City*, April 1913, 402–403; C. E. Terry, "The Public Dump and the Public Health," *American Journal of Public Health* 3 (April 1913): 341.

12. See Wilson G. Smillie, *Public Health Administration in the United States*, 3d ed. (New York: Macmillan Co., 1947), 4, 255; Smillie, *Public Health: Its Promise*, 351–52.

13. American Child Health Association, Research Division, *A Health Survey of 86 Cities* (New York: American Child Health Association, 1925), 214.

14. Edwin T. Layton Jr., *The Revolt of the Engineers: Social Responsibility and the American Engineering Profession* (Cleveland, OH: Press of Case Western Reserve University, 1971), viii, 53–69.

15. David F. Noble, *America by Design: Science, Technology, and the Rise of Corporate Capitalism* (New York: Alfred A. Knopf, 1977), 44.

16. Ibid., 35–36. See also Layton, *Revolt of the Engineers*, 3–6, 53–71. On the growth of the engineering profession in the United States, see also Terry S. Reynolds, ed., *The Engineer in America* (Chicago: University of Chicago Press, 1991).

17. Schultz, *Constructing Urban Culture*, 187–89.

18. *A Biographical Dictionary of American Civil Engineers* (New York: American Society of Civil Engineers, 1972), 58–59; FitzSimons, "Pollution Fighter;" *Civil Engineering ASCE* (October 1971): 100; Jacqueline Wilkie, "Rudolph Hering," *APWA Reporter* 48 (April 1981): 4–5.

19. *Biographical Dictionary*, 23–24. See also Louis P. Cain, "Raising and Watering a City: Ellis Sylvester Chesbrough and Chicago's First Sanitation System," *Technology and Culture* 13 (July 1972): 353–72.

20. "Sanitary Engineer," 286–87.

21. "A Veteran in Garbage Disposal," *Municipal Journal and Engineer* 19 (September 1905): 124–25; "The Disposal of Municipal Waste," *Municipal Journal and Engineer* 20 (January 31, 1906): 102.

22. Michael Robinson, "William Mulholland," *APWA Reporter* 43 (September 1976): 12–13.

23. Ibid.; William L. Kahrl, "The Politics of California Water: Owens Valley and the Los Angeles Aqueduct, 1900–1927," *California Historical Quarterly* 55 (Spring 1976): 98–119; Remi Nadeau, "The Water War," *American Heritage*, December, 1961, 31–35.

24. Schultz and McShane, "To Engineer the Metropolis," 399. See also "Needed Reforms in the Collection and Disposal of City Refuse," *Engineering News* 59 (April 23, 1908): 462–63; Raymond H. Merritt, *Engineering in American Society, 1850–1875* (Lexington: University of Kentucky Press, 1969), 157–76.

25. See Christopher Hamlin, "Edwin Chadwick and the Engineers, 1842–1854: Systems and Antisystems in the Pipe-and-Brick Sewers War," *Technology and Culture* 33 (October, 1992): 680–706; Hamlin, *Public Health and Social Justice*; Anthony S. Wohl, *Endangered Lives: Public Health in Victorian Britain* (Cambridge: J. M. Dent, 1983); Luckin, *Pollution and Control*; M.W. Flinn, *Public Health Reform in Britain* (London: Macmillan, 1968); Jon A. Peterson, "The Impact of Sanitary Reform upon American Urban Planning, 1840–1890," *Journal of Social History* 13 (Fall 1979): 83–103; R. Winthrop Pratt, "The Industrial Need of Technically Trained Men: Sanitary Engineering," *Scientific American Supplement* (March 7, 1914): 150.

26. Ellen H. Richards, *Conservation by Sanitation* (New York: John Wiley and Sons, 1911), ix; Pratt, "Industrial Need," 150.

27. Stuart Galishoff, "Triumph and Failure: The American Response to the Urban Water Supply Problem, 1860–1923," in Melosi, *Pollution and Reform*, 35–57; Schultz and McShane, "To Engineer the Metropolis," 393. See also Melosi, *Sanitary City*, 73–89, 117–48.

28. Joel A. Tarr and Francis Clay McMichael, "Historical Decisions about Wastewater Technology: 1800–1932," *Journal of the Water Resources Planning and Management Division, ASCE* 103 (May 1977): 48–50.

29. Ibid., 50–61. For more on sewerage development prior to 1920, see Melosi, *Sanitary City*, 90–99, 149–74.

30. Joel A. Tarr, James McClurley, and Terry F. Yosie, "The Development and Impact of Urban Wastewater Technology: Changing Concepts of Water Quality Control, 1850–1930," in Melosi, *Pollution and Reform*, 68–69.

31. "Needed Reforms ," 462. See also M. N. Baker, "Condition of Garbage Disposal in United States," *Municipal Journal and Engineer* 11 (October 1901): 147; William T. Sedgwick, *Principles of Sanitary Science and the Public Health* (New York: Macmillan Co., 1918), 117–18; Rich, "State Health Departments," 135; Herman G. James, *Municipal Functions* (New York: D. Appleton and Co., 1917), 85; Earl B. Phelps, *The Principles of Public Health Engineering* (New York: Macmillan Co., 1925), 236.

32. *Proceedings of the American Society for Municipal Improvements, Eighth Annual Convention* (1901): 183 (hereafter cited as *Proceedings of the ASMI).*

33. Mansfield Merriman, *Elements of Sanitary Engineering* (New York: John Wiley and Co., 1898; 2d ed. 1899), 7.

34. William Paul Gerhard, *Sanitation and Sanitary Engineering* (New York: John Wiley and Co., 1898; 2d ed. 1899), 56. See also "Place of the Engineer in Public Health," *American Journal of Public Health* 4 (July 1914): 589. The persistent emphasis upon "men" as engineers reflected the gender-specific character of the profession in this period.

35. Gerhard, *Sanitation and Sanitary Engineering*, 56–59; Merriman, *Elements of Sanitary Engineering*, 8; Richards, *Conservation by Sanitation*, 216–19.

36. A. Prescott Folwell, *Municipal Engineering Practice* (New York: John Wiley and Sons, 1916), 1–2.

37. Richards, *Conservation by Sanitation*, v. See also "The 'Civil Engineer,'" *Engineering Record* 67 (May 10, 1913): 509.

38. Richards, *Conservation by Sanitation*, 222. According to Schultz and McShane, "Sanitarians, landscape architects, and engineers formed a troika that tried to pull critics and officials alike from the mire of governmental inaction to the higher ground of municipal planning and administration." Schultz and McShane, "To Engineer the Metropolis," 396.

39. Gerhard, *Sanitation and Sanitary Engineering*, 58. In the 1920s, sanitary engineering led to the formation of yet another new engineering subspecialty—public health engineering, a combination of sanitary engineering and sanitary inspection. Later the term "environmental engineer" became a popular title for modern engineers with duties similar to those of the sanitary engineer. See Smillie, *Public Health Administration*, 260; Phelps, *Principles of Public Health Engineering*, 336–43. See also Robert S. Dorney, *The Professional Practice of Environmental Management* (New York: Springer-Verlag, 1989); Jeffrey K. Stine, "Engineering a Better Environment," (Paper presented at the SHOT/HSS Critical Problems and Research Frontiers Conference, Madison, Wisconsin, 1991).

40. Rudolph Hering, "Report of the Committee on Disposal of Garbage and Refuse," *Public Health: Papers and Reports* (APHA) 29 (October 1903): 129; APHA, *A Half Century of Public Health*, 190–91.

41. *Proceedings of the ASMI, Twenty-fourth Annual Convention* (1918): 296–313.

42. "Society for Street Cleaning and Refuse Disposal of the United States and Canada," *Municipal Journal and Engineer* 41(November 23, 1916): 646–47.

43. *Engineering News* 48 (November 30, 1902): 359; *Municipal Journal and Engineer* 17 (November 1904): 214–15; International Association of Public Works Officials, *Report of Proceedings—Conference of Street Cleaning Officials* (1919): 5–26.

44. Rudolph Hering, "How to Attack the Sewage and Garbage Problems," *American City*, August 1913, 111.

45. Rudolph Hering, "The Need for More Accurate Data in Refuse Disposal Work," *American Journal of Public Health* 2 (December 1912): 909–10. See also APHA, Sanitary Engineering Section, "Report of the Committee on Street Cleaning" (Report presented to the APHA, Jacksonville, FL, November-December, 1914); Rudolph Hering, "Report of the Committee on Disposal of Refuse Materials" *Public Health: Papers and Reports* (APHA) 27 (September 1901): 184–85.

46. "Report of the Committee on the Disposal of Garbage and Refuse," *Public Health: Papers and Reports* (APHA) 23 (October 1897): 206–18. See also William F. Morse, "The Next Step in the Work of Refuse and Garbage Disposal," *Public Health: Papers and Reports* (APHA) 25 (October-November 1899): 312; APHA, *A Half Century of Public Health,* 190–91; *Proceedings of the ASMI, Seventeenth Annual Convention,* (1910): 67; Rudolph Hering, "Vexed Question of Garbage Disposal," *Engineering Magazine*, June 1897, 392–98; Hering, "Modern Practice in the Disposal of Refuse," *American Journal of Public Health* 1 (December 1911): 910; "Refuse Disposal Investigations," *Municipal Journal and Engineer* 25 (December 2, 1908): 787; Hering and Greeley, *Collection and Disposal,* 12–20.

47. "Report of the Committee on Refuse Collection and Disposal," *American Journal of Public Health* 5 (September 1915): 933–34.

48. Samuel A. Greeley, "A Standard Form for Statistics of Municipal Refuse," *American Journal of Public Health* 2 (June 1912): 403. See also Louis L. Tribus, "Refuse and Garbage Disposal—A General Survey," *American Journal of Public Health* 6 (December 1916): 1307–14.

49. "Refuse Disposal in America," *Engineering News* 58 (July 25, 1908): 85. See also S. Whinery, "Recent Progress in Methods and Character of Street Cleaning," *American Journal of Public Health* 4 (August 1914): 680; S. Whinery, "How to Keep the Streets Clean," *American City*, January 1914, 23–24; M. N. Baker, "Condition of Garbage Disposal," 147; F. C. Bamman, "Analysis of Cost Keeping as Applied to Municipal Management of Street Cleaning," *American Journal of Public Health* 4 (August 1914): 674–78; P. M. Hall, "Methods of Accounting in the Collection of City Waste," *American Journal of Public Health* 2 (June 1912): 399–402; Hering, "How to Attack," 111.

50. Hering and Greeley, *Collection and Disposal*, 4.

51. Tribus, "Refuse and Garbage Disposal," 1311.

52. M. N. Baker, "Condition of Garbage Disposal," 148. See also James, *Municipal Functions,* 234–44; League of American Municipalities, *Proceedings of the Fifth Annual Convention,* (1901): 21; *Proceedings of the ASMI, Sixth Annual Convention* (1899): 232–35; "Municipal Ownership of Refuse Plants," *Municipal Journal and Engineer* 32 (June 27, 1912): 987.

53. Gleaned from US Department of Labor, "Statistics of Cities," Bulletin 36 (September 1901): 880–85; *Municipal Journal and Engineer* 37 (December 10, 1914): 836–44.

54. Gleaned from "Refuse Collection and Disposal," *Municipal Journal and Engineer* 39 (November 11, 1915): 723–27; B. F. Miller, "Garbage Collection and Disposal," *Proceedings of the ASMI, Twenty-second Annual Convention* (1915): 10–11.

55. Sedgwick, *Principles of Sanitary Science,* 115.

56. "The Sanitary Disposal of Municipal Refuse," *Transactions of the American Society of Civic Engineers* 50 (1903): 104 (hereafter cited as *Transactions of the ASCE).*

57. Hering, "Disposal of City Refuse," 398–406; "Report of the Committee on the Disposal of Garbage and Refuse" (October 1897), 207; Harry R. Crohurst, "Municipal Wastes: Their Character, Collection, Disposal," *US Public Health Service Bulletin* 107 (October 1920): 79ff; *Proceedings of the ASMI, Twenty-second Annual Convention* (1916): 244–45; M. N. Baker, "Condition of Garbage Disposal," 147–48; "Report of the Committee on Street Cleaning," *American Journal of Public Health* 5 (March 1915): 255–59.

58. "Report of the Committee on the Disposal of Garbage and Refuse," *Public Health: Papers and Reports* (APHA) 22 (April 1897): 108.

59. William F. Morse, "The Disposal of Municipal Wastes," *Municipal Journal and Engineer* 22 (March 6, 1907): 232–35. For an in-depth discussion of the failure to adapt the British destructor to American cities, see Melosi, "Technology Diffusion and Refuse Disposal," in Tarr and Dupuy, *Technology and the Rise of the Networked City,* 207–26.

60. William F. Morse, "The Sanitary Disposal of Municipal Refuse," *ASCE Transactions* 50 (1903): 114ff; "Refuse Disposal in America," *Engineering Record* 58 (July 25, 1908): 85; "Why American Garbage Crematories Fall," *Municipal Journal and Engineer* 13 (November 1902): 234; *Proceedings of the ASMI, Eighteenth Annual Convention* (1911): 8–11.

61. "Report of the Committee on the Disposal of Garbage and Refuse," (October, 1897), 215–18. See also *Proceedings of the ASMI, Fourth Annual Convention* (1897): 223.

62. "Garbage Collection and Disposal," *City Government* 7 (September 1899): 50. See also Joseph G. Branch, *Heat and Light from Municipal and Other Waste* (Saint Louis: W. H. O'Brien, 1906), 7–8.

63. "Report of Committee on Refuse Disposal and Street Cleaning," *Proceedings of the ASMI, Twenty-second Annual Convention* (1916): 245.

64. Samuel Haber, *Efficiency and Uplift: Scientific Management in the Progressive Era. 1890–1920,* ix. The "efficiency" movement was a central feature of reform efforts of the late nineteenth and early twentieth centuries. For a discussion of this issue from various standpoints see Martin J. Schiesl, *The Politics of Efficiency: Municipal Administration and Reform in America: 1880–1920* (Berkeley: University of California Press, 1977); Samuel P. Hays, *Conservation and the Gospel of Efficiency: The Progressive Conservation Movement. 1890–1920* (Cambridge, MA: Harvard University Press, 1959; reprint New York: Atheneum, 1972); Layton, *Revolt of the Engineers;* Klein and Kantor, *Prisoners of Progress* 27–32.

FOUR. Refuse as an Aesthetic Problem: Voluntary Citizens' Organizations and Sanitation

1. Venable, *Garbage Crematories in America,* 1.

2. Thomas C. Devlin, *Municipal Reform in the United States* (New York: G. P. Putnam's Sons, 1896), 39–40. For additional references on the emergence of municipal reform in the 1890s and beyond, see William M. Leary Jr., and Arthur S. Link, eds., *The Progressive Era and the Great War, 1896–1920,* 2d ed. (Arlington Heights, IL: AHM Publishing Corp., 1978); Adam W. Rome, "Coming to Terms with Pollution: The Language of Environmental Reform, 1865–1915," *Environmental History* 1 (July 1996): 6–28; Martin V. Melosi, "Battling Pollution in the Progressive Era," *Landscape* 26 (1982): 35–41; Melosi, *Effluent America,* 207–10.

3. Roy Lubove, "The Twentieth Century City: The Progressive as Municipal Reformer," *Mid-America* 41 (October 1959): 200.

4. Melvin G. Holli, "Urban Reform in the Progressive Era," in *The Progressive Era*, ed. Lewis L. Gould, (Syracuse, NY: Syracuse University Press, 1974), 133–41. Griffith, *History of American City Government: The Progressive Years*, 118–19.

5. See William Howe Tolman, *Municipal Reform Movements in the United States* (New York: Fleming H. Revell Co., 1895). See also Kenneth Feingold, *Experts and Politicians: Reform Challenges to Machine Politics in New York, Cleveland, and Chicago* (Princeton: Princeton University Press, 1995). According to Feingold, there were three patterns of urban reform politics that emerged in the Progressive Era: traditional reform led by predominantly native-stock and elite forces; municipal populists drawn from new immigrants and the upper-working class; and progressive coalitions that combined elements of the other two. See pp. 13–22.

6. Devlin *Municipal Reform in the United States,* 14–15.

7. Tolman, *Municipal Reform Movements,* 47–136. See also Holli, "Urban Reform," 135–36; Griffith, *History of American City Government: The Progressive Years*, 117–19.

8. For a discussion of the Commission Plan, see Bradley Robert Rice, *Progressive Cities: The Commission Government Movement in America, 1901–1920* (Austin: University of Texas Press, 1977); see Rice's bibliography for citations on the City Manager Plan.

9. Griffith, *History of American City Government: The Progressive Years*, 156–60.

10. Clifford Patton, *Battle for Municipal Reform: Mobilization and Attack, 1875–1900* (Washington, DC: American Council on Public Affairs, 1940), 34–35.

11. Griffith, *History of American City Government: The Progressive Years*, 159–60.

12. See Melosi, *Pollution and Reform.* For additional references, see Carolyn Merchant, *The Columbia Guide to American Environmental History* (New York: Columbia University Press, 2002).

13. Holli, "Urban Reform," 133.

14. Goldfield and Brownell, *Urban America*, 214. See also David F. Burg, *Chicago's White City of 1893* (Lexington, KY: University Press of Kentucky, 1976).

15. Goldfield and Brownell, *Urban America,* 214.

16. Ibid., 214–17.

17. Jon A. Peterson, "The City Beautiful Movement: Forgotten Origins and Lost Meanings," *Journal of Urban History* 2 (August 1976): 416–28. See also William H. Wilson, *The City Beautiful Movement* (Baltimore: Johns Hopkins University Press, 1989).

18. Ibid., 421–30.

19. Ibid., 424. See also "Clean Streets and Smokeless Chimneys Demanded," *California Municipalities* 5 (November 1901): 114; Civic League of Saint Louis, *Yearbook* (1907), 9–11, 16–17, 36–37; Saint Louis Civic Improvement League, *Keep Our City Clean* (1902).

20. Caroline Bartlett Crane, "The Work for Clean Streets," New York City Women's Municipal League, *Bulletin* 5 (August 1906): 1.

21. Civic Club of Philadelphia, *Civic Club Bulletin* 5 (October 1911): 15.

22. See such periodicals as *City Government; Pacific Municipalities; City Hall: Bulletin of the League of American Municipalities;* and *American Municipalities.* See also bulletins of local civic groups and proceedings of national civic organizations.

23. "The Advantages of Municipal Construction over the Contract System," *Proceedings of the League of American Municipalities* (1903): 21. See also City Club of Philadelphia, *City*

Club Bulletin 5 (May 13, 1912): 199–200; "Street Cleaning without Contract," *City Government* 6 (January 1899): 16.

24. "The Advantages of Municipal Construction over the Contract System," 21ff. The Cleveland Board of Control offered a unique alternative to the contract and the municipal systems. In 1900, it adopted a resolution stating that if a citizen took charge of his own street cleaning the city would turn over to him from the street fund the amount allocated to the particular street. See "Cleveland Citizens Keep Streets Clean," *City Government* 9 (July 1900): 19.

25. "Garbage Contracts," *City and State* 4 (20 January 1898): 259; Civic League of Saint Louis, *Yearbook* (1912), 5; Saint Louis Civic League, "The Collection and Disposal of Rubbish and Ashes," *Civic Bulletin* 2 (January 8, 1912): 2–4. See also Frederick C. Wilkes, "Should the City Remove Rubbish?" *City Hall* 4 (October 1905): 134–36; "Wants No Garbage Contract," *City Government* 6 (March 1899): 57; New York American, Public Welfare Department, *Considerably More Than Too Much!* (New York: McConnell Printing Co.), 3–9; Pittsburgh Civic Club of Allegheny County, *Fifty Years of Civic History, 1896–1954*, comp. H. Marie Dermitt, 1945, 7–8.

26. Michigan Citizen's Research Council, *Report on Street Cleaning and Refuse Collecting, Department of Public Works, City of Detroit* (February 1917).

27. City Club of Chicago, "Household Pests and Their Relation to Public Health," *City Club Bulletin* 4 (May 1, 1911): 77.

28. San Francisco Citizen's Health Committee, *Eradicating Plague from San Francisco*, prepared by Frank Morton Todd, March 31, 1909, 122–25. For information about other civic lobbying for better conditions in San Francisco, see "San Francisco Street-cleaning Department Assailed," *Engineering Record* 69 (March 28, 1914): 368.

29. Saint Louis Civic Improvement League, Public Sanitation Committee, *Disposal of Municipal Waste* (1906), 1–19; Saint Louis, Civic League, *Yearbook* (1907), 36–37; *Yearbook* (1908), 12; *Yearbook* (1909), 39–40; Saint Louis Civic League, *Civic Bulletin* 1 (March 6, 1911): 2.

30. See Saint Louis City Improvement League, *Keep Our City Clean* (1902), 2–3; Philadelphia Civic Club, *Annual Report* (1911), 56; "Street Cleaning," *City Hall* 4 (September 1905): 61–62; Merchants Association of San Francisco, *Street Cleaning Problem in San Francisco* (1909) 10–11, 28–29; "Street Cleaning," *City Hall* 8 (July 1907): 55–58; Douglas Sutherland, *Fifty Years on the Civic Front: A History of the Civic Federation's Dynamic Activities* (Chicago: Chicago Federation, 1943), 16–17; Henry G. Selfridge, *Suggestions on the Problem of Cleaning the Streets of Chicago* (Chicago: City Homes Association, 1901); Carol Aronovici, "Municipal Street Cleaning and Its Problems," *National Municipal Review* 1 (April 1912): 225.

31. Frederick C. Wilkes, "Cleanliness and Economy: How Pittsburgh Might Be Cleaned," *City Hall* 4 (November 1905): 154. See also "Citizen Co-operation in Street Cleaning," *Municipal Journal and Engineer* 29 (November 30, 1910): 746.

32. Lois W. Banner, *Women in Modern America: A Brief History* (New York: Harcourt Brace Jovanovich, 1974), 87–89. See Zane L. Miller, *The Urbanization of Modern America: A Brief History* (New York: Harcourt Brace Jovanovich, 1973), 104–5. See also Maureen Flanagan, "The City Profitable, the City Livable: Environmental Policy, Gender, and Power in Chicago in the 1910s," *Journal of Urban History* 22 (January 1996): 163–90. For additional references on women and the environment, see Merchant, *Columbia Guide*.

33. Raymond W. Smilor, "Toward an Environmental Perspective: The Anti-Noise Campaign, 1893–1932," in Melosi, *Pollution and Reform*, 141–46.

34. R. Dale Grinder, "The Battle for Clean Air: The Smoke Problem in Post-Civil War America," in Melosi, *Pollution and Reform*, 89.

35. See Hoy, *Chasing Dirt*.

36. Mildred Chadsey, "Municipal Housekeeping," *Journal of Home Economics* 7 (February 1915): 53.

37. Crane, "Work for Clean Streets," 2.

38. Susan Strasser, *Waste and Want: A Social History of Trash* (New York: Henry Holt and Company, 1999), 136. See also 121–24, 137–40; Daniel Eli Burnstein, "Progressivism and Urban Crisis: The New York City Garbage Workers' Strike of 1907," *Journal of Urban History* 16 (August 1990): 403.

39. Edith Parker Thomson, "What Women Have Done for the Public Health," *Forum* 24 (September 1897): 54–55.

40. Hoy, "'Municipal Housekeeping,'" 174.

41. See, for example, Elizabeth D. Blum, "Pink and Green: A Comparative Study of Black and White Women's Environmental Activism in the Twentieth Century" (PhD diss., University of Houston, 2000).

42. Banner, *Women in Modern America*, 47.

43. Samuel A. Greeley, "The Work of Women in City Cleaning," *American City*, June 1912, 874.

44. Imogen B. Oakley, "The More Civic Work, the Less Need of Philanthropy," *American City*, June 1912, 807–808.

45. Ibid., 808–11; Wagner, "What the Women Are Doing," 35.

46. "Cleaning Up American Cities," *Survey* 25 (October 1910): 85; Oakley, "More Civic Work," 810–11; Mrs. Lee Bernheim, "A Campaign for Sanitary Collection and Disposal of Garbage," *American City*, August 1916, 135–36; Hester M. McClung, "Women's Work in Indianapolis," *Municipal Affairs* 2 (September 1898): 523–24; Mary Ritter Beard, *Women's Work in Municipalities*, 89–90. See also Tolman, *Municipal Reform Movements in the United States*, I67–82.

47. *New York City, Woman's Municipal League, Campaign Bulletin*, November 1903, 12.

48. See the following publications by New York City, Woman's Municipal League: *Bulletin* 5 (April 1907): 8–9; 5 (June 1907): 3–4; 6 (June 1908): 7–8; *Yearbook* (1911), 45–50; *Yearbook* (1912), 26–46; *Yearbook* (1913), 24–25; *Yearbook* (1914), 22–25. See also William H. Edwards, "Four Kinds of Cooperation Needed by Street Cleaning Departments," *American City*, July 1913, 65; Mrs. Julius Henry Cohen, *What We Should All Know about Our Streets: Prepared for the Use of Our Young Citizens in the City Schools* (New York: Women's Municipal League of the City of New York, 1916).

49. New York City, Woman's Municipal League, *Bulletin* 5 (July-August 1907): 1.

50. Hoy, "'Municipal Housekeeping,'" 181–88; "Cleaning Up American Cities," 83–84; Mary Ritter Beard, *Woman's Work in Municipalities*, 86–87; Crane, "Work for Clean Streets," 1–10; Suellen M. Hoy, "Caroline Bartlett Crane," *APWA Reporter* 45 (June 1978): 4–5.

51. Hoy, "'Municipal Housekeeping,'" 188–93; Mary Ritter Beard, *Woman's Work in Municipalities*, 88–89; "Chicago's Struggle for Scientific Garbage Collection and Disposal," *Survey* 31 (March 21, 1914): 776–77. For more on the role of women in sanitation and urban environmental reform in general, see Hoy, *Chasing Dirt*; Robert Gottlieb, *Forcing the*

Spring: The Transformation of the American Environmental Movement (Washington, DC: Island Press, 1993); Gottlieb, "Reconstructing Environmentalism: Complex Movements, Diverse Roots," *Environmental History Review* 17 (Winter, 1993): 1–19.

52. Hoy, "'Municipal Housekeeping,'" 190; Mary Ritter Beard, *Woman's Work in Municipalities,* 88; Addams, *Twenty Years at Hull-House,* 200–205. See also Harold L. Platt, "Jane Addams and the Ward Boss Revisited: Class, Politics, and Public Health in Chicago, 1890–1930," *Environmental History* 5 (April 2000): 194–222.

53. See Robert Clarke, *Ellen Swallow: The Woman Who Founded Ecology* (Chicago: Follett Publishing Co., 1973).

54. Lawrence A. Cremin, *The Transformation of the School: Progressivism in American Education, 1876-1957* (New York: Alfred A. Knopf, 1961), viii–ix. See also Burnstein, "Clean Streets."

55. See Raymond E. Callahan, *Education and the Cult of Efficiency: A Study of the Social Forms that Have Shaped the Administration of the Public Schools* (Chicago: University of Chicago Press, 1962), 5–6.

56. Wilcox, *American City*, 91.

57. Ibid., 92. See also "Junior Improvements," *American City*, October 1910, 196; "The Junior Civic Leagues of Binghamton, N.Y.," *American City*, November 1911, 297; Charles Dwight Willard, *City Government for Young People* (New York: Macmillan Co., 1906); Julia Richman and Isabel Richman Wallach, *Good Citizenship* (New York: American Book Co., 1908); Cohen, *What We Should All Know.*

58. Cremin, *Transformation of the School,* 58–61; Callahan, *Education and the Cult of Efficiency,* 8ff.; Wilcox, *American City,* 102–20; Samuel H. Ziegler, "Practical Citizenship Taught to High School Boys," *American City*, July 1914, 20–23.

59. "The Children's Responsibility," *American City*, July 1911, 41; William H. Allen, "Teaching Civics by Giving Pupils Civic Work to Do," *American City,* February 1916, 154–55; Ethel Rogers, "Playing at Citizenship," *American City,* November 1913, 445–48.

60. "Junior Improvements," 196.

61. Simons, "Juvenile Street Cleaning Leagues," 163–66; Pt. 2, November 1910, 239–43. "New York Street Cleaning," *Municipal Journal and Engineer* 39 (July 13, 1910): 49. During the years between the termination of the league and its reestablishment, the Woman's Municipal League of New York kept the idea alive with the Juvenile City League. See *Bulletin of the Woman's Municipal League* 2 (October 1903): 3–4; 2 (December 1903): 5; 2 (February 1904): 3–4; 3 (September 1904): 4; 3 (October 1904): 3; 3 (December 1904): 4; 5 (November 1906): 1–3; 5 (June 1907): 4.

62. "A Street Cleaning Nurse," *Literary Digest* 52 (March 18, 1916): 709–10; "Junior Civic Leagues," 297–98; "Health Leagues in the Schools," *American City*, August 1917, 152–53; William W. B. Seymour, "Junior Deputy Sanitary Inspectors, *American City*, December 1916, 696–98; "Young Boosters for a Chicago Beautiful," *American City*, June 1911, 290; "Anti-Litter League," *American City* 15, July 1916, 67; Bernheim, "Campaign for Sanitary Collection," 136; "Street Cleaning," *City Hall* 8 (July 1907): 55–56; Oakley, "More Civic Work," 811; Saint Louis Civic League, *Civic Bulletin* 1 (December 12, 1910): 4; Cleveland Department of Public Service, Subdivision of Street Cleaning, *Annual Report* (1914), 3: Philadelphia Department of Public Works, Bureau of Highways and Street Cleaning, *Annual Report* (1913), 53.

63. For a good example of the effort to "civilize" the immigrants, see David Willard, "The Juvenile Street-Cleaning Leagues," in Waring, *Street Cleaning*, 177–86.

64. William Parr Capes and Jeanne Daniels Carpenter, *Municipal Housecleaning* (New York: E. P. Dutton and Co., 1918), 213–14.

65. Mary Ritter Beard, *Woman's Work in Municipalities,* 85.

66. Mrs. C. J. Baxter, "A Women's League That Keeps the Streets Clean," *American City*, June 1912, 898–99, 901. See also Mrs. George E. Bird, "The Parade Inaugurated a Village Clean-up Campaign," *American City*, February 1916, 162–64; "Recognizing the Work of the Children," *American City* 3, July 1910, 42. See also Capes and Carpenter, *Municipal Housecleaning,* 214–32; Baltimore Department of Street Cleaning, *Annual Report* (1912), 4–5; *Annual Report* (1915), 6–7.

67. Gustavus A. Weber, "A 'Clean-up' Campaign Which Resulted in a 'Keep-Clean' Ordinance," *American City*, March 1914, 231–34; Ewing Galloway, "How Sherman Cleans Up," *American City*, July 1913, 40–41; Mary Ritter Beard, *Woman's Work in Municipalities,* 84.

68. "Children in City Clean-up Work," *American City*, February 1916, 156–61.

69. See R. P. Crawford, "Training the Young in Civic Duties," *American City*, April 1917, 359–61; *American City*, March 1911, 146; "A Plan for Interesting Children in Civic Betterment," *American City*, April 1913, 415.

70. "Philadelphia's Second Annual Clean-up Week," *Municipal Journal and Engineer* 37 (September 10, 1914): 348–49; "Annual Municipal Clean-up Week in Philadelphia," *Engineering News* 73 (April 1, 1915): 620–21; R. Robinson Barett, "Philadelphia's Second Annual Clean-up Week," *American City*, April 1915, 299–304; "Get the Habit," *American City*, February 1917, 154–55. See also Philadelphia Department of Public Works, Bureau of Highways and Street Cleaning, *Annual Report* (1914), 63–75; *Annual Report* (1915), 32–35; Philadelphia Bureau of Street Cleaning, *Annual Report* (1917), 9–12.

71. Zueblin, *American Municipal Progress,* 78.

72. For more details of environmental reform efforts in the late nineteenth and early twentieth centuries, see Melosi, ed., *Pollution and Reform;* Melosi, *Effluent America.*

FIVE. Street-Cleaning Practices in the Early Twentieth Century

1. "Disposition of Municipal Refuse," *Municipal Journal and Engineer* 19 (October 1905): 169. See also *Proceedings of the ASMI, Twenty-first Annual Convention, Boston,* (1914): 1–3, 49–58; Rudolph Hering, "Disposal of Municipal Refuse; Review of General Practice," *Transactions of the ASCE* 54, pt. E (1904): 265–66.

2. Venable, *Garbage Crematories in America,* 2.

3. M. N. Baker, *Municipal Engineering and Sanitation* (New York: Macmillan Co., 1902), 5–6; see also 151–56; Crohurst, "Municipal Wastes," 7–8; William F. Morse, *The Collection and Disposal of Municipal Waste* (New York: Municipal Journal and Engineer, 1908), 1–95.

4. Aronovici, "Municipal Street Cleaning," 218–19. See also Edward T. Hartman, "The Social Significance of Clean Streets," *American City*, October 1910, 173.

5. McShane, "Transforming the Use of Urban Space," 279–307. See also "Street Cleaning Standards," *Municipal Journal and Engineer* 35 (December 11, 1913): 794–96; "Street Cleaning and Pavement Economy," *Municipal Journal and Engineer* 40 (January 6, 1916): 8–9; Charles A. Beard, *American City Government: A Survey of Newer Tendencies* (New York: Century Co., 1912), 242–48.

6. See "Tables of Street Cleaning Statistics," *Municipal Journal and Engineer* 37 (December 10, 1914): 833–44; Capes and Carpenter, *Municipal Housecleaning,* 64–65.

7. See L. M. King, "Street Cleaning in San Francisco," *Engineering News* 50 (August 20, 1903): 169; "Indianapolis Discards Contract Street Cleaning," *Municipal Journal and Engineer* 18 (June 1905): 31; "Street Cleaning without Contractors in Washington, D.C.," *Engineering News* 70 (October 20, 1913): 681–82; J. W. Paxton, "Municipal versus Contract Street Cleaning in Washington, D.C.," *American Journal of Public Health* 4 (November 1914): 1032–34; "Philadelphia Street Cleaning by Contract," *Engineering News* 74 (July 1, 1915): 6; Philadelphia Bureau of Municipal Research, *Municipal Street Cleaning in Philadelphia* (June 1924), 9.

8. Paxton, "Municipal versus Contract Street Cleaning," 1033. See also Philadelphia Bureau of Municipal Research, *Municipal Street Cleaning in Philadelphia,* 10–12.

9. An important exception to anticontract sentiment caused a change in policy in New York City. Until 1909, the city force had been completely responsible for street cleaning. In that year, the governor of New York signed a bill giving the city the authority to enter into five-year contracts for machine street cleaning and sprinkling. The city did not abandon its sweeping patrols but added machine sweeping by contract to its operations. The argument in favor of this move was two-fold: municipal operation was more successful in service branches with small numbers of employees than in large departments where politics influenced decisions, and mechanical substitutes for hand sweeping had come of age and should be substituted for brooms for some tasks. The example of New York City was not followed by many other cities, however. Only 1 or 2 percent of the cities surveyed in this period combined municipal and contract operations. See "Street Cleaning by Contract or Day Labor," *Municipal Journal and Engineer* 26 (May 26, 1909): 931.

10. In 1916, the following comment was made about street cleaning in suburban Philadelphia: "Another advance in the department [of street cleaning] was the provision for cleaning suburban streets and country roads. Prior to 1915, street cleaning work covered only the paved streets. Now a force of uniformed men are at work on the suburban roads, resulting in great improvement in appearance at a slight cost." See "Street Cleaning in Philadelphia," *Municipal Journal and Engineer* 40 (June 29, 1916): 898.

11. William L. Riordan, *Plunkitt of Tammany Hall* (New York: E. P. Dutton and Co., 1963), 6.

12. Soper, *Modern Methods of Street Cleaning,* 11.

13. Stuart Stevens Scott, "The Street Cleaning Department of Baltimore," *American City*, December 1913, 546–48.

14. Philadelphia Department of Public Works, Bureau of Highways and Street Cleaning, *Annual Report* (1914), 49–53.

15. "New York City Will Try Innovation in Modern Street Cleaning," *Better Roads and Streets*, July 1915, 30.

16. Whinery, "Recent Progress in Methods and Character," 680.

17. Ibid., 679.

18. F. C. Bamman, "Analysis of Cost Keeping as Applied to Municipal Management of Street Cleaning," *American Journal of Public Health* 4 (August 1914): 677–78. See also J. W. Paxton, "Street Cleaning Methods and Costs at Washington, D.C.," *Engineering News* 72 (July 9, 1914): 58–66.

19. Statistics gleaned from US Department of Labor, *Bulletin* 24 (September 1899): 662–64; *Bulletin* 30 (September 1900): 958–62; *Bulletin* 36 (September 1901): 880–84; *Bulletin* 42 (September 1902): 958–62; US Bureau of the Census, *Statistics of Cities Having a Population of over 25,000: 1902–1903, Bulletin* 20 (1905): 121–29; US Bureau of the Census, *Statistics of Having a Population of over 3,000: 1905, Special Reports* (1907), 338–40. According to the 1905 census, sweeping machines were used by all surveyed cities with populations of 100,000 or more, by 90 percent of cities with populations of 50,000 to 100,000, and by 75 percent of cities with populations of 30,000 to 50,000.

20. US Bureau of the Census, *Statistics of Cities Having a Population of 8,000 to 25,000: 1903, Bulletin* 45 (1906): 92–97. For statistics on regional variations in cleaning methods, see "Street Cleaning Methods," *Municipal Journal and Engineer* 26 (March 24, 1909): 485–89.

21. Whinery, "How to Keep the Streets Clean," 21.

22. "Hand vs. Machine Street Cleaning," *Municipal Journal and Engineer* 17 (August 1904): 95; "Street Cleaning Here and Abroad," *Outlook* (August 1, 1914), 774–75; Gerhard, *Sanitary Engineering*, 58–59; Soper, *Modern Methods of Street Cleaning*, 2–3; Frank Hagerdorn, "Sweeping City Streets by Machine," *American City* 12 (February 1915): 147–48; "Some Notes on the Development of Street Methods," *Engineering and Contracting* 42 (October 21, 1914): 394–96; Edward D. Very, "Modern Methods of Street Cleaning," *American City*, November 1912, 435–39.

23. See *Municipal Engineering* 50 (January 1916): 26, 44; *Municipal Journal and Engineer* 33 (July 4, 1912): 2, 9, 14.

24. "The Street Cleaning Problem," *California [Pacific] Municipalities* 8 (May 1903): 109–10; "A New Street Cleaning and Disinfecting Method," *Public Improvements* 1(1 July 1899): 85; "Vacuum Street-Cleanings," *Literary Digest* 48 (14 March 1914): 548; William H. Connell, "Organization and Method of Street Cleaning Departments," *Canadian Engineer* 26 (26 March 1914): 503–505; "Street Sweeping," *Pacific (formerly California) Municipalities* 9 (December 1903): 164–68.

25. "A New Street Cleaning and Disinfecting Method," 85.

26. "Street Flushing Practice in American Cities," *American City* 16 (February 1917): 117–21; Raymond W. Parlin, "Flushing and Street Cleaning," *American Municipalities* 33 (May 1917): 49-53; Raymond W. Parlin, "Hand Flushing—Its Place in the Street Cleaning Field," *American City* 14 (May 1916): 441-48; "Sanitary Street Flushing," *Municipal Journal and Engineer* 18 (June 1905): 295; "Sanitary Street Flushing Machine," *Municipal Journal and Engineer* 19 (July 1905): 46; "Street Cleaning," *Municipal Journal and Engineer* 29 (23 November 1910): 713-14; Gus H. Hanna, "Economy in Street Cleaning," *American City* 14 (January 1916): 22; "Street Cleaning Methods and Costs in Several Ohio Cities," *Engineering and Contracting* 38 (18 September 1912): 319; "Methods and Costs of Street Cleaning at Washington, D.C., During 1911–1912," *Engineering and Contracting* 38 (18 December 1912): 684.

27. "Street Cleaning and Refuse Disposal in New York," *Engineering Record* 57 (22 February 1908): 208; New York City, Department of Street Cleaning, *Report of the Commission on Street Cleaning and Waste Disposal, 1907*, p. 138. See also Parlin, "Hand Flushing-Its Place in the Street Cleaning Field," 491ff; Whinery, "How to Keep the Streets Clean," 22–23.

28. For example, see Very, "Modern Methods of Street Cleaning," 435–36.

29. "Motor-driven Squeegees in Street Cleaning Service," *Scientific American*, July 15, 1916): 66, 68; "Motor Wagons for Municipal Work," *Municipal Journal and Engineer* 16 (February 1904): 81; Thomas Finegan, "The Comparative Cost of Sweeping Pavements by Horse-drawn Sweepers and by Motor Sweepers," *American City* February 1915, 148–49; "Street Cleaning by Motor Apparatus," *Municipal Journal and Engineer* (January 14, 1915): 33–34.

30. "Clean Streets and Motor Traffic," *Literary Digest* (September 5, 1914): 413–14.

31. Joseph J. Norton, "The Public and Clean Streets," *American Municipalities* 28 (January 1915): 136.

32. Edward D. Very, "Street Cleaning Methods and Results," *Municipal Engineering* 47 (September 1914): 176.

33. Michigan Citizens' Research Council, *Report on Street Cleaning and Refuse Collection, Department of Public Works, City of Detroit* (February 1917), 74–75.

34. Guy C. Emerson, "Individual Responsibility for Clean Streets," *New Boston*, August 1910, 2.

35. "Street Cleaning in Washington," *City Hall: Bulletin of the League of American Municipalities* 10 (May 1909): 385; "New Haven Fights Street Littering," *Municipal Journal and Engineer* 40 (January 13, 1916): 37–38; "To Prevent Littering Parks," *Municipal Journal and Engineer* 39 (July 1, 1915): 8; "Street Cleaning and Pavement Economy," *Municipal Journal and Engineer* 40 (January 6, 1916): 7–9. See also "Street Cleaning in New York," *Municipal Journal and Engineer* 35 (December 11, 1913): 789; New York City Department of Street Cleaning, *Unsightly Streets and Careless People: Control of the Loose Paper Nuisance* (1914); "Street Cleaning," *Municipal Journal and Engineer* 29 (November 23, 1910): 713.

36. "Amount and Cost of Street Cleaning," *Municipal Engineering* 30 (April 1906): 280. See also "Comparative Cost of Street Cleaning," *Municipal Journal and Engineer* 12 (April 1902): 160.

37. Capes and Carpenter, *Municipal Housecleaning*, 39; Baltimore Department of Street Cleaning, *Annual Report* (1900), 6. See also Paul Iglehart, "Street Cleaning in Baltimore," *Municipal Journal and Engineering* 10 (March 1901): 89; "Amount and Cost of Street Cleaning," 280.

38. "Some Hazards of City Housecleaning," *Survey* (April 14, 1917): 42; "Strike of New York Street Cleaners," *Survey* (November 25, 1911): 1243–49; "Street Cleaning and Refuse Collection Methods," *Municipal Journal and Engineer* 42 (May 17, 1917): 688; William H. Edwards, "The Work of the Street Cleaning Department of New York City," City Club of Philadelphia, *City Club Bulletin* 2 (March 24, 1910).

SIX. Collection and Disposal Practices in the Early Twentieth Century

1. Morse, *Collection and Disposal of Municipal Waste*, 1.

2. Hering and Greeley, *Collection and Disposal of Municipal Refuse*, 3; Samuel A. Greeley, "Refuse Disposal and Street Cleaning, *Engineering Record* 69 (January 3, 1914): 15; "Report of Committee on Refuse Collection and Disposal" *American Journal of Public Health* 7 (April 1917): 412–13, Refuse Disposal in America *Engineering Record* 58 (July 25, 1908): 85. See also M. L. Davis, "The Disposal of Garbage " *Journal of the American Medical Association* 31 (July 2, 1898): 23–26; "Street Cleaning and Refuse Disposal," *Municipal Journal and Engineer* 30 (January 4, 1911): 15; *Proceedings of the ASMI, Seventeenth Annual Meeting, Erie, Pa.* (1910): 57-58.

3. M. N. Baker, *Municipal Engineering and Sanitation,* 5–6; M. N. Baker, "Condition of Garbage Disposal," 147.

4. Luther E. Lovejoy, "Garbage and Rubbish," *Proceedings of the Academy of Political Science* 2 (1911–12): 300; "Need of Sanitary Garbage Disposal," *Municipal Journal and Engineer* 18 (April 1905): 173–74; John H. Simon, "Municipal Waste," *Municipal Journal and Engineer* 18 (January 1905): 10; William P. Munn, "Collection and Disposal of Garbage," *City Government* 2 (January 1897): 6; "The Unsatisfactory Condition of Refuse Disposal in America," *Sanitary Record* 21 (February 11, 1898): 146; *Proceedings of the ASMI, Eighth Annual Meeting, Niagara Falls, N.Y.* (1901): 183; "The Unsatisfactory Condition of Garbage Disposal in America," *Sanitarian* 40 (January 1898): 20.

5. See Hering and Greeley, *Collection and Disposal of Municipal Refuse,* 155–56.

6. M. N. Baker, "Condition of Garbage Disposal," 148. See also "Garbage Collection and Disposal of Philadelphia," *Engineering News* 45 (January 17, 1901): 41; Philadelphia Bureau of Health, *Annual Report* (1898), xix.

7. In *Collection and Disposal of Municipal Refuse,* p. 156, Hering and Greeley enumerated the advantages and disadvantages of the contract system. Among the advantages they listed a more effective application of business principles, the elimination of politics from the operations, a simplification of the work of cities and especially of small towns, a fixed expenditure for service, and limitations in capital expenses for the city. This list assumes an institutional vulnerability of public but not private service; however, Hering and Greeley were not myopic in their evaluation of the contract system. They recognized that contractors were motivated by profit, could ignore contract provisions that were not spelled out clearly, kept inadequate records (at least for municipal consumption), and bore no direct responsibility to the citizenry. See also I. S. Osborn, *Disposal of Garbage in the District of Columbia,* US Congress, House, 64th Cong., 1st sess., 1915, Doc. 661; "How Not to Award a Contract," *Municipal Journal and Engineer* 11 (August 1901): 69–70; Boston Special Commissions on Collection and Disposal of Refuse, *Reports of the First and Second Special Commissions to Investigate the Subject of the Collection and Disposal of Refuse in the City of Boston* (1908, 1910), 24–25; Washington DC Department of Street Cleaning, *Report of the Superintendent* (1899), 550; *Engineering News* 46 (October 24, 1901): 308; Cleveland Board of Public Service, Division of Engineering, *Annual Report* (1906), 93–95; J. W. Paxton, "Collection and Disposal of City Refuse, Washington, D.C.," *Engineering News* 72 (October 1, 1914): 671–74; "The Latest Garbage-Disposal Contract of Los Angeles," *Engineering News* 70 (August 28, 1913): 422.

8. Hering and Greeley, *Collection and Disposal of Municipal Refuse,* 104–5; Parsons, *Disposal of Municipal Refuse,* 43.

9. Morse, *Collection and Disposal of Municipal Waste,* 36; John H. Gregory, "Collection and Disposal of Municipal Refuse," *American Journal of Public Health* 2 (December 1912): 919; Saint Louis Civic League, *Civic Bulletin* 2 (January 8, 1912): 2; "Refuse Disposal in Ohio," *Municipal Journal and Engineer* 25 (December 2, 1908): 776. See also William F. Morse, "The Collection of Municipal Waste," *American Journal of Public Health* 4 (July 1914): 564–69.

10. Robert H. Wyld, "Modern Methods of Municipal Refuse Disposal," *American City,* October 1911, 205–207; Parsons, *Disposal of Municipal Refuse,* 44. For a discussion of the controversy over separation in New Orleans, see *Engineering News* 39 (March 10, 1898): 160.

11. See C. E. A. Winslow and P. Hansen, "Some Statistics of Garbage Disposal for the Large American Cities in 1902," *Public Health: Papers and Reports* (APHA) 29 (October 1903): 141–52; *Proceedings of the ASMI* (1915): 10–11; "Refuse Collection and Disposal," (*Municipal Journal and Engineer*), 723–25; Pittsburgh Commission on Garbage and Rubbish Collection and Disposal, *Report on Methods of Garbage and Rubbish Collection and Disposal in American Cities* (1918), 14–15; Capes and Carpenter, *Municipal Housecleaning,* table vi; "Garbage Collection and Disposal: A Compilation from Questionnaires Returned by 101 City Manager Cities in the US and Canada," *City Manager Magazine,* July 1924, 12–14.

12. Winslow and Hansen, "Some Statistics of Garbage Disposal," 141–53.

13. See B. F. Miller Jr., "Horse or Motor for Collecting City Garbage," *Engineering News* 76 (November 23, 1916): 1006–1007; Baltimore Department of Street Cleaning, *Annual Report* (1913), 6–7; "Collection of Refuse and Disposal in Chicago," *Engineering Record* 69 (April 11, 1914): 424–25; Samuel A. Greeley, "Motor Trucks for Refuse Collection," *American City,* March 1916, 239–43; "Garbage Collection Studies in Chicago Justify Continued Use of Horses," *Engineering Record* 72 (July 10, 1915): 52–53.

14. These figures are primarily for summer months; collections were less frequent in the winter months, especially in northern cities. Collections for ashes and rubbish or combined waste with garbage were also less frequent. See "Disposal of Municipal Refuse," *Municipal Journal and Engineer* 35 (November 6, 1913) 632–33; "Uniform Statistics of Refuse Collection and Disposal," *Engineering News* 70 (October 2, 1913): 678; "Refuse Disposal and Street Cleaning," *Municipal Journal and Engineer* 36 (March 12, 1914): 361; "Refuse Collection and Disposal," (*Municipal Journal and Engineer*), 725–27; Pittsburgh Commission on Garbage and Rubbish Collection and Disposal *Report,* 14–15; Capes and Carpenter, *Municipal Housecleaning,* table vi; "Garbage Collection and Disposal," 12–13.

15. Hering and Greeley, *Collection and Disposal of Municipal Refuse,* 70.

16. Rudolph Hering, "Disposal of Municipal Refuse; Review of General Practice," *Transactions of the ASCE* 54 (1904): 278–79; US Bureau of the Census, *Statistics of Cities Having a Population of over 30,000: 1905,* 337–41; H. de B. Parsons, "Disposal of Municipal Refuse and Rubbish Incineration," *Transactions of the ASCE* 57 (December 1906): 56–65; Parsons, "City Refuse and Its Disposal," *Journal of the Society of Chemical Industry* 27 (April 30, 1908): 376–79; "City Waste Studies in Ohio Cities," *Engineering News* 67 (March 28, 1912): 608; "Disposal of Municipal Refuse," *Municipal Journal and Engineer 35* (November 6, 1913): 627ff.; "Uniform Statistics of Refuse Collection and Disposal," 678; Morse, "Collection of Municipal Waste," 569–70; "Refuse Disposal and Street Cleaning," *Municipal Journal and Engineer* 36 (March 12, 1914): 361; "Refuse Collection and Disposal," (*Municipal Journal and Engineer*), 725–27; Morris Irwin Evinger and Daniel C. Faber, "The Collection and Disposal of City Refuse," *Bulletin of Iowa State College of Agriculture and Mechanical Arts* 14 (January 1, 1915): 6–8; Folwell, *Municipal Engineering Practice,* 320–21; Capes and Carpenter, *Municipal Housecleaning,* 167–68; Pittsburgh Commission on Garbage and Rubbish Collection and Disposal, *Report,* 14–15; Hering and Greeley, *Collection and Disposal of Municipal Refuse,* 28, 37, 40; Parsons, *Disposal of Municipal Refuse,* 56; Crohurst, "Municipal Wastes," 11, 17, 20.

17. Parsons, *Disposal of Municipal Refuse,* 56; *Statistics of Cities Having a Population of over 30,000: 1905,* 337; US Bureau of the Census, *Statistics of Cities Having a Population of over 30,000: 1907,* 452. Total tonnage figures did not always increase since certain kinds of refuse varied each year and changed over time. For instance, as electrical heating became more

popular, coal and wood consumption dropped markedly, and consequently the volume of ash residue declined sharply.

18. A survey conducted in ten New England cities in 1909 demonstrated that cities with large residential populations (such as Cambridge and Somerville, Massachusetts) produced more than twice as much garbage per capita as manufacturing cities (such as New Bedford, Lawrence, and Lynn, Massachusetts, and Manchester, New Hampshire) produced, and about one-third more ashes and rubbish. See Hering and Greeley, *Collection and Disposal of Municipal Refuse,* 37.

19. Ibid., 38–39. Figures were calculated by wards, which is an inexact measure for evaluating volume of waste per ethnic group. Another variable, of course, is the efficiency and completeness of collections in each ward.

20. Frederick L. Stearns, *The Work of the Department of Street Cleaning* (New York: Municipal Engineers of the City of New York, 1913), 210. See also *Engineering News* 48 (July 17, 1902): 48; "The Sanitary Disposal of Municipal Refuse," *Transactions of the ASCE* 50 (1903): 104; Charles A. Meade, "City Cleansing in the City," *Municipal Affairs* 4 (December 1900): 738.

21. Crohurst, "Municipal Wastes," 42–43; Parsons, *Disposal of Municipal Refuse*, 93; William P. Munn, "Collection and Disposal of Garbage," *City Government* 2 (1897): 6–7; Capes and Carpenter, *Municipal Housecleaning,* 175. See also US Committee on Interstate and Foreign Commerce, *Hearings on Bill to Prevent the Dumping of Refuse Material in Lake Michigan or Near Chicago* (Washington, DC: Government Printing Office, 1910).

22. Louisiana Board of Health, *Annual Report* (1898–99): 177.

23. See George W. Fuller, *Sewage Disposal* (New York: McGraw Hill, 1912), 201; Langdon Pearse, "The Dilution Factor," *Transactions of the ASCE* 85 (1922): 451.

24. C. E. Terry, "The Public Dump and the Public Health," *American Journal of Public Health* 3 (April 1913): 338–39.

25. Cleveland Chamber of Commerce, Committee on Housing and Sanitation, *Report on Collection and Disposal of Cleveland's Waste* (1917), 7. See also Boston Health Department, *Annual Report* (1905), 40–41; *Annual Report* (1909), 12–13; *Annual Report* (1916), 89–90; Detroit Board of Health, *Annual Report* (1907), 17; *Annual Report* (1908), 16; *Annual Report* (1909), 23–24; *Annual Report* (1910) 27–28.

26. Terry, "Public Dump and the Public Heath," 338.

27. Hering and Greeley, *Collection and Disposal of Municipal Refuse,* 257. See also Capes and Carpenter, *Municipal Housecleaning,* 174–75; Schneider, "Disposal of a City's Waste," 24–25; "Dumping Garbage Unsanitary," *Municipal Journal and Engineer* 24 (April 22, 1908): 493.

28. "The Land Disposal of Garbage: An Opportunity for Engineers and Contractors," *Engineering News* 53 (April 6, 1905): 367–69.

29. "Dumping City Refuse," *Municipal Journal and Engineer* 42 (January 25, 1917): 103.

30. Crohurst, "Municipal Wastes," 43–45; Parsons, *Disposal of Municipal Refuse,* 78–80; D. C. Faber, "Collection and Disposal of Refuse," *American Municipalities* 30 (February 1916): 185–86; Wyld, "Modern Methods of Municipal Refuse Disposal," 207–208; A. M. Compton, "The Disposal of Municipal Waste by the Burial Method," *American Journal of Public Health* 2 (December 1912): 925–29; "Refuse Disposal in California," *Municipal Journal and Engineer* 42 (January 25, 1917): 100–1.

31. Charles A. Meade, "City Cleansing in New York City," *Municipal Affairs* 4 (December 1900): 735–36; New York City Department of Street Cleaning, *Annual Report* (1902–1905), 74; City Club of Philadelphia, *City Club Bulletin* 2 (March 24, 1910): 116; Charles W. Staniford, *Report on the Disposal of City Wastes* (New York: New York City Department of Docks and Ferries, 1913), 1–19; "Waste-Material Disposal of New York," *Engineering News* 77 (January 18, 1917): 119. Burial of garbage, which faced many of the same complaints as dumping and filling, was reconsidered by some cities. See Faber, "Collection and Disposal of Refuse," 186; Crohurst, "Municipal Wastes," 45–47; Parsons, *Disposal of Municipal Refuse*, 93–94.

32. See F. G. Ashbrook and A. Wilson, "Feeding Garbage to Hogs," *Farmer's Bulletin* 1133, 3–16; Charles V. Chapin, "Disposal of Garbage by Hog Feeding," *American Journal of Public Health* 7 (March 1918): 234–35; and US Food Administration, *Garbage Utilization, with Particular Reference to Utilization by Feeding* (Washington, DC, 1918) 3–11.

33. Parsons, *Disposal of Municipal Refuse*, 94; Capes and Carpenter, *Municipal Housecleaning*, 173; Charles V. Chapin, "The Collection and Disposal of Garbage in Providence, R.I.," *Public Health: Papers and Reports* 28 (1903): 48–50; D. C. Faber, "Collection and Disposal of Refuse," *American Municipalities* 30 (February 1916): 184–85.

34. Capes and Carpenter, *Municipal Housecleaning*, 174.

35. X. H. Goodnough, "The Collection and Disposal of Municipal Waste and Refuse," *Journal of the Association of Engineering Societies* 40 (May 1908): 246–47.

36. Hering and Greeley, *Collection and Disposal of Municipal Refuse*, 258–59. See also Alvah W. Brown, "Garbage Piggeries," *American Journal of Public Health* 2 (December 1912): 930–36.

37. Wyld, "Modern Methods of Municipal Refuse Disposal," 208; "Garbage Disposal in St. Louis," *Municipal Journal and Engineer* 19 (November 1905): 220; E. N. Stacy, "Refuse Collection and Disposal," *Journal of the Association of Engineering Societies* 54 (January 1915): 15; J. J. Jessup. "Refuse Incineration," *Pacific Municipalities* 27 (May 1913): 258; Cleveland Chamber of Commerce, *Report on Collection*, 9; American Garbage Cremation Co. *Cremation of Garbage* (Boston: n.d), 13.

38. "Refuse Disposal in California," *Municipal Journal and Engineer* 42 (January 25, 1917): 101; Chamber of Commerce of the United States, *Refuse Disposal*, 15–16.

39. W. F. Goodrich suggested that, of the more than 250 plants constructed in Great Britain in the thirty years before 1908, fewer than 10 were dismantled or abandoned. Walter Francis Goodrich, *Modern Destructor Practice* (London: C. Griffin and Co., 1912), 15–30; Morse, "Utilization and Disposal of Municipal Waste," 420–21; "British Refuse Destructors and American Garbage Furnaces," *Engineering News* 53 (April 13, 1905): 388–89; Morse, *Collection and Disposal of Municipal Waste*, 216–79.

40. Chamber of Commerce of the United States, *Refuse Disposal*, 16; Joseph G. Branch, "Garbage Disposal," *Municipal Journal and Engineer* 20 (January 1906): 4–5; Morse, "Disposal of the City's Waste," 224–27; "Garbage Disposal in St. Louis," *Municipal Journal and Engineer* 19 (November 1905): 220.

41. Joseph B. Rider, "Public Refuse Destruction a Municipal Asset, Not a Liability," *Fire and Water Engineering* 54 (October 15, 1913): 311.

42. See Martin V. Melosi, "The Viability of Incineration as a Disposal Option: The Evolution of a Niche Technology, 1885–1995," *Public Works Management & Policy* 1 (July, 1996): 33. See also Chamber of Commerce of the United States, *Refuse Disposal*, 15–16;

"British Refuse Destructors," 388–89; Hering and Greeley, *Collection and Disposal of Municipal Refuse*, 311–12; Wyld, "Modern Methods of Municipal Refuse Disposal," 208; "Sanitary Disposal of Municipal Refuse," 106; Morse, *Collection and Disposal of Municipal Waste*, 137; M. N. Baker, "Condition of Garbage Disposal," 147; "Refuse Disposal in America," 85; Morse, "Utilization and Disposal of Municipal Waste" (July 1904), 28; *Proceedings of the ASMI* (1910): 58

43. See Hering and Greeley, *Collection and Disposal of Municipal Refuse*, 311–12.

44. Morse, *Collection and Disposal of Municipal Waste*, 98. See also Morse, "Utilization and Disposal of Municipal Waste," 28–30.

45. Morse, *Collection and Disposal of Municipal Waste*, 148–49, 161–63, 191–93.

46. "Report of the Committee on Disposal of Refuse Materials," *Public Health* 27, 184.

47. Rudolph Hering, "Disposal of Municipal Refuse: Review of General Practice," *Transactions of the ASCE* 54 (1904): 266.

48. *Proceedings of the ASMI, Twenty-third Annual Convention* (1916): 245.

49. *Engineering News* 64 (August 11, 1910): 153. See also Morse, "Disposal of Municipal Waste," 235.

50. Stacy, "Refuse Collection and Disposal," 16.

51. M. N. Baker, "Condition of Garbage Disposal," 148. See also Jessup, "Refuse Incineration," 258; "British Refuse Destructors," 388–89; "Refuse Incineration and Engineering Problems," *Engineering News* 67 (February 15, 1912): 311; William F. Morse, "Garbage Disposal Work in America," *Municipal Journal and Engineer* 17 (October 1904): 158; Howard G. Bayles, "The Incineration of Municipal Waste," *Municipal Engineering* 29 (October 1905): 255.

52. Morse, *Collection and Disposal of Municipal Waste*, 98–99; J. T. Fetherston, "Incineration of Refuse," *American Journal of Public Health* 2 (December 1912): 943–45.

53. "Modern Refuse Disposal Plants," *Municipal Journal and Engineer* 32 (May 30, 1912): 832; Chamber of Commerce of the United States, *Refuse Disposal*, 16. For the most detailed chronology of incinerators in America between 1885 and 1908, see Morse, *Collection and Disposal of Municipal Waste*, 114–19.

54. *Engineering News* 63 (June 23, 1910): 729; Wyld, "Modern Methods of Municipal Refuse Disposal," 208; Hering and Greeley, *Collection and Disposal of Municipal Refuse*, 314.

55. "Recent Refuse Disposal Practice," 849–50. For additional statistics on refuse incinerating plants and their operations in the early twentieth century, see Crohurst, "Municipal Wastes," 62–63; "Refuse Collection and Disposal," (*Municipal Journal and Engineer*), 731–34.

56. Melosi, "Viability of Incineration," 34–35.

57. "Disposal of Garbage," *City Government* 5 (August 1898): 67; Morse, "Disposal of the City's Waste," 271.

58. "Recent Refuse Disposal Practice," 848–49. See also Morse, "Disposal of the City's Waste," 272; Folwell, *Municipal Engineering Practice*, 334–35; "Refuse Collection and Disposal," (*Municipal Journal and Engineer*) ,730–32; US Bureau of the Census, *General Statistics of Cities: 1909*, 49.

59. Schneider, "Disposal of a City's Waste," 25; Chicago City Waste Commission, *Report of the City Waste Commission of the City of Chicago* (1914), 25; Morse, "Disposal of the City's Waste," 271.

60. Hering and Greeley, *Collection and Disposal of Municipal Refuse*, 444.

61. Folwell, *Municipal Engineering Practice,* 333.

62. Morse, "Disposal of the City's Waste," 274; "British Refuse Destructors," 389; H. de B. Parsons, "City Refuse and Its Disposal," *Scientific American Supplement* July 4, 1908, 8; Walter Francis Goodrich, *The Economic Disposal of Towns' Refuse* (New York: John Wiley and Co., 1901), 236; Baltimore Department of Street Cleaning, *Annual Report* (1906): 5.

63. Merriman, *Elements of Sanitary Engineering,* 229; Crohurst, "Municipal Wastes," 77; "Street Cleaning and Refuse Disposal," 16; Parsons, "City Refuse and Its Disposal," 8–9; Goodrich, *Economic Disposal,* 236–38.

64. B. F. Miller; "Garbage Collection and Disposal," 8; "Street Cleaning and Refuse Disposal," 16; "Recent Refuse Disposal Practice," 850.

65. Frederick P. Smith, "Final Disposition of Garbage and Rubbish," *Proceedings of the ASMI, Fourteenth Annual Convention* (1907): 169. See also F. Allen Phillips, "Selling Garbage for Reduction at Los Angeles," *Engineering News* 70 (August 28, 1913): 429–30; Merriman, *Elements of Sanitary Engineering,* 230–32.

66. "The Unsatisfactory Condition of Garbage Disposal in the United States," *Proceedings of the League of American Municipalities* (1901): 20; "Four Garbage-Disposal Contracts," *Engineering News* 70 (December 9, 1913): 718–19; M. N. Baker, "Condition of Garbage Disposal," 147; Folwell, *Municipal Engineering Practice,* 330–31; Irwin S. Osborn, "Disposal of Garbage by the Reduction Method," *American Journal of Public Health* 2 (December 1912): 939.

67. "The Success of Two Municipal Garbage-Reduction Plants," *Engineering News* 73 (May 27, 1915): 1042; "Operating Results of the Garbage-Reduction Works of Cleveland and Columbus, Ohio," *Engineering News* 66 (November 30, 1911): 633–65; "Two Years' Operations of the Municipal Garbage Reduction Works, Cleveland, O.," *Engineering News* 57 (May 2, 1907): 487–90; "Garbage Reduction in Cleveland, Ohio," *Municipal Journal and Engineer* 19 (December 1905): 274–75; "Cleveland's Garbage Reduction Plant," *Municipal Journal and Engineer* 22 (February 13, 1907): 147–48; "Municipal Garbage Reduction," *Municipal Journal and Engineer* 22 (February 13, 1907): 149–51; "Street Cleaning Methods in Cleveland," *Municipal Engineering* 31 (December 1906): 437–38. See also Morse, "Disposal of the City's Waste," 272–73; "Recent Refuse Disposal Practice," 848; Osborn, "Disposal of Garbage," 940–41.

68. Merriman, *Elements of Sanitary Engineering,* 230; Folwell, *Municipal Engineering Practice,* 336; "The Garbage Disposal Problem in Boston and Elsewhere," *Engineering News* 48 (August 7, 1902): 96–97; Stacy, "Refuse Collection and Disposal," 13; Rider, "Public Refuse Destruction," 312. An additional problem with reduction was the volatility of the naphtha used in the process. Explosions were common. In May 1908 a large tank of naphtha exploded at the Chicago Reduction Company plant, killing one man and seriously injuring five, with eight reported missing. Morse, *Collection and Disposal of Municipal Waste,* 307–308.

69. "Garbage Collection and Disposal," *City Government* 7 (September 1899): 50.

70. "Garbage Destruction," *Pacific Municipalities* 9 (October 1903): 81; Schneider, "The Disposal of a City's Waste," 25.

71. "Garbage Reduction Plant for New York City," *Municipal Journal and Engineer* 41 (November 9, 1916): 568–69; "New York Garbage-Reduction Works Controversy May Be Over," *Engineering News* 77 (January 18, 1917): 125; "New York's Garbage-Reduction Fuss," *Engineering News* 77 (February 1, 1917): 201–2; Richard Fenton to Martin Melosi, October

27, 1980. The Barren Island reduction plant in New York, the world's largest reduction plant, was also the subject of controversy and complaint because of the noxious odors it emitted. See New York City Board of Health, *A Report as to the Existing Conditions on Barren Island* (1899), 1–31; "Collection of Garbage," *Municipal Journal and Engineer* 39 (July 8, 1915): 43; "Reduction of New York's Garbage," *Municipal Journal and Engineer* 39 (July 8, 1915): 37; Charles F. Bolduan, *Over a Century of Health Administration in New York City* (New York: New York City, Department of Health, 1916), 27; Morse, *Collection and Disposal of Municipal Waste,* 350.

72. Parsons, "City Refuse and Its Disposal," 9; Hering and Greeley, *Collection and Disposal of Municipal Refuse,* 501–2; Morse, "Disposal of the City's Waste," 273; Howard G. Bayles, "Incineration of Municipal Waste," *Municipal Engineering* 29 (October 1905): 255; Goodrich, *Economic Disposal,* 238–39 "Recent Refuse Disposal Practice," 848–50; Crohurst, "Municipal Wastes," 76, Osborn, "Disposal of Garbage," 939–42; Sterling H. Bunnell, "Municipal Refuse Sorting and Utilization Plant Pittsburgh, Penn.," *Engineering News* 71 (April 30, 1914) 980–84; "Reduction of New York's Garbage," 37–38; "Collection of Garbage," 43.

73. George E. Dyck, *The Treatment of Garbage* (Chicago: George B. Harmer Co., 1916), 4.

74. Theodore Waters, "The Chemical House That Jack Built," *Cosmopolitan Magazine* (July 1907): 290–93.

75. Zueblin, *American Municipal Progress,* 78–79. See also Strasser, *Waste and Want,* for an excellent discussion of home-based recycling in the late nineteenth to mid-twentieth centuries.

76. William F. Morse, *The Disposal of Refuse and Garbage,* (New York: J. J. O'Brien and Sons, 1899), 3. See also "A Chance to Save Money from the Refuse of New York City," *Engineering News* 67 (February 8, 1912): 265.

77. Albert C. Day, *The Garbage Question: A Profitable Solution* (N.p., 1902), 1. See also Edgar L. Culver, *Value of City Waste: Treating on the Subject of Sanitation, Incineration, Reduction, Gathering, and Commercial Value* (Kansas City, MO: Co-Coal-Co., 1916).

78. Ervin E. Ewell, *The Fertilizing Value of Street Sweepings,* US Department of Agriculture, Bulletin 55 (1898): 7–19, "Fertilizer from City Refuse," *Municipal Journal and Engineer* 30 (June 28, 1911): 918; Arturo Bruttini *Uses of Waste Materials: The Collection of Waste Materials and Their Uses for Human and Animal Food, in Fertilizers, and in Certain Inidustries, 1914-1922* (London: A. S. King and Sons, 1923), 1–5; Edward A. Oldham, "Value of Street Sweepings," *Municipal Engineering 16* (February 1899) 80–84; "Concrete from Refuse Clinker," *Municipal Journal and Engineer* 31 (August 9, 1911): 175; "Value of Coal Ashes," *City Government* 4 (January 1898): 31.

79. Thomas J. Keenan, "How Waste Paper Is Treated to Make New Paper," *Scientific American* (December 23, 1916): 574–75; "Where Waste Newspapers Go," *Scientific American* (December 5, 1914): 471; "Sale of Waste Material," *Municipal Journal and Engineer* 41 (August 31, 1916): 261–62; Morse, *Collection and Disposal of Municipal Waste,* 421–25.

80. "Alcohol from Garbage," *Municipal Journal and Engineer* 42 (May 24, 1917): 707–709; Robert H. Moulton, "Turning Garbage into Fuel," *Independent* 89 (February 5, 1917): 222; Morse, *Collection and Disposal of Municipal Waste,* 444–46.

81. William F. Morse, "Refuse Disposal and Power Production" *Municipal Journal and Engineer* 17 (September 1904): 107–109; Goodrich, *Refuse Disposal and Power Production,* v–vii, 1–26; Joseph G Branch *Heat and Light,* 1ff.; Morse, "Utilization and Disposal of

Municipal Waste," 420–21; "City Refuse Destructors as Power Plants," *California (Pacific) Municipalities* 2 (April 1900): 85–86; Louis L. Tribus, "Disposal of Garbage, a Large City's Problem," *Proceedings of the ASMI* (1917): 250; William F. Morse, "Steam Power from City Waste," *Municipal Journal and Engineer* 10 (March 1901): 90–91; Morse, *Collection and Disposal of Municipal Waste,* 216–79; Fetherston, "Municipal Refuse Disposal," 384–86; "Running Municipal Trolley Cars with Garbage and Refuse," *Scientific American*, June 1, 1907, 446.

82. It should be noted that the development of industrial waste heat boilers to generate steam antedated municipal incinerators in the United States.

83. "New York Light from Rubbish" *Bulletin of the League of American Municipalities* 4 (December 1905): 190; *Engineering News* 53 (January 5, 1905): 17; *Municipal Journal and Engineer* 19 (July 1905): 40–41; Samuel A. Greeley, "Refuse Disposal and Street Cleaning," 15; Zueblin, *American Municipal Progress*, 80; Tribus, "Report of the Committee on Street Cleaning," 61–62; Wyld, "Modern Methods of Municipal Refuse Disposal," 208–9; Goodrich, *Modern Destructor Practices,* 194.

84. Chamber of Commerce of the United States, *Refuse Disposal,* 17.

85. Morse, *Collection and Disposal of Municipal Refuse,* 431–32; P. M. Hall, "Report of the Committee on City Wastes: The Economics of Waste Collection and Disposal," *American Journal of Public Health* 5 (November 1915): 1164–67.

86. Charles V. Chapin, "Profit from Garbage," *American Journal of Public Health* 1 (April 1911): 288.

87. It was widely believed that American waste had much higher water content than European waste and a higher percentage of organic materials. Contemporary statistics—at least those for the major cities—do not bear out this belief. For example, in a table reproduced in Hering's "Disposal of Municipal Refuse," 271, the average percentage of moisture in American garbage was 70 percent; in English garbage, 65 percent, and in Berlin garbage, 60 percent. Since many American incinerators primarily burned organic material, unlike the mixed refuse burned in Europe, water content was a legitimate issue, but couched in different terms. See also Venable, *Garbage Crematories in America,* 271–74; Parsons, *Disposal of Municipal Refuse,* 20; Morse, *Collection and Disposal of Municipal Waste,* 37; "Some Financial, Political, and Sanitary Phases of Garbage Disposal," *Engineering News* 45 (February 14, 1901): 120–21.

SEVEN. Solid Waste as Pollution in Twentieth-Century America

1. See William E. Small, *Third Pollution: The National Problem of Solid Waste Disposal* (New York: Praeger Publishers, 1970).

2. US Bureau of the Census, *Historical Statistics of the United States: Colonial Times to 1970* (Washington, DC: Government Printing Office, 1975), 8, 11.

3. John C. Bollens and Henry J. Schmandt, *The Metropolis: Its People, Politics, and Economic Life*, 2d ed. (New York: Harper and Row, 1970), 17, 19; Gil A. Stelter, "Metropolis," in *Encyclopedia of Urban America: The Cities and Suburbs*, ed. Neil Larry Shumsky (Santa Barbara, CA: ABC-CLIO, 1998), 455–56; Chudacoff and Smith, *Evolution of American Urban Society*, 207, 216–17; Carl Abbott, *Urban America in the Modern Age: 1920 to the Present* (Arlington Heights, IL: Harlan Davidson, 1987), 2, 4–5, 7; Amos Hawley, *The Changing Shape of Metropolitan America: Deconcentration Since 1920* (Glencoe, IL: Free Press, 1956), 2; US Bureau of the Census, *Historical Statistics*, 8, 11.

4. William H. Whyte Jr., "Urban Sprawl," in *The Exploding Metropolis*, ed. William H. Whyte Jr. (Berkeley, CA: University of California Press, 1993; orig. pub., 1957), 134.

5. Marcus Felson, "Urban Sprawl," Shumsky, *Encyclopedia of Urban America*, 833–34.

6. Joseph Interrante, "The Road to Autopia: The Automobile and the Spatial Transformation of American Culture," *Michigan Quarterly Review* 19–20 (Fall-Winter 1980–81): 502–17; Kenneth T. Jackson, *Crabgrass Frontier: The Suburbanization of the United States* (New York: Oxford University Press, 1985), 139–40, 162–63; Jon Teaford, *The Twentieth-Century American City* (Baltimore: Johns Hopkins University Press, 1986), 63; Michael Ebner, "Suburbanization," Shumsky, *Encyclopedia of Urban America*, 759.

7. For instance, see Bollens and Schmandt, *Metropolis*, 102–04.

8. The article suggested that in the New York City metropolitan area there was no uniform disposal pattern, while in Washington DC, because of its peculiar metropolitan status, "...the disposal of trash quickly becomes one of awe-inspiring importance, involving interstate commerce regulations as well as the problem of maintaining friendly relations with other states." See "Metropolitan Refuse Disposal Problems," *American City*, February, 1952, 104–05.

9. See Donald C. Stone, *The Management of Municipal Public Works* (Chicago: Municipal Administration Service, 1939), 241.

10. Samuel A. Greeley, "Administrative and Engineering Work in the Collection and Disposal of Garbage: A Review of the Problem," *ASCE Transactions* 89 (1926): 800.

11. "Financial Statistics of Cities: 1926–1934 Sanitation Service," *American City* 51 (December, 1936): 11, 13, 15, 17, 19.

12. See Bernard Baum, et al., *Solid Waste Disposal* (Ann Arbor, MI: Science Pub. Inc., 1974), 1:vi; D. Joseph Hagerty, Joseph L. Pavoni, and John E. Heer Jr., *Solid Waste Management* (New York: Van Nostrand Reinhold, 1973), 13–14; APWA, *Municipal Refuse Disposal*, 3d ed. (Chicago: APWA, 1970;), viii–ix; National League of Cities and United States Conference of Mayors, Solid Waste Management Task Force, *Cities and the Nation's Disposal* Crisis (Washington, DC: National League of Cities-United States Conference of Mayors, 1973), 3, 32.

13. Harrison P. Eddy Jr., "Why Not Make Garbage Collection and Disposal Self-Sustaining?" *American City*, October, 1932, 52–53; Stone, *Management of Municipal Public Works*, 241.

14. Institute for Training in Municipal Administration, *Municipal Public Works Administration* (Chicago, IL: International City Managers' Association, 1957), 372–74; Edward Scott Hopkins, W. McLean Bingley, and George Wayne Schucker, *The Practice of Sanitation in Its Relation to the Environment* (Baltimore: Williams and Wilkins, Co., 1970), 226. Hopkins, Bingley, and Schucker reported that in 1950, 321 cities with populations exceeding 10,000 had service charges. See p. 226.

15. Peter Kemper and John M. Quigley, *The Economics of Refuse Collection* (Cambridge, MA: Ballinger Pub. Co., 1976), 109–11; Hagerty, Pavoni, and Heer, *Solid Waste Management*, 13.

16. Kemper and Quigley, *Economics of Refuse Collection*, 111.

17. Stone, *Management of Municipal Public Works*, 240.

18. Harold Crooks, *Dirty Business: The Inside Story of the New Garbage Agglomerates* (Toronto: James Lorimer & Co., 1983), 7.

19. Ibid., 8.

20. Ibid., 8–9; Peter Reuter, "Regulating Rackets," *Regulation* (December 1984): 33–34; "What We Must Do—Part 2: 'God Bless Chem Waste,'" *Rachel's Hazardous Waste News* 89, August 8, 1988, http://www.ejnet.org/rachel/rhwn089.htm.

21. In New York City the garbage collection business has often been referred to as the "carting" industry.

22. Reuter, "Regulating Rackets," 29–30. See also John D. Hanrahan, *Government for Sale: Contracting-Out the New Patronage* (Washington, DC: American Federation of State and County Municipal Employees, 1977), 44.

23. Stewart E. Perry, *San Francisco Scavengers: Dirty Work and the Pride of Ownership* (Berkeley, CA: University of California Press, 1978), 15–21.

24. G. P. Gordon, "Water Works Financing," *Journal of the American Water Works Association* 26 (April 1934): 519.

25. Abbott, *Urban America in the Modern Age*, 15, 47–48; Goldfield and Brownell, *Urban America* 323–24; Teaford, *Twentieth-Century American City*, 74–80; Chudacoff and Smith, *Evolution of American Urban Society*, 233–34.

26. Goldfield and Brownell, *Urban America*, 325–26.

27. Roger Daniels, "Public Works in the 1930s: A Preliminary Reconnaissance," in *The Relevancy of Public Works History: 1930s—A Case Study* (Washington, DC: Public Works Historical Society, 1975), 3.

28. Ibid., 3–4; L. Evans Walker, comp., *Preliminary Inventory of the Records of the Public Works Administration* (Washington, DC: National Archives, 1960), 1; J. Kerwin Williams, *Grants-in-Aid Under the Public Works Administration* (New York: AMS Press, 1939), 22–28; Charles Trout, "The New Deal and the Cities," in *Fifty Years Later: The New Deal Evaluated*, ed. Harvard Sitkoff (New York: Knopf, 1985), 134; Goldfield and Brownell, *Urban America*, 326.

29. John H. Mollenkopf, *The Contested City* (Princeton: Princeton University Press, 1983), 55.

30. Ibid., 65.

31. Trout, "New Deal," 136; "Sanitary Landfill and the Decline of Recycling as a Solid Waste Management Strategy in American Cities" (Unpublished paper), 15.

32. Public Works Administration, *America Builds: The Record of PWA* (Washington, DC: Public Works Administration, 1939), 279. See also Harold Ickes, *Back to Work: The Story of PWA* (New York: Macmillan, 1935), 170; US Public Works Administration, *The First 3 Years* (Washington, DC: Government Printing Office, 1936), 10.

33. "Sanitary Landfill and the Decline of Recycling," 15; Trout, "New Deal," 144.

34. Harrison P. Eddy Jr., "Refuse Disposal: A Review," *Municipal Sanitation* 8 (January 1937): 86.

35. Trout, "New Deal," 137, 139, 141, 144–45; Mark Gelfand, *A Nation of Cities: The Federal Government and Urban America, 1933–1965* (New York: Oxford University Press, 1975), 48–49; Mollenkopf, *Contested City*, 71–72; Eric H. Monkkonen, *America Becomes Urban: The Development of U.S. Cities & Towns, 1780–1980* (Berkeley, CA.: University of California Press, 1988), 134–35.

36. Teaford, *Twentieth-Century American City*, 90–91; Goldfield and Brownell, *Urban America*, 336–41.

37. Gelfand, *Nation of Cities*, 148–51, 242–45; Walker, *Preliminary Inventory*, 3.

38. Martin V. Melosi, *Coping with Abundance: Energy and Environment in Industrial America* (New York: Alfred Knopf, 1985), 103–05.

39. Ariel Parkinson, "Responsible Waste Management in a Shrinking World," *Environment* 25 (December 1983): 61.

40. APWA, *Solid Waste Collection Practice*, 4th ed. (Chicago: Public Administration Service, 1975), 22; Hopkins, Bingley, Schucker, *Practice of Sanitation*, 227; Kemper and Quigley, *Economics of Refuse Collection*, 6; Peter Steinhart, "Down in the Dumps," *Audubon*, May 19, 1986, 104.

41. US Council of Environmental Quality, *Environmental Quality* (Washington, DC: Government Printing Office, 1993), 505.

42. US Environmental Protection Agency, *Solid Waste Dilemma: Background Documents* (Washington, DC: Environmental Protection Agency, September, 1988), 1–19 (hereafter cited as US EPA).

43. In 1966, officials of the Office of Solid Waste—then affiliated with the Department of Health, Education, and Welfare—identified polyethylene containers as possibly the "biggest problem" in solid waste management at the time. Aside from being nonbiodegrable, polyethylene burns at temperatures high enough to melt conventional grates in incinerators. See US Department of Health, Education, and Welfare, Public Health Service, Environmental Health Service, Bureau of Solid Waste Management, Solid Waste Management, *Abstracts and Excerpts from the Literature,* Pub. no. 2038, 2 vols. (Washington, DC: GPO, 1970), 35.

44. John A. Burns and Michael J. Seaman, "Some Aspects of Solid Waste Disposal," in *Our Environment: The Outlook of 1980,* ed. Alfred J. Van Tassel (Lexington, Mass.: Lexington Books, 1973), 457–58; Baum, et al., *Solid Waste Disposal*, 1:vi; Laurent Hodges, *Environment Pollution: A Survey Emphasizing Physical and Chemical Principles* (New York: Holt, Rinehart, and Winston, 1973), 219; C. L. Mantell, *Solid Wastes: Origins, Collection, Processing, and Disposal* (NY: John Wiley and Sons, 1975), 32, 35.

45. Strasser, *Waste and Want*, 199–200.

46. Samuel A. Greeley, "Modern Methods of Disposal of Garbage and Some of the Troubles Experienced in Their Use," *Proceedings of the ASMI* 28 (1922): 242; George B. Gascoigne, "A Year's Progress in Refuse Disposal and Street Cleaning," *American Society of Municipal Engineers, Official Proceedings of the 38th Annual Convention* 38 (January 1933): 191–92. For cost figures, by city, see *Municipal Index* (1926): 162–83; *Municipal Index and Atlas* (1930): 618–35.

47. APWA, *Solid Waste Collection Practice*, 36–39; Hagerty, Pavoni, and Heer, *Solid Waste Management*, 10; Korbitz, *Urban Public Works Administration*, 439.

48. Samuel A. Greeley, "Street Cleaning and the Collection and Disposal of Refuse," *ASCE Transactions* 92 (1928): 1245. See also "Refuse Collection in 28 Cities in the United States," *American City*, July 1938, 57–60; Archer M. Soby and John H. Nuttall, "Time Studies of Refuse Collection Activities," *American City*, August 1929, 112–16.

49. Kemper and Quigley, *Economics of Refuse Collection*, 109–11; Hagerty, Pavoni, and Heer, *Solid Waste Management*, 13.

50. Armstrong, Robinson, and Hoy, *History of Public Works*, 438–40; Institute for Training in Municipal Administration, *Municipal Public Works Administration*, 197–209. For information on snow removal, see pp. 209–17.

51. US EPA, *The Private Sector in Solid Waste Management* (Washington, DC: EPA, 1973), 1:4.

52. Armstrong, Robinson, and Hoy, *History of Public Works*, 441–42, 444–45. See also Hopkins, Bingley, and Schucker, *Practice of Sanitation*, 230–31; C. G. Gillespie and E. A. Reinke, "Municipal Refuse Problems and Procedures," *Civil Engineering* 4 (September 1934): 488.

53. Armstrong, Robinson, and Hoy, *History of Public Works*, 441–42, 445; Gascoigne, "Year's Progress," 188–89; Lent D. Upson, *Practice of Municipal Administration* (New York: Century Co., 1926) 459–60; J. E. Doran, "The Economical Collection of Municipal Wastes," *American City*, October 1928, 98; Joseph J. Butler, "Refusé Collection and Disposal in Chicago," *American City*, October 1937, 81; Greeley, "Modern Methods of Disposal of Garbage," 242.

54. Kemper and Quigley, *Economics of Refuse Collection*, 109.

55. See Robert B. Brooks, "Contract Collection," *American City*, April 1934, 66; Armstrong, Robinson, and Hoy, *History of Public Works*, 446.

56. "Combined Treatment of Sewage and Garbage," *National Municipal Review* 13 (August 1924): 450–51; C. E. Keefer, "The Disposal of Garbage with Sewage," *Civil Engineering* 6 (March 1936): 178–80; "Disposal of Ground Garbage into Sewers Arouses Interest," *Municipal Sanitation* 7 (March 1936): 94; Charles Gilman Hyde, "Recent Trends in Sewerage and Sewage Treatment," *Municipal Sanitation* 7 (February 1936): 46–47; "Send out the Garbage with the Sewage from the Home," *American City* 50 (September 1935): 13.

57. Suellen Hoy, "The Garbage Disposer: The Public Health and the Good Life," *Technology and Culture* 26 (October 1985): 761. See also Hoy, *Chasing Dirt*, 170–71.

58. Susan Strasser, "Leftovers and Litter: Food Waste in the Late Twentieth Century" (Paper presented at the Organization of American Historians meeting, Atlanta, Georgia, April 1994), 6–8. See also Strasser, *Waste and Want*, 273–74; "Home Garbage Grinder," *Scientific American*, September, 1935, 145.

59. "The Garbage Grinder in Municipal Sanitation," *American City*, April, 1950, 106; Hoy, "Garbage Disposer," 758ff; Hoy, "Public Health and Sanitation in an Indiana Community: The Garbage Disposer and Jasper," *Indiana Magazine of History* 82 (June 1986): 139–60; Strasser, "Leftovers and Litter," 7. For a more positive assessment of the impact of grinders on sewage treatment, see Mark Owen, "The Future of Domestic Garbage Grinders in the Municipal Sanitation Field," *American City*, March 1949, 96–97; Harry P. Croft, "Kitchen Garbage Disposal Means Sewage-Plant Expansion," *American City*, November, 1948, 110–11.

60. George W. Schusler, "The Disposal of Municipal Wastes," *American City*, August 1936, 86.

61. Martin V. Melosi, "Waste Management: The Cleaning of America," *Environment* 23 (October 1981): 12; Eddy, "Refuse Disposal: A Review," 79; Roger J. Bounds, "Refuse Disposal in American Cities," *Municipal Sanitation* 2 (September 1931): 431; Gillespie and Reinke, "Municipal Refuse Problems," 490.

62. Melosi, "Waste Management," 12; Richard Fenton, "Current Trends in Municipal Solid Waste Disposal in New York City," *Resource Recovery and Conservation* 1 (1975): 170; Hodges, *Environment Pollution*, 260; Esber I. Shaheen, *Environmental Pollution: Awareness and Control* (Mahomet, IL.: Engineering Technology, 1974), 260.

63. Bounds, "Refuse Disposal in American Cities," 432; "Sanitary Landfill and the Decline of Recycling," 18; Upson, *Practice of Municipal Administration*, 462.

64. Ralph Stone and Francis R. Bowerman, "Incineration and Alternative Refuse Disposal Processes," *ASCE Transactions* 121 (1956): 310.

65. Martin V. Melosi, "Historic Development of Sanitary Landfills and Subtitle D," *Energy Laboratory Newsletter* 31 (1994): 20; Armstrong, Robinson, and Hoy, *History of Public Works*, 449–50; "An Interview with Jean Vincenz," *Public Works Historical Society Oral History Interview* (Chicago: Public Works Historical Society, 1980),1:9–10; John J. Casey, "A Disposal of Mixed Refuse by Sanitary Fill Method at San Francisco," *Civil Engineering* 9 (October 1939): 590–92.

66. The initial sanitary landfill in Fresno—an experimental fill—was opened on October 15, 1934, at the City Sewer Farm. A 90-acre parcel three miles from city hall was purchased in 1937, which became the Fresno Sanitary Landfill. GeoSyntec Consultants, *Final Remedial Action Work Plan: Construction and Operations Activities for Operable Unit 1: Fresno Sanitary Landfill, Fresno, California*, 1-1–1-2; GeoSyntec Consultants, *Design of Source Control Operable Unit: Fresno Sanitary Landfill, Fresno, California*, vol. 1, *Design Report* (Walnut Creek, CA: GeoSyntec Consultants, March 24, 1997), 1–2; Camp Dresser & McKee, Inc., *City of Fresno: Fresno Sanitary Landfill: Remedial Investigation Report, Final*, 1–3, 1–8; Camp Dresser & McKee, Inc., *Final Administrative Order No. 90-24: Sampling and Analysis Plan for Fresno Sanitary Landfill, Fresno, California*, 2-1; Camp Dresser & McKee, Inc., *Investigation and Feasibility Study for Fresno Sanitary Landfill, 2–3; U.S. Environmental Protection Agency, Fresno Sanitary Landfill: Superfund Site, Removal Action No. 1*, 3; Camp Dresser & McKee, Inc., *Investigation and Feasibility Study for Fresno Sanitary Landfill*, 2. For another view of the origins of the sanitary landfill, see Corey, "King Garbage," 207–08, 242–43.

67. "Interview with Jean Vincenz," 1.

68. "Sanitary Landfill and the Decline of Recycling," 19–21; "Interview with Jean Vincenz;" Jean L. Vincenz, "Sanitary Fill at Fresno," *Engineering News-Record* 123 (October 26, 1939): 539–40; Vincenz, "The Sanitary Fill Method of Refuse Disposal," *Public Works Engineers' Yearbook* (1940): 187–201; Vincenz, "Refuse Disposal by the Sanitary Fill Method," *Public Works Engineers' Yearbook* (1944): 88–96; Vincenz, "Sanitary Fill as Used in Fresno," *American City*, February 1940, 42–43. The Fresno Landfill contained no liners, containment structures, leachate collection systems, or leak detection systems upon its original construction. In subsequent investigations, at least 20 hazardous substances were found in the groundwater at the site, including volatile organic compounds (VOCs) such as vinyl chloride and trans-1, 2-dichloroethene. There also was a migration of methane off-site indicating high concentrations of methane gas production within the landfill. The landfill was first evaluated by the Superfund program as a result of a notification filed by the City of Fresno's Solid Waste Management Division on May 27, 1981, and the city subsequently began the process of closing it, which was completed in 1987. On October 4, 1989, the site was placed on the National Priorities List of Superfund as defined in Section 105 of the Comprehensive Environmental Response, Compensation and Liability Act. *Administrative Consent Order in the Matter of Fresno Sanitary Landfill, City of Fresno*, US EPA Docket No. 90–22 (n.d.), 5– 6; US EPA, *Fresno Sanitary Landfill: Superfund Site, Removal Action No. 1*, 2– 3. GeoSyntec Consultants, *Final Remedial Action Work Plan*, 1–2 to 1–3.

69. "Sanitary-Fill Refuse Disposal at San Francisco," *Engineering News-Record* 116 (February 27, 1936): 314–17; "Fill Disposal of Refuse Successful in San Francisco," *Engineering News-Record* 116 (July 6, 1939): 27–28; J. C. Geiger, "Sanitary Fill Method," *Civil Engineering* 10 (January 1940): 42; Casey, "Disposal of Mixed Refuse," 590–92.

70. "Sanitary Landfill and the Decline of Recycling," 22–25; Vincenz, "Sanitary Fill Method of Refuse Disposal," 199. See also Desmond P. Tynan, "Modern Garbage Disposal," *American City*, June 1939, 100–1; Rolf Eliassen and Albert J. Lizee, "Sanitary Land Fills in New York City," *Civil Engineering* 12 (September 1942): 483–86.

71. W. Rayner Straus, "Use of Sanitary Fill in Baltimore to Continue After the War," *American City*, January 1945, 82–83.

72. "Interview with Jean Vincenz," 17–19; Vincenz, "Refuse Disposal by the Sanitary Fill Method," 88–89; Melosi, "Historic Development," 20.

73. Melosi, "Historic Development," 20.

74. Melosi, "Waste Management," 13.

75. Joel A. Tarr, "Risk Perception in Waste Disposal: A Historical Review," (Unpublished paper, Pittsburgh, PA), 20–22.

76. Crohurst, "Municipal Wastes," 48–49.

77. Cyril E. Marshall, "Incinerator Knocks Out Garbage Dump in Long Island Town," *American City*, June 1929, 129; Hering and Greeley, *Collection and Disposal of Municipal Refuse*, 313; *Municipal Index* (1924), 68; "Garbage Collection and Disposal," *City Manager Magazine*, July 1924, 12–13.

78. Michael R. Greenberg, et al., *Solid Waste Planning in Metropolitan Regions* (New Brunswick, NJ: Center for Urban Policy Research, Rutgers University, 1976), 8; G. C. Holbrook, "The Modern Refuse Incinerator—A Sanitary Municipal Utility," *American City*, December, 1936, 59.

79. Henry W. Taylor, "Incineration of Municipal Refuse: Part 1—Municipal Wastes," *Municipal Sanitation* 6 (May 1935): 142.

80. Henry W. Taylor, "Incineration of Municipal Refuse: Part 6—Past and Present," *Municipal Sanitation* 6 (October 1935): 300; Taylor, "Incineration of Municipal Refuse: Part 4," *Municipal Sanitation* 6 (August 1935): 239; George L. Watson, "What Constitutes a Low Bid on an Incinerator?" *American City*, October 1934, 66.

81. H. S. Hersey, "Incinerators for Garbage and Refuse Disposal—Part I," *American City* February 1938, 61, 63. Henry W. Taylor, "Incineration of Municipal Refuse: Part 6—Past and Present," 300. See also H. S. Hersey, "Incinerators for Garbage and Refuse Disposal—Part II," *American City*, March 1938, 89.

82. "Public Still Wants No Incinerator as a Next Door Neighbor," *Municipal Sanitation* 8 (November 1937): 585. See also Schusler, "Disposal of Municipal Wastes," 86; Alden E. Stilson, "Incinerators for Garbage and Refuse Disposal," *American City*, April 1938, 109–17; Michael W. Wipfler, "Modern Incineration Solves Another Garbage Problem," *American City*, May 1930, 101–05.

83. R. H. Stellwagen, "Incinerators and How to Use Them," *American City*, March 1948, 113.

84. Rolf Eliassen, "Incinerator Mechanization Wins Increasing Favor," *Civil Engineering* 19 (April 1949): 17–19; Morris M. Cohn, "Highlights of Incinerator Construction—1941," *Sewage Works Engineering* 13 (February 1942): 87. See also "National Census of Refuse Incinerator Construction—1940," *Sewage Works Engineering* 12 (February 1941): 88; Fenton, "Current Trends," 171, 173–74; T. E. Maxson, "New Incinerator Solves a Wartime Problem," *American City*, May 1945, 81–82.

85. Of the five largest cities in the country, only Chicago abandoned incineration for sanitary fills. F. R. Bowerman, "What Cities Use Incinerators—and Why?" *American City*, March 1952, 100.

86. Casimir A. Rogus, "Refuse Incineration—Trends and Developments," *American City*, July 1959, 94–97.

87. "The Incinerator—'A Machine of Beauty,'" *American City*, August 1954, 85; "Sanitary Landfill or Incineration?" *American City*, March 1951, 98–99.

88. See Rodney R. Fleming, "Solid-Waste Disposal: Part II—Incineration and Composting," *American City*, February 1966, 95; Mantell, *Solid Wastes*, 21.

89. "Incinerator-residue Study Under Way," *American City*, March 1965, 20; Junius W. Stephenson, "Planning for Incineration," *Civil Engineering* 34 (September 1964): 38, 40; APWA, *Municipal Refuse Disposal*, viii.

90. This included salvage of scrap metals, recovery of other saleable items, and recycling/reuse of food wastes.

91. Armstrong, Robinson, and Hoy, *History of Public Works*, 448; Eddy, "Refuse Disposal: A Review," 80; Bounds, "Refuse Disposal in American Cities," 433–34.

92. Melosi, "Waste Management," 12.

93. See Brian J. L. Berry and Frank E. Horton, *Urban Environmental Management*, (Englewood Cliffs, NJ: Prentice-Hall, 1974), 267, 269; Joseph A. Salvato Jr., *Environmental Engineering and Sanitation* (New York: Wiley-Interscience, 1969), 402; Hopkins, Bingley, and Schucker, *Practice of Sanitation*, 234–35; Casimir A. Rogus, "Sanitary Fills and Incinerators," *American City* 70, March 1955, 114–15.

94. Unfortunately, not a great deal is known about the reduction plants because about half of them were run by private contractors. Nevertheless, see Eddy, "Refuse Disposal: A Review," 79; Bounds, "Refuse Disposal in American Cities," 433–34; Greeley, "Street Cleaning," 1246; Upson, *Practice of Municipal Administration*, 463–64; Harry A. Mount, "A Garbage Crisis: Must We Solve Anew the Problem of the Disposition of Domestic Wastes?" *Scientific American*, January 1922, 38; "Trends in Refuse Disposal," *American City*, May 1939, 13.

95. As early as 1920, interest in composting as a means of disposal and treatment of municipal waste was noted in Europe. The first significant research with composting took place in India in 1925. Sir Albert Howard's process relied simply on placing organic material in a pit, turning the compost periodically, and draining off the moisture. Dr. Giovanni Beccari of Florence, Italy received a patent in 1922 for a fermentation process leading to decomposition of waste. Other patents soon followed. The first full-scale composting plant was built in the Netherlands in 1932 by a non-profit utility company established by the government for the disposal of city refuse. See APWA, *Municipal Refuse Disposal*, 293–94; Bounds, "Refuse Disposal in American Cities," 432.

96. Shaheen, *Environmental Pollution*, 271; Hopkins, Bingley, and Schucker, *Practice of Sanitation*, 247; Max L. Panzer and Harvey F. Ludwig, "Should We Reconsider Composting of Organic Refuse?" *Civil Engineering* 21 (February 1951): 40–41; APWA, *Municipal Refuse Disposal*, 296–98.

97. Upson, *Practice of Municipal Administration*, 465–66; APWA, *Municipal Refuse Disposal*, 337; Nathan B. Jacobs, "What Future for Municipal Refuse Disposal?" *Municipal Sanitation* 1 (July 1930): 384. See also Strasser, *Waste and Want*, 222–27 on salvage charities such as Goodwill and the Salvation Army.

98. APWA, *Solid Waste Collection Practice*, 25 For statistics on recovery from recycling in 1960 and 1965, see US Council for Environmental Quality, *Environmental Quality* (Washington, DC: Council for Environmental Quality, 1993), 505.

99. Hoy and Robinson, *Recovering the Past*, 20–22. See also Strasser, *Waste and Want*, 233–63; APWA, *Solid Waste Collection Practice*, 25.

EIGHT. The Garbage Crisis in the Late Twentieth Century

1. Goldfield and Brownell, *Urban America*, 375. See also John J. Harrigan, *Political Change in the Metropolis*, 5th ed. (New York: Harper Collins College Publishers, 1993) 39–40; Joel Garreau, *Edge Cities: Life on the New Frontiers* (New York: Anchor Books, 1992). For more on the Northeast versus the Southwest, see Mollenkopf, *Contested City* 213–53.

2. US Bureau of the Census, *Statistical Abstract of the United States* (Washington, DC: Department of Commerce, 1995), 39; US Bureau of the Census, *1990 Census of Population and Housing, Supplemental Reports: Urbanized Areas of the United States and Puerto Rico* Section 1 (Washington, DC: Department of Commerce, 1993), II–2. See also Peter Mieszkowski and Mahlon Straszheim, eds., *Current Issues in Urban Economics* (Baltimore: Johns Hopkins University Press, 1979), 4; Richard Stren, Rodney White, and Joseph Whitney, eds., *Sustainable Cities: Urbanization and the Environment in International Perspective* (Boulder: Westview Press, 1992), 173; Chudacoff and Smith, *Evolution of American Urban Society*, 292; Goldfield and Brownell, *Urban America*, 376.

3. US Bureau of the Census, *Statistical Abstract of the United States: 2001* (Washington, DC: Department of Commerce, 2001), 8, 29.

4. See Robert Fishman, "America's New City: Megalopolis Unbound," in *America's Cities: Problems and Prospects*, ed. Roger L. Kemp (Aldershot: Avebury, 1995), 128.

5. Teaford, *Twentieth-Century American City*, 153–54; Chudacoff and Smith, *Evolution of American Urban Society*, 288–89, 301; Abbott, *Urban America in the Modern Age*, 111, 113–15, 132; Miller and Melvin, *Urbanization of Modern America*, 213; Goldfield and Brownell, *Urban America*, 435–48.

6. A 1966 act expanded the National Register of protected sites to include buildings and districts in the central city that might be of architectural and historical significance. In 1976, federal tax credits were available to those intending to restore structures in historic districts. Restorations and a renewed interest in older established neighborhoods attracted some wealthier groups back to the central cities. This process of gentrification seemed to slow the outmigration and even reverse it for a time in the mid-1970s. In the long run, however, decentralization continued throughout the country. See Chudacoff and Smith, *Evolution of American Urban Society*, 289–90; Abbott, *Urban America in the Modern Age*, 136; Goldfield and Brownell, *Urban America*, 380–81; 414–23.

7. In the 1970s, there was a significant acceleration in black suburbanization, with the proportion of African Americans in suburbia increasing in 44 of the 50 largest SMSAs. Minority suburbanization continued to increase in the 1980s and 1990s. However, the suburban population of blacks and Hispanics still tended to be concentrated in a relatively small number of fringe communities, and thus integration was uneven and severe racial imbalance continued to exist between core cities and the fringe. See Christopher Silver, "Housing Policy and Suburbanization: An Analysis of the Changing Quality and Quantity of Black Housing in Suburbia since 1950," in *Race, Ethnicity, and Minority Housing in the United States*, ed. Jamshid A. Momeni (New York: Greenwood Press, 1986), 71; Bernard H. Ross and Myron A. Levine, *Urban Politics: Power in Metropolitan America*, 5th ed. (Itasca, IL: F. E. Peacock Pub., Inc., 1996), 59, 285–89.

8. Teaford, *Twentieth-Century American City,* 153. See also Marian Lief Palley and Howard A. Palley, *Urban America and Public Policies* 2d ed. (Lexington, MA: D.C. Heath and Co., 1981), 19; Stren, White, and Whitney, *Sustainable Cities,* 173; Chudacoff and Smith, *Evolution of American Urban Society,* 289, 292; Kenneth Fox, *Metropolitan America: Urban Life and Urban Policy in the United States, 1940–1980* (New Brunswick, NJ: Rutgers University Press, 1986), 51.

9. See US Council on Environmental Quality, *Environmental Quality: Twenty-Fourth Annual Report of the Council of Environmental Quality* (Washington, DC: Government Printing Office, 1993), 385; US Bureau of the Census, *Statistical Abstract* (1995), 43.

10. See Chudacoff and Smith, *Evolution of American Urban Society,* 294.

11. Howard Chernick and Andrew Reschovsky, "Urban Fiscal Problems: Coordinating Actions among Governments," in *The Urban Crisis: Linking Research to Action,* ed. Burton A. Weisbrod and James C. Worthy (Evanston, IL: Northwestern University Press, 1997), 132, 135–41.

12. Teaford, *Twentieth-Century American City,* 142; Jon C. Teaford, *The Rough Road to Renaissance: Urban Revitalization in America, 1940–1985* (Baltimore: Johns Hopkins University Press, 1990), 218, 225, 262, 265; Chudacoff and Smith, *Evolution of American Urban Society,* 294–95; Lawrence J. R. Herson and John M. Bolland, *The Urban Web: Politics, Policy, and Theory* (Chicago: Nelson-Hall Pubs., 1990), 347.

13. Palley and Palley, *Urban America and Public Policies,* 24, 59; Teaford, *Twentieth-Century American City,* 140–41; Chudacoff and Smith, *Evolution of American Urban Society,* 293; Herson and Bolland, *Urban Web,* 335–36.

14. Benjamin Kleinberg, *Urban America in Transformation: Perspectives on Urban Policy and Development* (Thousand Oaks, CA.: Sage Pub., 1995), 187–88; Miller and Melvin, *Urbanization of Modern America,* 210–11; Teaford, *Twentieth-Century American City,* 140.

15. See Kleinberg, *Urban America in Transformation,* 210–13; Goldfield and Brownell, *Urban America,* 392–94.

16. George E. Peterson and Carol W. Lewis, eds., *Reagan and the Cities* (Washington, DC: Urban Institute Press, 1986), 1.

17. An enterprise zone was an area in a low-income neighborhood in which the federal government would offer tax concessions and other inducements to business in exchange for the promise of economic development. Some were established in the 1980s, but few were successful. Several economists have voiced skepticism about the ability of enterprise zones to encourage economic development. See Chernick and Reschovsky, "Urban Fiscal Problems," 146.

18. Ibid., 141. See also Peterson and Lewis, *Reagan and the Cities,* 1–10; Kleinberg, *Urban America in Transformation,* 226–36, 242–44; Miller and Melvin, *Urbanization of Modern America,* 228–39; Abbott, *Urban America in the Modern Age,* 131; Chudacoff and Smith, *Evolution of American Urban Society,* 294, 296–97; Goldfield and Brownell, *Urban America,* 433–35. See also Paul Kantor, *The Dependent City Revisited: The Political Economy of Urban Development and Social Policy* (Boulder, CO.: Westview Press, 1995), 99, 101, 103, 106.

19. Committee on Infrastructure Innovation, National Research Council, *Infrastructure for the 21st Century: Framework for a Research Agenda* (Washington, DC: National Academy Press, 1987), 9; National Council on Public Works Improvement, *The Nation's Public Works: Defining the Issues* (Washington, DC: NCPWI, September 1986), 57; Council on National

Public Works Improvement, *Fragile Foundations: A Report on America's Public Works* (Washington, DC: Government Printing Office, February, 1988), 12–13. See also Government Finance Research Center, Municipal Finance Officers Association, *Building Prosperity: Financing Public Infrastructure for Economic Development* (Washington, DC: Municipal Finance Officers Association, October 1983), 8; George E. Peterson and Mary John Miller, *Financing Public Infrastructure: Policy Options* (Washington, DC: Community and Economic Development Task Force, HUD, 1982), 5–7, 9.

20. Pat Choate and Susan Walter, *America in Ruins: The Decaying Infrastructure* (Durham, NC: Duke Press Paperbacks, 1981), xi–xii, 1–4. See also CONSAD Research Corporation, *A Study of Public Works Investment in the United States* (Pittsburgh: CONSAD, March, 1980), upon which Choate and Walker drew a substantial portion of their information.

21. Council on National Public Works Improvement, *Fragile Foundations*, 1, 6.

22. See *Renewing America's Infrastructure: A Citizen's Guide National* (Reston, VA: ASCE, 2001); League of Cities and United States Conference of Mayors, *Cities and the Nation's Disposal Crisis* (Washington, DC: National League of Cities and United States Conference of Mayors, March 1973), 1. See also Greenberg, et al., *Solid Waste Planning*, 11–22; Berry and Horton, *Urban Environmental Management*, 376–77. Examples of "garbage crises" can be found much earlier than the 1970s, however. For instance, see Mount, "Garbage Crisis," 38. Mount pleaded for the coming of "a Moses of the Garbage Can" to solve the garbage crisis.

23. Homer A. Neal and J. R. Schubel, *Solid Waste Management and the Environment: The Mounting Garbage and Trash Crisis* (Englewood Cliffs, NJ: Prentice-Hall, 1987), 5.

24. Robert Emmet Long, ed., *The Problem of Waste Disposal* (New York: H. W. Wilson Co., 1989), 9. See also Newsday, *Rush to Burn: Solving America's Garbage Crisis?* (Washington, DC: Island Press, 1989).

25. James Cook, "Not in Anybody's Backyard," *Forbes*, November 28, 1988, 172; "New Ways to Keep a Lid on America's Garbage Problem," *Wall Street Journal*, April 15, 1986; National Center for Policy Analysis, "A Consumer's Guide to Environmental Myths and Realities," October 31, 2001, http://www.ncpa.org/studies/s165/s165.html.

26. "The Trash Mess Won't Be Easily Disposed Of," *Wall Street Journal*, December 15, 1988.

27. The same case can be made for street cleaning—less significant than in the days of the horse—but necessary to help keep cities clean.

28. APWA, *Solid Waste Collection Practice*, 1. See also Arthur J. Warner, Charles H. Parker and Barnard Baum, *Solid Waste Management of Plastics* (Washington, DC: Manufacturing Chemists Association, December 1970), 4–5. Total public spending in 1985 for construction and operation of solid waste facilities alone was estimated to be $5.1 billion. See National Council on Public Works Improvement, *Fragile Foundations*, 195. Other studies indicated that the disparity between the budgeted amount for collection and the actual cost may be great. See E. S. Savas, "How Much Do Government Services Really Cost?" *Urban Affairs Quarterly* 15 (September 1979): 23–42.

29. Kemper and Quigley, *Economics of Refuse Collection*, 109–11. In the mid-1970s, compactor trucks represented over half the collection vehicles in operation. See Melosi, "Waste Management," 9.

30. Neal and Schubel, *Solid Waste Management*, 29. The technology of collection underwent substantial change in the 1970s and 1980s. In 1969–1970, Scottsdale, Arizona,

pioneered the "Godzilla," a mechanical arm that grabbed trashcans and deposited the waste into a sideloading truck. In 1975, Tempe, Arizona, was the first city to convert completely to sideloaders for residential collection. Also in the 1970s, the first co-collection trucks were introduced, which collected both trash and recyclables. In the 1980s, frontloading vehicles were used in some western states. See John T. Aquino, "MSW Collection: A History," *Waste Age* 30 (February 1999): 30.

31. E. S. Savas, "Solid Waste Collection in Metropolitan Areas," in *The Delivery of Urban Services*, ed. Elinor Ostrom (Beverly Hills, CA: Sage Pub., 1976), 220–21.

32. Laith D. Ezzet, "Collection: Who Handles the Trash in the 100 Largest U.S. Cities," *Waste Age* (April 1, 1998): 1–3, http://wasteage.com.

33. Quoted in *Waste Age* 26 (April 1995), 173 and also in John T. Aquino, "Privatization of Solid Waste Services: The Tie That Binds?" *Waste Age* 27 (September 1996): 86.

34. Ibid.

35. Kim A. O'Connell, "Back in the Game," *Waste Age* (March 1, 2001): 1–5, http://wasteage.com.

36. This approach most often represents a public-private partnership where private companies derive their authority and some degree of supervision from the city.

37. E. S. Savas, "Intracity Competition Between Public and Private Service Delivery," *Public Administration Review* 41 (January/February 1981): 47–48.

38. O'Connell, "Back in the Game," 4–5.

39. Bryan D. Jones, *Service Delivery in the City: Citizen Demand and Bureaucratic Rules* (New York: Longman, 1980), 128–32.

40. Armstrong, Robinson, and Hoy, *History of Public Works*, 446–47; E. S. Savas, "Policy Analysis for Local Government: Public vs. Private Refuse Collection," *Policy Analysis* 3 (Winter 1977): 54–58; APWA, *Solid Waste Collection Practice*, 236–37.

41. See National Solid Wastes Management Association, "Privatizing Municipal Waste Services: Saving Dollars and Making Sense" (Washington, DC, 1984), 1–2. See also Sarah Halsted, "All Ends of the Spectrum: The Public/Private Balance," *Waste Age* 28 (October 1997): 30–32, 34, 36, 38,40, 42.

42. See Eugene J. Wingerter, "The Role of Privatization," *Waste Age* 19 (September 1988): 210.

43. "Efficiency," however, might simply be measured as the cost to households for collection service and not the cost of disposal. Alternatively, some supporters of privatization claimed that private companies incurred costs not faced by municipalities. See E. S. Savas, "Intracity Competition Between Public and Private Service Delivery," *Public Administration Review* 41 (January/February 1981): 46, 50; Savas, "Policy Analysis" 69–71; James T. Bennett and Manuel H. Johnson, "Public versus Private Provision of Collective Goods and Services: Garbage Collection Revisited," *Public Choice* 34 (1979): 55, 60–61; Julia Marlowe, "Private Versus Public Provision of Refuse Removal Service: Measures of Citizen Satisfaction," *Urban Affairs Quarterly* 20 (March 1985): 356–57, 362; John Vickers and George Yarrow, *Privatization: An Economic Analysis* (Cambridge, MA: MIT Press, 1988), 41; Gordon Garner, "Cities and Contracting Out: How Public-Private Partnerships Can Work," *Current Municipal Problems* 12 (Winter 1986): 376–77; Bruce Johnson, "Privatization Solves Small City's Problems," *World Wastes* 29 (May 1986): 21–22. See also E. S. Savas, "How Much Do Government Services Really Cost?" *Urban Affairs Quarterly* 15 (September 1979): 25, 30–31; Franklin R. Edwards and Barbara J. Stevens, "The

Provision of Municipal Sanitation Services by Private Firms," *Journal of Industrial Economics* 27 (December 1978): 133.

44. John N. Collins and Bryan T. Downes, "The Effects of Size on the Provision of Public Services: The Case of Solid Waste Collection in Smaller Cities," *Urban Affairs Quarterly* 12 (March 1977): 345.

45. Ezzet, "Collection: Who Handles the Trash?" 1.

46. Bethany Barber, "Apples vs. Oranges: Managing Public/Private Competition," *Waste Age* 29 (October 1998): 36–38, 40, 42, 44, 46.

47. Anne Hartman, "The Solid Waste Control Industry," *Waste Age* 4 (July/August 1973): 54. See also Denny Paul, "The Solid Waste Agglomerates," *Waste Age* 5 (August 1974): 22.

48. In the 1950s, Waste Management was Ace Scavenger. It acquired several other disposal companies under Dean L. Buntrock, but really began its rapid rise when Buntrock joined forces with H. Wayne Huizenga, who at the time had his own collection company in Florida. In 1968 they incorporated under the name Waste Management. See Charles G. Burck, "There's Big Business in All That Garbage," *Fortune* 101 (April 7, 1980): 107–8. See also Michael A. Oberman, "Waste Management, Inc.: The Expanding Role of Private Industry," *Waste Age* 4 (March/April 1973): 4; "A Roundtable Discussion on the Newly Emerged Solid Waste Agglomerates," *Waste Age* 4 (July/August 1973): 10ff.

49. *Facing America's Trash: What Next for Municipal Solid Waste?* (New York: Van Nostrand Reinhold, 1992), 53.

50. "The Politics of Waste Disposal," *Wall Street Journal*, September 5, 1989.

51. See Nancy Shute, "The Selling of Waste Management," *Amicus Journal* 7 (Summer 1985): 8–15; James Cook, "Waste Management Cleans Up," *Forbes*, November 18, 1985; Bob Sablatura, "BFI, Waste Management Face Probe," *Houston Chronicle*, sec. 2, July 6, 1988, 2; Richard Asinof, "The Nation's Dumpster," *Environmental Action* 17 (May/June 1986): 13–16; Janet Novack, "A New Top Broom," *Forbes,* November 28, 1988, 200, 202.

52. However, consolidations in this fragmented industry had been going on for 25 years. For information about the internationalization of the solid waste industry, see John T. Aquino, "Yanks Abroad: U.S. Solid Waste Firms' International Experience," *Waste Age* 29 (April 1998): 84–86, 88–90, 92–93; Bethany Barber and John T. Aquino, "The Waste Age 100," *Waste Age* 28 (September 1997): 37.

53. John T. Aquino, "The Future is (Almost) Now," *Waste Age* 27 (December 1996): 52–53.

54. Scott Jones, "The Latest Moves in Waste Industry Consolidations," *Waste Age* 28 (May, 1997): 180, 184.

55. "USA Waste Services, Inc.," *Hoover's Company Capsules* (Austin: Hoover's, Inc., 1998); John T. Aquino, "The Waste Age 100," *Waste Age* 29 (September 1998): 83–84; John T. Aquino and Sarah Halsted, "A Mid-Year Review: Half-Way Through an Already-Full Year," *Waste Age* 29 (June 1998): 57–58, 60. See also John T. Aquino, "Mega-Mergers & Shakers, *Waste Age* 29 (December 1998): 38, 40, 42–43.

56. "Experts Predict Busier 1998," *Waste News* 3 (January 12, 1998): 1.

57. Cheryl L. Dunson, "Consolidation: Rearranging the Pieces," *WasteAge* (July 1, 1999), http://wasteage.com. See also Sarah Halsted, "Mid-Cap Companies: A Shrinking Middle Class?" *Waste Age* 29 (May 1998): 110–12, 114, 116.

58. Barnaby J. Feder, "'Mr. Clean' Takes on the Garbage Mess," *New York Times* sec. 3, March 11, 1990, 1, 6. See also Peter Reuter, "Regulating Rackets," *Regulation: AEI Journal on Government and Society* 8 (September/December 1984): 29–36.

59. Crooks, *Dirty Business*, 8.

60. Harold Crooks, *Giants of Garbage: The Rise of the Global Waste Industry and the Politics of Pollution Control* (Toronto: James Lorimer & Co., 1993), 55.

61. APWA, *Municipal Refuse Disposal*, x.

62. Frank P. Grad, "The Role of the Federal and State Governments," in Savas, *Organization and Efficiency*, 169–70.

63. Mantell, *Solid Wastes*, 3–7; Hagerty, Pavoni, and E. Heer, *Solid Waste Management*, 268–69; APWA, *Municipal Refuse Disposal*, 1–2; Shaheen, *Environmental Pollution*, 9.

64. J. Ernest Flack and Margaret C. Shipley, *Man and the Quality of His Environment* (Boulder: University of Colorado Press, 1968), 117–19.

65. The link between street cleaning and collection and disposal weakened after the advent of the automobile.

66. George Tchobanoglous, Hilary Theisen, and Rolf Eliassen, *Solid Wastes: Engineering Principles and Management Issues* (New York: McGraw-Hill Book Co., 1977) 40; Mantell, *Solid Wastes*, 11–12; Kemper and Quigley, *Economics of Refuse Collection*, 5. See also APWA, *Municipal Refuse Disposal*, 346–47.

67. Armstrong, Robinson, and Hoy, *History of Public Works*, 453; Tchobanoglous, Theisen, and Eliassen, *Solid Wastes*, 40–43; Hagerty, Pavoni, and Heer, *Solid Waste Management*, 269, 283–91; Douglas B. Cargo, *Solid Wastes: Factors Influencing Generation Rates* (Chicago: University of Chicago, Department of Geography, 1978), 74.

68. Melosi, "Waste Management," 7; Stanley E. Degler, *Federal Pollution Control Programs: Water, Air and Solid Wastes*, rev. ed. (Washington, DC: Bureau of National Affairs, 1971), 37–38.

69. Alfred Van Tassel, ed., *Our Environment: The Outlook of 1980* (Lexington, MA: Lexington Books, 1973), 468; Grad, "Role of the Federal and State Governments," 169–83; Armstrong, Robinson, and Hoy, *History of Public Works*, 453; Peter S. Menell, "Beyond the Throwaway Society: An Incentive Approach to Regulating Municipal Solid Waste," *Ecology Law Quarterly* 17 (1990): 671; William L. Kovacs, "Legislation and Involved Agencies," in *The Solid Waste Handbook*, ed. William D. Robinson (N.Y.: John Wiley and Sons, 1986), 9; Berry and Horton, *Urban Environmental Management*, 361–62. See also David Mafrici, "The Role of the Local Health Department in Solid Waste Management," *American Journal of Public Health* 61 (October 1971): 2010–14.

70. Bill Wolpin and Lourdes Dumke, "Former EPA Administrator: Regs Will Move Industry," *World Wastes* 32 (May 1989): 36.

71. James R. Pfafflin and Edward N. Ziegler, eds., *Encyclopedia of Environmental Science and Engineering* (Philadelphia: Gordon and Breach Science Pubs., 1992), 2:704–05.

72. EPA, *The Solid Waste Dilemma: An Agenda for Action* (Washington, DC: EPA, February 1989), 2. See also EPA, *Decision-Maker's Guide to Solid Waste Management* (November 1989); National Solid Waste Management Association, *Landfill Capacity in the Year 2000*, 2–3 (hereafter cited as NSWMA); Jonathan V. L. Kiser and Lisa M. Gills, "Integrated Waste Management: The Crisis Solution," *Waste Alternatives* 2 (March 1987): 40. In *Handbook of Solid Waste Management*, ed. George Tchobanoglous and Frank Kreith

(New York: McGraw-Hill, 2002), 1.8, integrated waste management is defined as "...the selection and application of suitable techniques, technologies, and management programs to achieve specific waste management objectives and goals."

73. The 1980 Solid Waste Amendments, however, were regarded as retrenchment from RCRA, including exceptions from Subtitle C for certain categories of special wastes—oil, gas, and thermal energy waste.

74. Sam M. Cristofano and William S. Foster, eds., *Management of Local Public Works* (Washington, DC: International City Management Association, 1986), 318. See also P. Aarne Vesilind and J. Jeffrey Peirce, *Environmental Pollution and Control* 2nd ed. (Ann Arbor, MI: Ann Arbor Science, 1983), 231–39; Duane A. Siler, "Resource Conservation and Recovery Act," Timothy A. Vanderver Jr., *Environmental Law Handbook* (Washington, DC: Bureau of National Affairs, Inc., 1994), 247–54.

75. National Council on Public Works Improvement, *The Nation's Public Works: Report on Solid Waste* (Washington, DC: National Council on Public Works Improvement, May 1987), 1.

76. Michael G. Malloy, "EPA in the 1980s: Scandals, Reform, and a Solid (Waste) Comeback," *Waste Age* 21 (December 1990): 41, 44. See also "From Cleanup to Prevention," *Waste Age* 21 (March 1990): 131–32, 134.

77. Marc K. Landy, Marc C. Roberts, Stephen R. Thomas, *The Environmental Protection Agency: Asking the Wrong Questions from Nixon to Clinton* (New York: Oxford University Press, 1994), 97–8; *Facing America's Trash*, 299; *The Environment Index*, 1973, 8.

78. Cathy Dombrowski, "Reilly Predicts More Regs and Higher Disposal Costs" *World Wastes* 32 (May 1989): 39. See also EPA, *Solid Waste Dilemma: An Agenda for Action*, 8–11.

79. Kovacs, "Legislation and Involved Agencies," 19.

80. Landy, Roberts, and Thomas, *Environmental Protection Agency*, 89.

81. Menell, "Beyond the Throwaway Society," 674.

82. US EPA, Office of Solid Waste and Emergency Response, *Municipal Solid Waste in the United States: 2000 Facts and Figures, Executive Summary* (Washington, DC: EPA, June, 2002), 2. EPA figures, although widely quoted, are in stark contrast with others. *BioCycle*'s 2001 nationwide survey found the total of 409 million tons of municipal solid waste was generated in the United States in 2000. The survey is based on detailed questionnaires sent out to state solid waste management and recycling officials. "The State of Garbage in America," *BioCycle* (December 2001): 42, http://www.jgpress.com?BCArticles/2001/SOG2001/120142.html. See also Van Tassel, *Our Environment: The Outlook of 1980* , 460; US EPA, *Characterization of Municipal Solid Waste in the United States: 1990 Update; Executive Summary* (Washington, DC: Government Printing Office, June 13, 1990), 3; National Council on Public Works Improvement, *The Nation's Public Works: Executive Summaries of Nine Studies* (Washington, DC: National Council on Public Works Improvement, May, 1987), 48.

83. Changes in consumption patterns include new working patterns, eating habits, technical innovations, and the level of affluence. Affluence alone, however, has not been the key variable in waste generation. See the "Comparative Data on National Solid Waste Generation and Economic Output" in *Recycling and Incineration: Evaluating the Choices*, ed. Richard A. Denison and John Ruston (Washington, DC: Island Press, 1990), 34–5. It shows, for example, that Switzerland has a higher annual per capita GNP than the United States, but less than half the annual per capita waste generation.

84. Franklin Associates, Ltd., *Analysis of Trends in Municipal Solid Waste Generation, 1972 to 1987: Final Report* (January 1992), ES-1 – ES-2, 1–4. The report was prepared for Procter & Gamble Company, Browning-Ferris Industries, General Mills, and Sears.

85. NSWMA, *Solid Waste Disposal Overview* (Washington, DC: NSWMA, 1987), 1. See also Stren, White, and Whitney, *Sustainable Cities,* 184.

86. In 1988, 14.4 million tons of all types of plastics were produced in the United States, and sales of plastic products in the 1980s exceeded $150 billion a year. See Debra L. Strong, *Recycling in America: A Reference Handbook* 2d ed. (Santa Barbara, CA: ABC-CLIO, 1997), 54–56; Stratford P. Sherman, "Trashing a $150 Billion Business," *Fortune*, August 28, 1989, 90.

87. Various studies completed in the late 1980s and early 1990s showed some wide variations in composition by region, especially in categories such as food waste, yard waste, and even paper discards. See Philip O'Leary and Patrick Walsh, "Introduction to Solid Waste Landfills," *Waste Age* 22 (January 1991): 44. See also Franklin Associates, Ltd., *Analysis of Trends*, 1–6.

88. EPA, *Solid Waste Dilemma: Background Documents*, 1–18 and 1–19.

89. Lewis Erwin and L. Hall Healy Jr., *Packaging and Solid Waste* (Washington, DC: AMA Membership Publications Division, American Management Assoc., 1990), 19; Burns and Seaman, "Some Aspects of Solid Waste Disposal," Van Tassel, *Our Environment*, 457–58; Hodges, *Environmental Pollution*, 219. One packaging item that did increase substantially was corrugated containers in food packaging. Franklin Associates stated that it was the largest single manufactured product in solid waste with 9.6 percent of total MSW by weight in 1972, and 8.5 percent in 1987. See Franklin Associates, Ltd., *Analysis of Trends*, 1-7–1-8.

90. US EPA, *The Solid Waste Dilemma: An Agenda for Action: Appendices A-B-C* (Washington, DC: EPA 1988), A.C-1 to A.C-15; EPA, *Solid Waste Dilemma: Background Documents*, 1–19 and 1–20; Robert E. Landreth and Paul A. Rebers, eds., *Municipal Solid Wastes: Problems and Solutions* (Boca Raton: LA: Lewis Publishers, 1997), 11.

91. Styrofoam in particular was targeted because it was made of CFCs that were believed to deplete the ozone layer. The pollution resulting from the dispersal of chemical precursors and waste byproducts used to make Styrofoam and the styrene leached from the packaging into the food were also of concern. See Herbert F. Lund, ed., *The McGraw-Hill Recycling Handbook* (New York: McGraw-Hill, 2001), 2.4–2.5; "Waste Reduction Efforts," www.mcdonalds.com / countries / usa / community / environ / info / waste / waste.html; "The McToxics Victory," www.mcspotlight.org / campaigns / countries / usa / usa_toxics.html; "Cutting-Edge Packaging Cuts Down on Waste," www.packaging-technology.com / informer / Manage / manage5; "Packaging and the Solid Waste Problem," *Packaging* (August 1989): 33–86; "Packaging in the 90's: The Environmental Impact," *Modern Materials Handling* (June 1990): 52–57; Sherman, "Trashing a $150 Billion Business," 90–91, 94–95, 98.

92. EPA, *Solid Waste Dilemma: Appendices A-B-C*, A.A-1–A.A-48; Franklin Associates, Ltd., *Analysis of Trends*, 1-10–1-11, 4-1–4-6, 4-14, 5-3–5-4, 5-11, 5-14, 6-1–6-2, 6-10, 7-3.

93. The name Fresh Kills comes from the tidewater creek originating with the first Dutch settlers on Staten Island.

94. Matthew Gandy, *Concrete and Clay: Reworking Nature in New York City* (Cambridge, MA: MIT Press, 2002), 192–93; Hans Tammemagi, *The Waste Crisis: Landfills, Incinerators, and the Search for a Sustainable Future* (New York: Oxford University Press, 1999), 194–95;

Newsday, *Rush to Burn*, 49–57; "Waste Disposal in New York City," *Waste Age* 12 (December 1981): 45; Bill Breen, "Landfills are #1," *Garbage* 2 (September/October 1990): 43; "What to Do With Our Waste," *Newsweek*, July 27, 1987, 51; J. Tevere MacFadyen, "Where Will All the Garbage Go?" *Atlantic* 255, March 1985, 29; *Renewing America's Infrastructure*, 32.

95. "Waste Disposal in New York City," 45; Tammemagi, *Waste Crisis*, 194–95.

96. Jim Johnson, "New York City 'Nightmare' Ends," *Waste News* 6 (March 26, 2001): 1, 35. See also Jim Johnson, "NYC Develops Competition for Fresh Kills Site," *Waste News* 5 (November 13, 2000): 11; Kim A. O'Connell, "The Closure of Fresh Kills: What's the Forecast?" *Waste Age* 29 (October 1998): 48–50.

97. "Attack Resurrects NYC's Fresh Kills," *Waste News* 6 (November 12, 2001): 13; "Debris Gone: Memories Remain," *Waste News* 7 (September 2, 2002): 1.

98. See J. J. Dunn Jr. and Penelope Hong, "Landfill Siting—An Old Skill in a New Setting," *APWA Reporter* 46 (June 1979): 12.

99. Quoted in Peter Steinhart, "Down in the Dumps," *Audubon* 88 (May 19, 1986): 106.

100. "Solid Waste Organizing Project," *Everyone's Backyard* 11 (February 1993): 8. See also Robert Gottlieb, *Forcing the Spring: the Transformation of the American Environmental Movement* (Washington, DC: Island Press, 1993), 168, 189–90, 261; Larry S. Luton, *The Politics of Garbage: A Community Perspective on Solid Waste Policy Making* (Pittsburgh: University of Pittsburgh Press, 1996), 204, 208, 210–11.

101. Robert D. Bullard, ed., *Confronting Environmental Racism: Voices from the Grassroots* (Boston: South End Press,1993), 3.

102. See Eileen Maura McGurty, "From NIMBY to Civil Rights: The Origins of the Environmental Justice Movement," *Environmental History* 2 (July 1997): 305–14. See also Andrew Szasz, *Ecopopulism: Toxic Waste and the Movement for Environmental Justice* (Minneapolis: University of Minnesota Press), 1994); Robert D. Bullard, *Dumping in Dixie: Race, Class, and Environmental Quality* 2d. ed. (Boulder, CO: Westview Press, 1994); Bunyan Bryant and Paul Mohai, eds., *Race and the Incidence of Environmental Hazards: A Time for Discourse* (Boulder, CO: Westview Press, 1992); Dana A. Alston, ed., *We Speak for Ourselves: Social Justice, Race and Environment* (Washington, DC, 1990); Christopher H. Foreman Jr., *The Promise and Perils of Environmental Justice* (Washington, DC: Brookings Institution, 1998); and Martin V. Melosi, "Environmental Justice, Political Agenda Setting, and the Myths of History," in Melosi, *Effluent America*, 238–262.

103. Martin V. Melosi, "Equity, Eco-Racism, and the Environmental Justice Movement, in *The Face of the Earth: Environment and World History*, ed. J. Donald Hughes (Armonk, New York: M.E. Sharpe, 2000), 65–66; David E. Newton, *Environmental Justice: A Reference Handbook* (Santa Barbara, CA: ABC-CLIO, 1996), 134–42; Blum, "Pink and Green." Sociologist Robert Bullard found that between the early 1920s and 1980, all of the public municipal landfills and six of the eight incinerators were located in African American neighborhoods in Houston. In the 1970s also three of four privately owned landfills were located in African American neighborhoods. While only 28 percent of the city's population was black, 82 percent of the solid waste disposal sites in the 1970s were located near them. See Robert D. Bullard, *Invisible Houston: The Black Experience in Boom and Bust* (College Station, TX: Texas A&M University Press, 1987).

104. Newton, *Environmental Justice*, 35–36. See also Bullard, *Dumping in Dixie*, xv; Bullard, "Race and Environmental Justice in the United States," *Yale Journal of International*

Law 18 (1993): 319–35. An interesting case study on the subject of "job blackmail" is Dan McGovern, *The Campo Indian Landfill War: The Fight for Gold in California's Garbage* (Norman: University of Oklahoma Press, 1995).

105. Michael C. Gross, "Is Justice Served?" *Waste Age* 30 (May 1999): 166.

106. David Naguib Pellow, *Garbage Wars: The Struggle for Environmental Justice in Chicago* (Cambridge, MA: MIT Press, 2002), 1–5, 9–12.

107. Neal and Schubel, *Solid Waste Management*, 116; Sue Darcey, "Landfill Crisis Prompts Action," *World Wastes* 32 (May 1989): 28; Joanna D. Underwood, Allen Hershkowitz, and Maarten de Kadt, *Garbage* (New York: Inform, 1988), 8–12; Richard A. Denison and John Ruston, eds., *Recycling and Incineration: Evaluating the Choices* (Washington, DC: Island Press, 1990), 4–5. See also National Council on Public Works Improvement, *Fragile Foundations*, 193.

108. William L. Rathje, "Rubbish!" *Atlantic Monthly* 264 (December 1989): 103.

109. "The State of Garbage in America," *BioCycle* 41 (March 2000): 30, http://www.jgpress.com/BCArticles/2000/040032.html.

110. Chaz Miller, "Garbage by the Numbers," *NSWMA Research Bulletin* (July 2002): 2; "State of Garbage in America," (1990): 49; "The State of Garbage in America," *BioCycle* 32 (April 1991): 34–36; "State of Garbage in America," (2000): 30; "Municipal Solid Waste Management: An Integrated Approach," *State Factor* 15 (June 1989): 2; Edward W. Repa and Allen Blakey, "Municipal Solid Waste Disposal Trends: 1996 Update," *Waste Age* 27 (January 1996): 43; NSWMA, *Landfill Capacity in the Year 2000* (Washington, DC, 1989), 1–3; Edward W. Repa, "Landfill Capacity: How Much Really Remains," *Waste Alternatives* 1 (December 1988): 32; Ishwar P. Murarka, *Solid Waste Disposal and Reuse in the United States* (Boca Raton, Fla., 1987), 1:5; "Land Disposal Survey," *Waste Age* 12 (January 1981): 65; National Council on Public Works Improvement, *Fragile Foundations*, 193; Conservation Foundation, *State of the Environment: A View Toward the Nineties* (Washington, DC: Conservation Foundation, 1987), 107. See also US EPA, "MSW Disposal," http://www.epa.gov/epaoswer/non-hw/muncpl/disposal/htm.

111. Kovacs, "Legislation and Involved Agencies," 10, 12–18. See also EPA, *Report to Congress: Solid Waste Disposal in the United States Executive Summary* (October 1988).

112. For more details on Subtitle D, see Melosi, "Historic Development of Sanitary Landfills and Subtitle D," 20–24; Repa and Blakey, "Municipal Solid Waste Disposal Trends," 46.

113. Casey Bukro, "Eastern Trash Being Dumped in America's Heartland," *Houston Chronicle*, November 24, 1989, 1F. See also Lori Gilmore, "The Export of Nonhazardous Waste," *Environmental Law* 19 (Summer 1989): 879–907; Jim Schwab, "Garbage In, Garbage Out," *Planning* 52 (October 1986): 5–6; F. Cairncross, "Waste and the Environment," *The Economist*, 327, no. 7813 (1993): 17.

114. Edward W. Repa, "Interstate Movement: 1995 Update," *Waste Age* 28 (June 1997), 41–44, 48, 50. See also Newsday, *Rush to Burn*, 61–77.

115. Susanna Duff, "Interstate Waste Keeps Crossing the Lines," *Waste News* 6 (August 6, 2001): 4.

116. Ibid.

117. Repa, "Interstate Movement," 52, 54, 56; Duff, "Interstate Waste Keeps Crossing the Lines," 4. As regional landfills increased, rail hauling of waste became more popular. See Bruce Geiselman, "Parallel Tracks," *Waste News* 6 (June 25, 2001): 19.

118. Deb Starkey and Kelly Hill, *A Legislator's Guide to Municipal Solid Waste Management* (Washington, DC: National Conference of State legislatures, August 1996), 20–21; US EPA, Solid Waste and Emergency Response, Office of Solid Waste, *Environmental Fact Sheet: Report to Congress on Flow Control and Municipal Solid Waste* EPA530-F-95-008 (March 1995) http: / / www.epaoswer / non-hw / muncpl / flowctrl / fsflow.txt; H. Lanier Hickman Jr., *Principles of Integrated Solid Waste Management* (New York: American Academy of Environmental Engineers, 1999), 2.6.3–2.6.7; Lund, *McGraw-Hill Recycling Handbook*, 2.3–2.4; Luton, *Politics of Garbage*, 29, 107–8, 117–18, 133–34; John Aquino, "The Tie That Binds?" *Waste Age* 27 (September 1996): 90; Deanna L. Ruffer, "Life After Flow Control," *Waste Age* 28 (January 1997): 73.

119. See Martin V. Melosi, "Viability of Incineration," 31–42.

120. Ibid., 38.

121. See T. Randall Curlee, et al., *Waste-to-Energy in the United States: A Social and Economic Assessment* (Westport, CT: Quorum Books, 1994), 37–8.

122. Neil Seldman, "Mass Burn is Dying," *Environment* 31 (September 1989): 42; US Congress, Office of Technology Assessment, *Facing America's Trash: What Next for Municipal Solid Waste?* (Washington, DC : Office of Technology Assessment, 1989), 222; William D. Robinson, *The Solid Waste Handbook* (New York: John Wiley & Sons, 1986), 111–12.

123. Curlee, et al., *Waste-to-Energy*, 4.

124. David A. Tillman, Amadeo J. Rossi, and Katherine M. Vick, *Incineration of Municipal and Hazardous Solid Wastes* (San Diego: Academic Press, Inc., 1989), 59, 113; Institute for Local Self-Reliance, *An Environmental Review of Incineration Technologies* (Washington, DC: Institute for Local Self-Reliance, 1986), 2. See also Curlee et al., *Waste-to-Energy*, 9, 11–13, 29–32. Dioxin is a generic term for some 75 chemical compounds with the technical name poly-chlorinated dibenzo-p-dioxins (PCDDs). PCDFs, or furans, are a related group of 135 chemicals often found in association with PCDDs. Both can be released into the air as a result of burning.

125. Cheryl A. McMullen, "EPA Links Incinerator Dioxins to Cancer," *Waste News* 6 (December 18, 2000): 10. See also R. E. Hester and R. M. Harrison, eds., *Waste Incineration and the Environment* (Letchworth: Royal Society of Chemistry, 1994), 2.

126. See James E. McCarthy, "Incinerating Trash: A Hot Issue, Getting Hotter," *Congressional Research Service Review* 7 (April 1986): 19–20. See also Tillman, Rossi, and Vick, *Incineration of Municipal and Hazardous Wastes*, x; Citizens Clearinghouse for Hazardous Wastes, Inc., *Incineration: The Burning Issue of the 80's* (July 1985), 11–13, 26–41; Janet Marinelli and Gail Robinson, "Garbage: No Room at the Bin," *The Progressive* (December 1981): 24–25; Institute for Local Self-Reliance, *An Environmental Review of Incineration Technologies* (Washington, DC: Institute for Local Self-Reliance, October 1986): 1–43; Holly A. Hattemer-Frey and Curtis Travis, eds., *Health Effects of Municipal Waste Incineration* (Baton Rouge, LA: CRC Press, 1991) Curtis C. Travis, ed., *Municipal Waste Incineration Risk Assessment* (New York: Plenum Press, 1991): E. Malone Steverson, "Provoking a Firestorm: Waste Incineration," *Environmental Science Technology* 25 (1991): 1808–9.

127. Dick Russell, "Environmental Racism," *Amicus Journal* 11 (Spring 1989): 23–24; Gottlieb, *Forcing the Spring*, 189–90. See also Institute for Local Self-Reliance, *An Environmental Review of Incineration Technologies*, 8.

128. Pellow, *Garbage Wars*, 9–10. See also 48–51, 89–99, 131–35; Gandy, *Concrete and Clay*, 193–213; Newsday, *Rush to Burn*, 95–99; Edward Walsh, Rex Warland, D. Clayton Smith, "Backyards, NIMBYs, and Incinertaor Sitings: Implications for Social Movement Theory," *Social Problems* 40 (February 1993): 25–49.

129. See Louis Blumberg and Robert Gottlieb, *War on Waste: Can America Win Its Battle with Garbage?* (Washington, DC: Island Press, 1989), 58–60; Russell, "Environmental Racism," 25–26.

130. K. A. Godfrey Jr., "Municipal Refuse: Is Burning Best?" *Civil Engineering* 55 (April 1985): 54–55; McCarthy, "Incinerating Trash: A Hot Issue, Getting Hotter," 19; Institute for Local Self-Reliance, *An Environmental Review of Incineration Technologies*, 8; Seldman, "Mass Burn is Dying," 42; Allen Hershkowitz, "Burning Trash: How It Could Work," *Technology Review* (July 1987): 26, 30; US Congress, Office of Technology Assessment, *Facing America's Trash*, 222.

131. Martin V. Melosi, "Down in the Dumps: Is There a Garbage Crisis in America?" in *Urban Public Policy: Historical Modes and Methods*, ed. Martin V. Melosi, (University Park, PA: Pennsylvania State University Press, 1993), 111. See also Tim Darnell, "The Ups and Downs of Waste-to-Energy," *American City* 106 (January 1991): 4, 6, 8.

132. Seldman, "Mass Burn is Dying," 42. See also Neal and Schubel, *Solid Waste Management and the Environment*, 90, 92; Neil Seldman and Jon Huls, "Beyond the Throw-away Ethic," *Environment* 23 (November 1981): 32–34.

133. *Facing America's Trash: What Next for Municipal Solid Waste?* 222. For more information on solid waste bonds, see Robert Lamb and Stephen P. Rappaport, *Municipal Bonds* 2d ed. (New York: McGraw-Hill Book Co., 1987), 156–60.

134. John H. Skinner, "The Consequences of New Environmental Requirements: Solid Waste Management as a Case Study," (unpublished paper presented at Centre Jacques Cartier, Lyon, France, December, 1993), 6–7; Margaret Ann Charles, "New Trends in Waste-to-Energy," *Waste Age* 24 (November 1993); 59–60.

135. "State of Garbage in America," (2000), 5. *Waste Age* estimated 140 incinerators/WTE in 2001. See Edward W. Repa, "The U.S. Solid Waste Industry: How Big is It?" *Waste Age* (December 1, 2001), http://wasteage.com/ar/waste__us__solid__waste/index.htm. See also Landreth and Rebers, *Municipal Solid Wastes*, 135–41.

136. Charles, "New Trends in Waste-to-Energy," 1993, 59–60; Berenyi and Gould, "Municipal Waste Combustion in 1993," 1993, 51–52.

137. See Melosi, "Equity, Eco-racism and Environmental History," 4–11.

138. Charles, "New Trends in Waste-to-Energy," 59–60.

139. Schwab, "Garbage In, Garbage Out," 7.

140. Quoted in Long, *Problem of Waste Disposal*, 17. See also Tillman, Rossi, and Vick, *Incineration of Municipal and Hazardous Wastes*, ix; Marinelli and Robinson, "No Room at the Bin," 26.

141. See Seldman, "Waste Management," 43–44; Schwab, "Garbage In, Garbage Out," 9; T. Randall Curlee, "Plastics Recycling: Economic and Institutional Issues," *Conservation and Recycling* 9 (1986): 335–50; US EPA, *Recycling Works!* (Washington, DC: Government Printing Office, January 1989); Nicholas Basta, "A Renaissance in Recycling," *High Technology* 5 (October 1985): 32–39; Barbara Goldoftas, "Recycling: Coming of Age," *Technology Review* (November/December 1987), 30–35, 71; Anne Magnuson, "Recycling Gains Ground," *American City & County/Resource Recovery* (1988), RR10; Strong, *Recycling*

in America, 1–20. Recycling was already a major component of waste programs in Europe and Japan. In fact, the Japanese are the world's leader in the field. See Hershkowitz, "Burning Trash," 28.

142. In 1995, 2 billion pounds of aluminum cans were diverted from landfills, and more than half of new aluminum cans are made from recyclables. See Strong, *Recycling in America*, 45.

143. "Recycling Timeline," http://members.aol.com/Ramola15/timeline.html; George R. Stewart, *Not So Rich as You Think* (Boston: Houghton Mifflin, Co., 1967), 125–29. Between 1971 and 1993, 10 states enacted some form of beverage container refund law. See James E. McCarthy, "Bottle Bills and Curbside Recycling: Are They Compatible?" *Congressional Research Service Report* (January 27, 1993), 3, http://www.cnie.org/nle/plgen-3.html.

144. Chaz Miller, "Source Separation Programs," *NCRR Bulletin: The Journal of Resource Recovery* 10 (December 1980): 82–83.

145. Jim Glenn, "Curbside Recycling Reaches 40 Million," *BioCycle* 31 (July 1990): 30–31; Susan J. Smith and Kathleen M. Hopkins, "Curbside Recycling in the Top 50 Cities," *Resource Recycling* 11 (March 1992): 101–102; "State of Garbage in America," (1991), 36–37; US EPA, "Reduce, Reuse, and Recycle" (December 12, 2000), http://www.epa.gov/epaoswer/non-hw/muncpl/reduce.htm; US Bureau of the Census, *Statistical Abstract* (2001), 218; Lund, *McGraw-Hill Recycling Handbook*, 2.2.

146. McCarthy, "Bottle Bills and Curbside Recycling," 9; US Bureau of the Census, *Statistical Abstract* (2001), 218. There are four basic models of curbside collection: complete citizen separation; truckside sorting conducted by the collector; site separation or commingling, where sorting takes place at the processing facility; and co-collection, where separately bagged or boxed recyclables are collected at the same time as garbage. See Steve Apotheker, "Curbside Collection: Complete Separation versus Commingled Collection," *Resource Recovery* 9 (October 1990): 58.

147. See NSWMA, *Solid Waste Disposal Overview* (Washington, DC: NSMWA, 1988), 2; Cynthia Pollock, "There's Gold in Garbage," *Across the Board* (March 1987): 37; Tchobanoglous and Kreith, *Handbook of Solid Waste Management*, 3.5; "Municipal Solid Waste Management," 7; "Newark Claims East Coast's Largest Recycling Program," *World Wastes* 31 (December 1988): 47–49; Debi Kimball, *Recycling in America* (Santa Barbara, CA: ABC-CLIO, 1992), 3, 5–6, 22–24. See also Pete Grogan, "Recycling in Transition," *Resource Recovery* 2 (Summer 1988): 5–7; Richard Hertzeberg, "New Directions in Solid Waste and Recycling," *BioCycle* 27 (January 1986): 22–27.

148. Waste minimization is a broader term than waste reduction. "Waste minimization" as it appears in the 1984 amendments to the Resource Conservation and Recovery Act of 1976 includes waste reduction, recycling, and the treatment of wastes after they are generated. According to Kirsten U. Oldenburg and Joel S. Hirschhorn, "[W]aste minimization combines the concepts of both prevention and control and its goal is generally understood to be the avoidance of land disposal of hazardous wastes regulated under RCRA." See Oldenburg and Hirschhorn, "Waste Reduction: A New Strategy to Avoid Pollution," *Environment* 29 (March 1987): 17–20, 39–45. Unlike the various disposal options, waste reduction and waste minimization moved into relatively uncharted waters. In a society conditioned to deal with its waste problems from the "back end," "front end" solutions go beyond technical fixes and management efficiencies toward lifestyle and

behavioral changes. Essentially, waste minimization is much more difficult to achieve than increased recycling. See Craig Colten, "Historical Development of Waste Minimization," *Environmental Professional* 11 (1989): 94–99; Masood Ghassemi, "Waste Reduction: An Overview," *Environmental Professional* 11 (1989): 100–116; EPA, *Waste Minimization: Environmental Quality with Economic Benefits* (Washington, DC: Government Printing Office, October 1987); Anne Magnuson, "What Has Happened to Waste Reduction?" *American City & County* 106 (April 1991): 30, 32, 34, 36–37; Douglas Wrenn, "Resource Recovery," *Environmental Comment* (March 1981): 3–15; Chaz Miller, "Source Reduction: Less of More," *Waste Age* 26 (February 1995): 79–84: Paula Comella and Robert W. Rittmeyer, "Waste Minimization / Pollution Prevention," *Pollution Engineering* 22 (April 1990): 71–80. For current statistics on specific state and local recycling programs, see *Waste News*, especially issues which include the municipal recycling survey.

149. In recent years about half of the states supported backyard composting programs, and the number of yard trimmings compost facilities have increased substantially. In 1988 there were 651 such facilities; in 1994 there were 3,202. Most of the facilities that handle organics are operated by local municipalities. See Landreth and Rebers, *Municipal Solid Wastes*, 86–88. For recent recycling rates by state and by municipalities, see "State of Garbage in America," (2001): 4; "Municipal Recycling Survey," *Waste News* 6 (February 19, 2001): 13–19.

150. For criticisms of recycling as an answer to waste disposal, see John Tierney, "Recycling is Garbage," *New York Times Magazine*, June 30, 1996, 24–29, 44, 48–49, 53; Cairncross, "Waste and the Environment," 7, 9. See also Oscar W. Albrecht, Ernest H. Manuel Jr., and Fritz W. Efaw, "Recycling in the USA: Vision and Reality," *Resources Policy* 7 (September 1981): 188–89, 194; Chaz Miller, "The Shape of Things to Come," *Waste News* 26 (September 1995): 60.

151. See Pollock, "There's Gold in Garbage," 37.

152. See Joe Truini, "No End in Sight for Recycling Downturn," *Waste News* 7 (November 26, 2001), 1, 27.

153. See Frank Ackerman, *Why Do We Recycle? Markets, Values and Public Policy* (Washington, DC: Island Press, 1997) 11–13, 23, 169; Jerry Powell, "The Anti-recyclers: What's Their Message?" *Resource Recycling* 11 (September 1992): 74–77.

154. Tierney, "Recycling is Garbage," 24, 26.

155. See John T. Aquino, "A Recycling *Pilgrim's Progress*: An Anatomy of 'Recycling is Garbage,'" *Waste Age* 28 (May 1997): 220, 222, 224, 226, 228, 230–32.

156. APWA, *Street Cleaning Practices*, 110–12, 122–29, 161; Korbitz, *Urban Public Works Administration*, 337–38; Stewart, *Not So Rich as You Think*, 125–29; Kenneth A Hammond, George Macinko, and Wilma B Fairchild, eds., *Sourcebook on the Environment* (Chicago: University of Chicago Press, 1978), 332–33; Lund, *McGraw-Hill Recycling Handbook*, 13.5.

157. Armstrong, Robinson, and Hoy, *History of Public Works*, 440; APWA, *Street Cleaning Practice*, 122–25; Stewart, *Not So Rich as You Think*, 114–15; Keep America Beautiful, *Clean Community System Bulletin* (March-April 1980): 1–3; "A Clean Sweep in Georgia," *Time*, May 19, 1980, 51; Arthur H. Purcell, *The Waste Watchers: A Citizen's Handbook for Conserving Energy and Resources* (Garden City, NY: Anchor Books, 1980), 121. Through direct grants and other means KAB inspired other national organizations to become involved in the antilitter campaign. Under a KAB grant the APWA developed a litter-measuring technique, the APWA Photometric Index, to evaluate the extent of litter in a given location. The

index measured the accumulation and spread of litter and provides a relatively simple means whereby cities identified their major problem areas and then devised ways of dealing with them. See APWA, *Street Cleaning Practice*, 112–13; Armstrong, Robinson, and Hoy, *History of Public Works*, 441.

Conclusion

1 Austin Bierbower, "American Wastefulness," *Overland Monthly*, April 1907, 358–59.

2. See Allen M. Wakstein, ed., *The Urbanization of America: A Historical Anthology*, (Boston: Houghton Mifflin, 1970), 115–16.

3. Many cities had smoke inspectors and even noise inspectors, but they were dealing with forms of pollution that defied more tangible means of control, such as refuse or sewerage.

4. See Jon A. Peterson, "Impact of Sanitary Reform," 83–103. Peterson argues that the establishment of elaborate sewer lines played a central role in determining the physical layout of cities and tended to dictate plans for city development. Planners, therefore, had to follow the lead established by the application of this form of technology.

5. For information on solid waste from an international perspective, see Melosi, "Waste Management," in *Encyclopedia of World Environmental History*, ed. Shepard Krech III, J. R. McNeill, and Carolyn Merchant, (New York: Routledge, 2004), 1291-99; L. Bonomo and A. E. Higginson, eds., *International Overview on Solid Waste Management* (London: Academic Press, 1988); J. S. Carra and R. Cossu, eds., *International Perspective on Municipal Solid Wastes and Sanitary Landfilling* (London: Academic Press, 1990); E. Pollock, "Wide World of Refuse," *Waste Age* 16 (December 1985):89–90; United Nations Environment Programme, Divison of Technology, Industry, and Economics, *Newsletter and Technical Publications: Municipal Solid Waste Management* (2002), http://www.unep.or.jp/ietc/ESTdir/Pub/MSW/RO.

INDEX